Foundations of
High-Energy Astrophysics

Other Theoretical Astrophysics titles available
from the University of Chicago Press

Inner Space / Outer Space
Edward Kolb, Michael Turner, Keith Olive, and David Seckel, editors (1985)

Theory of Neutron Star Magnetospheres
Curtis F. Michel (1990)

High-Energy Radiation from Magnetized Neutron Stars
Peter Meszaros (1992)

Cauldron in the Cosmos
Claus E. Rolf and William S. Rodney (1988)

Stars as Laboratories for Fundamental Physics
Georg G. Raffelt (1996)

Foundations of High-Energy Astrophysics

Mario Vietri

THE UNIVERSITY OF CHICAGO PRESS
Chicago and London

MARIO VIETRI is professor of astrophysics at the Scuola
Normale Superiore in Pisa, Italy.

The University of Chicago Press, Chicago 60637
The University of Chicago Press, Ltd., London
© 2008 by The University of Chicago
All rights reserved. Published 2008
Printed in the United States of America

17 16 15 14 13 12 11 10 09 08   1 2 3 4 5

ISBN-13: 978-0-226-85569-1 (cloth)
ISBN-10: 0-226-85569-4 (cloth)

Library of Congress Cataloging-in-Publication Data
Vietri, Mario.
Foundations of high-energy astrophysics / Mario Vietri.
      p.   cm.
Includes bibliographical references and index.
ISBN-13: 978-0-226-85569-1 (cloth : alk. paper)
ISBN-10: 0-226-85569-4 (cloth : alk. paper)
1. Astrophysics—Textbooks.   2. Infrared astronomy—Textbooks.
3. X-ray astronomy—Textbooks.   4. Gamma ray
astronomy—Textbooks.   5. Particles (Nuclear physics)—
Textbooks.   I. Title.
QB468.V54 2008
523.01′976—dc22

                                        2007030145

A Luisa, che aveva ragione

Contents

Appendix

Preface

When we bring a new scientific text to the attention of colleagues and students, it is necessary to provide a justification for this novelty. The main reason that drove me to write this book is the desire to present all the basic notions useful to the study of the current scientific literature in one single volume.

The choice of topics included in this book has been motivated primarily by this statement. Chapters 1 through 3 present those basic notions (hydrodynamics, magnetohydrodynamics, and radiative processes) that are indispensable for successive developments. Chapters 4 through 7 treat the classical topics of high-energy astrophysics: the acceleration of nonthermal particles, astrophysical accretion flows, both spherical and disk-shaped, and explosive motions (supernovae and gamma-ray bursts). Chapter 8 presents the basic results of electrodynamics around compact objects.

The second half of chapter 2, as well as the whole of chapters 7 and 8, present material not usually included in textbooks on high-energy astrophysics. The second half of chapter 2 contains a discussion of magnetic fields in astrophysics: methods of measurement and their results, and the fundamental ideas of batteries (generating the seeds of astrophysical magnetic fields) as well as of dynamos (amplifying these seeds). These topics are hardly ever found in high-energy astrophysics textbooks, even though they are crucial for an understanding of various phenomena, such as the acceleration of non-thermal particles, accretion disks on magnetized objects, and pulsars. The discussion is kept at a basic level and must be considered a simplified introduction to a very complex subject. Chapter 7 contains further developments in the theory of accretion disks which have acquired a certain relevance only in the last few years, namely, pair plasmas in the hot coronae of accretion disks, accretion flows dominated by advection, thick disks, and accretion on strongly magnetized objects. In chapter 8 I discuss those processes in which the coupling between magnetic field and rotation plays a fundamental role: here I introduce the main results on the magnetospheres of pulsars and black holes, as well as the coupling between disk and magnetic field, with the ensuing collimation of jets.

The choice of topics omitted from this book also requires an explanation. I decided to neglect all phenomena connected with the interiors of neutron stars and white dwarfs. Apart from the fact that these topics are outside my expertise, this choice is due to the presence of the beautiful book by Shapiro and Teukolsky, *Black Holes, White Dwarfs, and Neutron Stars* (1983), which, though a bit old, still provides a wonderful introduction to the subject. Furthermore, there is no discussion of the observational properties of astrophysical objects, nor of the match (or mismatch!) between theory and observations. This neglect is certainly more unusual and requires an explanation. At first, I wanted to include these topics as well and present a more complete textbook. However, the increased dimensions of this volume, as I added chapter after chapter, convinced me that it was impossible to illustrate, in a satisfactory way, both observations and their comparison with theory, within the reasonable boundaries of a textbook. Moreover, while there are many complete and timely review articles about observational data, it is more difficult to find a systematic introduction to theoretical topics. That is why I chose, even if with many regrets, to leave out these other subjects. The only comfort I can find is that, inevitably, the observations (thanks to continuous technological innovations, the construction of new telescopes, and the launch of new space missions) are destined to become obsolete more quickly than theory.

The contents of this book are obviously excessive for any university course, but one can pick out the material to be covered by taking into account what follows. Chapters 1 through 6 are written so that the initial parts cover the fundamental notions, while further developments of lesser generality and/or interest are discussed in the last few paragraphs of each chapter. On the other hand, chapter 6 is fundamental, and only the paragraphs about the analysis of the stability of disks and the Lense-Thirring effect can be considered optional. Chapters 7 and 8 are certainly complementary and can be treated according to the interests of lecturer and students.

The problems at the end of each chapter are generally easy. The exceptions to this rule contain a reference to where the solution is accurately derived. The exception to the exception is the exercise in appendix 2 (problem 1), which is rather difficult; since it has never been proposed as an exercise, I cannot suggest a reference for the solution, which, however, will certainly not be missed by the experts on the gravitomagnetic field. Problem 2 in chapter 5 may appear deprived of the information necessary for the solution, but this is not so: it actually proposes the real-life situation of astronomers, who sometimes manage to find ingenious solutions to (apparently) unsolvable problems.

Pasquale Blasi (who also wrote section 4.3 of this book), Giorgio Matt, and Luigi Stella gave me support and suggestions during the preparation of this work, and Lorenzo Sironi carried out and corrected most exercises; I would like to thank them from my heart. Of course, any errors are my exclusive responsibility. I would never have thought of writing a book without the help of my father Dario and of my wife Luisa. This book is dedicated to them, as well as to those friends (Renato Finocchi, Vincenzo Guidetti, and Antonio Navarra) who, even though they will never read it, have contributed so much to its realization.

Chapter 1

Hydrodynamics

Most astrophysical phenomena involve the release of energy either inside or outside of stars, within the interstellar medium. In both cases, the medium where the energy emission takes place is a fluid, which, because of this energy release, starts moving, expanding or contracting, becoming warmer or colder. The properties of the radiation emitted (photons, neutrinos, or nonthermal particles), measured by us on Earth, depend—in detail—on the thermodynamic state and on the speed of the fluid where the emission takes place. Therefore, we can say that a necessary prerequisite for high-energy astrophysics is the study of hydrodynamics and magnetohydrodynamics.

In this chapter, I shall briefly explain the fundamental principles of hydrodynamics, as well as some key results, which will often be used in the following chapters. In particular, I shall first discuss the fundamental equations of hydrodynamics, and their rewording as conservation laws for the energy-momentum tensor. I shall derive Bernoulli's theorem, the properties of small perturbations (or sound waves), tangential discontinuities, and, last but not least, shock waves. Second, I shall discuss the properties of self-similar solutions of the first and the second kind. Finally, I shall discuss the

general principles of relativistic hydrodynamics and shock
waves, and de Laval's nozzle.[1]

Hydrodynamics considers a fluid as a macroscopic object,
therefore, ideally, as a continuous medium. Even when we
consider infinitesimal volumes of fluids, elements, or particles
of fluid, we will always assume them to be made up of a very
large number of molecules.

The description of a fluid in a state of rest requires knowledge of its local thermodynamical properties. We shall assume
here knowledge of its equation of state, so that it is necessary
to provide only two of the three fundamental thermodynamical quantities (namely, pressure, density, and temperature).
A generic fluid not in a state of rest will be therefore described
also by the instantaneous speed of motion. In the following,
we shall suppose that all these quantities $(P, \rho, T, \vec{v})$ are continuous functions of space and time, in accordance with the
above.

1.1   The Mass Conservation Equation

The first fundamental law of fluids expresses mass conservation. Let us consider a volume V containing a fluid mass ρV.
The law of mass conservation states that mass can neither be
created nor destroyed, so that the mass inside V can change
only when a certain amount crosses the surface of the volume
in question. The quantity of mass crossing an infinitesimal el-
ement $d\vec{A}$ of area, per unit time, is $\rho \vec{v} \cdot d\vec{A}$. Let us take, as
a convention, the outward area element $d\vec{A}$; in this way, a
positive mass flow $\rho \vec{v} \cdot d\vec{A}$ corresponds to an amount of mass
leaving the volume V. The total mass flow through the sur-
face will therefore be

$$\oint \rho \vec{v} \cdot d\vec{A} \qquad (1.1)$$

---

[1] These sections can be left out in a first reading.

corresponding to the total mass lost by the volume V, per time unit:

$$-\frac{d}{dt}\int \rho dV \qquad (1.2)$$

By comparison, we find that

$$-\frac{d}{dt}\int \rho dV = \oint \rho\vec{v}\cdot d\vec{A} \qquad (1.3)$$

which is the expression of mass conservation in integral form; please notice that this law applies to any quantity that can be neither created nor destroyed, such as, for instance, electric charge, as well as baryonic or leptonic numbers. However, it is convenient to write the above equation in differential form: since the volume V does not change as time passes, we have

$$-\frac{d}{dt}\int \rho dV = -\int \frac{\partial \rho}{\partial t}\,dV \qquad (1.4)$$

while the right-hand side can be rewritten by means of Gauss's theorem:

$$\oint \rho\vec{v}\cdot d\vec{A} = \int \nabla\cdot(\rho\vec{v})\,dV \qquad (1.5)$$

Thus we find

$$\frac{\partial \rho}{\partial t} + \nabla\cdot(\rho\vec{v}) = 0 \qquad (1.6)$$

which is the equation we were looking for. This equation expresses, in differential form, the fact that mass can be neither created nor destroyed: mass changes within a volume only if some is added or taken away by a mass flow $(\rho\vec{v})$ crossing the surface of the given volume.

1.2 The Momentum Conservation Equation

It is well known that a fluid exerts a force $pd\vec{A}$ on an area element $d\vec{A}$, where p is the pressure. Therefore there is a pressure $-\oint pd\vec{A}$ on a fluid within a volume dV. This can be rewritten as

$$-\oint pd\vec{A} = -\int \nabla p dV \qquad (1.7)$$

which, in its turn, can be interpreted as follows: on each element of mass ρdV contained in an element of volume dV a pressure is exerted that equals $-\nabla p dV$. The equation of motion of this infinitesimal mass is thus

$$\rho \frac{D\vec{v}}{Dt} = -\nabla p \qquad (1.8)$$

Here the derivative $D\vec{v}/Dt$ represents the acceleration of a fixed element of mass and is therefore different from $\partial\vec{v}/\partial t$, which is the variation of speed at a fixed point. In order to see the connection between the two, let us consider an element of mass placed at $\vec{x}_0$ at the time t_0; this moves with speed $\vec{v}_0 = \vec{v}(\vec{x}_0, t_0)$. A time dt later, the same element is in a new position, $\approx \vec{x}_0 + \vec{v}_0 dt$, where it moves with speed $\vec{v}_1 \approx \vec{v}(\vec{x}_0 + \vec{v}_0 dt, t_0 + dt)$. Its acceleration is therefore

$$\frac{D\vec{v}}{Dt} = \lim_{dt\to 0} \frac{\vec{v}_1 - \vec{v}_0}{dt} = \frac{\partial\vec{v}}{\partial t} + (\vec{v}\cdot\nabla)\vec{v} \qquad (1.9)$$

which eventually allows us to write the equation of fluid motion as

$$\frac{\partial\vec{v}}{\partial t} + \vec{v}\cdot\vec{\nabla}\vec{v} = -\frac{1}{\rho}\nabla p + \frac{1}{\rho}\vec{f}_{\text{ext}} \qquad (1.10)$$

This is exactly the equation we were looking for, which is called Euler's equation, or the equation of momentum conservation, even though, in its current formulation, it does not appear formally similar to equation 1.6; we shall illustrate later on (sect. 1.5) how it is possible to rewrite it so that its nature as an equation of conservation is clearer.

The term $\vec{f}_{\text{ext}}$ includes all forces of a nonhydrodynamic nature, for example, within a gravitational field,

$$\vec{f}_{\text{ext}} = -\rho\nabla\phi \qquad (1.11)$$

where ϕ is the gravitational potential. Friction forces are also included in $\vec{f}_{\text{ext}}$; these forces are not always negligible in astrophysics; that is why we shall soon discuss viscous fluids. In the meantime, let us simply assume that all fluids are ideal.

The operator

$$DX/Dt \equiv \partial X/\partial t + \vec{v} \cdot \vec{\nabla} X \qquad (1.12)$$

expresses the time variation of any physical quantity X, when we consider its variation not at a fixed position, but for a fixed element of mass. In other words, if we concentrate on a certain element of mass and follow it in its motion, DX/Dt expresses the variation of X with time, just as we would see it if we straddled the element of mass in question. Because of this, the operator D/Dt is called the *convective derivative*, from the Latin *convehere*, which means "to carry."

1.3   The Energy Conservation Equation

We have already assumed that our fluid is ideal and therefore not subject to dissipative phenomena, which are often so important for laboratory fluids. In particular, we have neglected friction and thermal conduction. Supposing that there are no further dissipative mechanisms, the equation of energy conservation must simply establish that the entropy of a given element of mass does not change as time varies. Therefore, as far as the specific entropy s is concerned, namely, entropy measured per unit mass—not unit volume, its convective derivative (namely, for a fixed element of mass), vanishes:

$$\frac{Ds}{Dt} = \frac{\partial s}{\partial t} + \vec{v} \cdot \nabla s = 0 \qquad (1.13)$$

In astrophysics, the fundamental dissipation process is radiation, since fluids are so rarefied that the chances that a photon, once emitted, is rapidly reabsorbed, are often (but not always!) negligible. In this case, the specific entropy cannot be conserved, and the above equation must be modified, as we shall explain in a minute. However, even in radiative cases, this equation is valid to a certain extent. This is due to the fact that radiative processes take place on definite time scales (which will be discussed in chapter 3), and therefore all phenomena taking place on shorter time scales are essentially adiabatic, namely, without heat loss.

Assuming the fluid to be ideal, and that therefore equation 1.13 holds, we can rewrite it in a more suggestive form, using equation 1.6. It is actually easy to see that

$$\frac{\partial(\rho s)}{\partial t} + \nabla \cdot (\rho s \vec{v}) = 0 \qquad (1.14)$$

which has the same form of the equation 1.6. ρs is the entropy density per unit volume, and $\rho s \vec{v}$ is the entropy flux. We can see that this equation can be easily interpreted as follows: in an ideal fluid (namely, an adiabatic fluid) the entropy within a given volume can be neither created nor destroyed but changes only because of a flux through the surfaces of the given volume, which is the same as the interpretation of the equation of mass conservation.

The most common reason in astrophysics for changes of entropy is that the fluid is heated and cooled by a certain number of radiative processes. We define with Γ and Λ the heating and cooling coefficients per unit mass and time; the equation of energy conservation can therefore be rewritten as

$$\frac{Ds}{Dt} = \Gamma - \Lambda \qquad (1.15)$$

or, alternatively,

$$\frac{\partial(\rho s)}{\partial t} + \nabla \cdot (\rho s \vec{v}) = \rho \Gamma - \rho \Lambda \qquad (1.16)$$

The coefficients $\rho \Gamma$ and $\rho \Lambda$ are, respectively, heating and cooling coefficients per unit volume and time. Please note that in the literature there is often some confusion: the terms Γ and Λ can indicate either terms per unit mass or per unit volume.

Assuming the fluid to be adiabatic, we can rewrite Euler's equation 1.10 in a useful form. If ϵ is the internal energy of the gas per mass unit, the specific enthalpy (that is to say, for each mass unit) is given by

$$w \equiv \epsilon + \frac{p}{\rho} \tag{1.17}$$

The second principle of thermodynamics for reversible transformations can be written as

$$dw = Tds + \frac{dp}{\rho} \tag{1.18}$$

which shows that, for adiabatic transformations only,

$$dw = \frac{dp}{\rho} \tag{1.19}$$

which can be introduced into equation 1.10 to get

$$\frac{\partial \vec{v}}{\partial t} + \vec{v} \cdot \vec{\nabla}\vec{v} = -\nabla w \tag{1.20}$$

which will often be useful in the following.

1.4 Bernoulli's Theorem

The equation above leads to a useful result when motion is stationary. With this term, we do not indicate a situation where the fluid is still (i.e., static), but a situation where the fluid moves, while its speed and all other quantities do not depend on time. In other words, pictures of the flow at arbitrary times t_1 and t_2 would be identical. This situation is often indicated with a significant term $-\partial/\partial t = 0$. This means that the partial derivative of *any* quantity vanishes. Equation 1.20 can be rewritten thanks to a mathematical identity, which is easy to prove

$$\frac{1}{2}\nabla v^2 = \vec{v} \wedge (\nabla \wedge \vec{v}) + (\vec{v} \cdot \nabla)\vec{v} \tag{1.21}$$

thus obtaining

$$-\vec{v} \wedge (\nabla \wedge \vec{v}) = -\nabla \left(\frac{1}{2}v^2 + w \right) \tag{1.22}$$

Let us now take the component of the equation along the trajectory of motion of a fluid particle (we call the trajectory a *flux line*). In order to do this, we multiply it by $\vec{v}$. Note that the term on the left-hand side disappears because it is orthogonal to $\vec{v}$ by construction, and therefore,

$$\vec{v} \cdot \nabla \left(\frac{1}{2}v^2 + w \right) = 0 \qquad (1.23)$$

In other words, the quantity $v^2 + w$ is constant along each flux line,

$$\frac{1}{2}v^2 + w = \text{constant} \qquad (1.24)$$

even though, in general, the constant varies from flux line to flux line. This is due to the fact that the gradient of $v^2/2 + w$ vanishes only along a flux line, not everywhere, as we can see from equation 1.22.

In order to guarantee that the constant remains the same along all flux lines, it is necessary that $\nabla(v^2/2 + w) = 0$ in the whole space, namely, that

$$\nabla \wedge \vec{v} = 0 \qquad (1.25)$$

Flows with this property are called irrotational. It is easy to check that purely radial flows, in spherical and cylindrical symmetry, are irrotational.

It is perfectly easy to see that when the motion takes place in a gravitational field with a potential ϕ, Bernoulli's theorem guarantees the conservation of the quantity

$$\frac{1}{2}v^2 + w + \phi = \text{constant} \qquad (1.26)$$

1.5   The Equations of Hydrodynamics in Conservative Form

As we said above, equation 1.6 directly and clearly expresses the fact that mass can be neither created nor destroyed. Since the same applies also to momentum and energy, it seems reasonable to expect that Euler's equation and that of energy conservation can be rewritten in the same form. In this paragraph, we shall carry out this task; the reason for this apparently academic exercise is that this new formulation allows us to handle both conditions at shock, and the generalization to relativistic hydrodynamics, in a very simple way. Let us start from the equation of energy conservation. A fluid has internal (or thermal) energy, described by a density per unit volume $\rho\epsilon$, and kinetic (or bulk) energy $\rho v^2/2$, which give the total energy density of the fluid, η:

$$\eta = \rho \left(\frac{v^2}{2} + \epsilon \right) \tag{1.27}$$

It seems reasonable to associate to this energy density an energy flux $\vec{j}_{\mathrm{E}} = \rho(v^2/2+\epsilon)\vec{v}$, and to connect the two elements through an equation just like equation 1.6, namely,

$$\frac{\partial \eta}{\partial t} + \nabla \cdot \vec{j}_{\mathrm{E}} = 0 \tag{1.28}$$

but this is wrong: the expression for j_{E} is incomplete. Indeed, it is well known from elementary courses on thermodynamics that the internal energy of a gas can either increase or decrease by a compression or an expansion, respectively, so that under adiabatic conditions we have

$$dE = -pdV \tag{1.29}$$

This compression heating must be included in the law of energy conservation of a fluid in motion. Of course, the heating rate per unit time is given by

$$-\frac{d}{dt} \int pdV = - \int p\frac{dV}{dt} = - \oint p\vec{v} \cdot d\vec{A} \tag{1.30}$$

We have therefore

$$\frac{d}{dt} \int \rho \left(\frac{v^2}{2} + \epsilon \right) dV = - \oint (\vec{j}_{\mathrm{E}} + p\vec{v}) \cdot d\vec{A}$$

$$= - \oint \rho \left(\frac{v^2}{2} + w \right) \vec{v} \cdot d\vec{A} \tag{1.31}$$

where, once again, $w = \epsilon + p/\rho$ is the specific enthalpy. Thus
we see that the true energy flux is given by

$$\vec{j}_{\mathrm{E}} = \rho \left(\frac{v^2}{2} + w \right) \vec{v} \qquad (1.32)$$

and that with this definition of energy flux, the equation of
energy conservation is actually given by equation 1.28. It is,
however, possible to derive equation 1.28 formally, directly
from equations 1.6, 1.10, and 1.13 (see problem 1).

We can now consider Euler's equation, which we want to
rewrite in a form similar to equation 1.6, apart from the fact
that mass density must be replaced by momentum density,
and mass flux by momentum flux. Of course, the momentum
density is $\rho \vec{v}$. We use

$$\frac{\partial}{\partial t} \rho \vec{v} = \frac{\partial \rho}{\partial t} \vec{v} + \rho \frac{\partial \vec{v}}{\partial t} \qquad (1.33)$$

and rewrite each of the two time derivatives on the right-
hand side using equations 1.6 and 1.10, thus obtaining the
following equation for the ith component of the vector $\rho \vec{v}$:

$$\frac{\partial}{\partial t} (\rho v_i) = -\frac{\partial p}{\partial x_i} - \frac{\partial}{\partial x_k} (\rho v_i v_k) \qquad (1.34)$$

This immediately suggests the following definition of a tensor
R_{ik}:

$$R_{ik} \equiv p \delta_{ik} + \rho v_i v_k \qquad (1.35)$$

which is called *Reynolds's stress tensor*. Thanks to this,
Euler's equation can be rewritten as

$$\frac{\partial}{\partial t} (\rho v_i) = -\frac{\partial}{\partial x_k} R_{ik} = -\nabla \cdot \hat{R} \qquad (1.36)$$

where the symbol over R reminds us that R is a tensor, not a
vector. Since it is clearly symmetric, it does not matter over
which index we sum.

The physical meaning of R emerges when we integrate equation 1.36 on a finite volume. We thus obtain

$$\int \frac{\partial}{\partial t}(\rho v_i)dV = \frac{d}{dt}\int \rho v_i dV = -\int \frac{\partial R_{ik}}{\partial x_k}dV = -\oint R_{ik}dA_k$$

(1.37)

The latter expression is the flux of R_{ik} through the surface, whereas the left-hand side is the variation of the ith component of the momentum contained in the volume. It follows that R_{ik} is the flux through a surface with its normal in the k direction, of the ith component of the momentum.

The existence of this quantity seems quite natural when one considers an element of fluid passing through a surface with a speed that is not parallel to the surface normal. In this case, there will be a variation of total momentum on both sides of the surface (but with different signs!) for both the parallel and the perpendicular component. We have seen that both the energy and the mass flux are vectors, carrying energy and mass densities, which—on the other hand—are scalar values. In the same way, since momentum density is a vector, it must be carried by R, a tensor.

We should notice from equation 1.35 that in the direction of $\vec{v}$, $R = p + \rho v^2$, whereas in the direction perpendicular to $\vec{v}$, $R = p$.

1.6 Viscous Fluids

We have so far considered only ideal fluids, where all internal dissipative processes—such as viscosity, heat conduction, and so on—have been neglected. In general this treatment is adequate, except for one important exception: when we take into consideration disk accretion flows, viscosity plays an absolutely fundamental role. Thus I now introduce the equations of motion for viscous fluids.

Viscosity can be found in the presence of speed gradients: particles starting from a point $\vec{x}$, with an average speed

$\vec{v}$, spread to nearby areas with average speeds that differ from those of the local particles. As a consequence, the inevitable collisions among particles tend to smooth out speed differences. From a macrophysical point of view, we call viscosity the smoothing out of speed gradients, which is due, from a microphysical point of view, to collisions among particles having different average speeds. Viscosity appears when $\partial v_i/\partial x_j \neq 0$.

We can therefore imagine that this phenomenon depends on speed gradients, so that we can do an expansion in the speed gradient, keeping only linear terms in the speed gradient because we assume them to be small: if this approximation were to fail, it would not even be clear whether local thermal equilibrium could exist at all. Since the speed gradient $\partial v_i/\partial x_j$ is not a vector, we may hope to change the equations of motion for the ideal fluid to the form

$$\frac{\partial}{\partial t}(\rho v_i) + \frac{\partial}{\partial x_l}(R_{il} + V_{il}) = 0 \qquad (1.38)$$

where V_{il} is the *viscous stress tensor*. It must be built from terms linear in $\partial v_i/\partial x_j$, which do not constitute a true tensor, just a dyad. There are three distinct tensors that can be built from linear combinations of $\partial v_i/\partial x_j$:

$$A = \frac{1}{3}\frac{\partial v_j}{\partial x_j}\delta_{il} \qquad (1.39)$$

$$B = \frac{1}{2}\left(\frac{\partial v_i}{\partial x_l} - \frac{\partial v_l}{\partial x_i}\right) \qquad (1.40)$$

$$C = \frac{1}{2}\left(\frac{\partial v_i}{\partial x_l} + \frac{\partial v_l}{\partial x_i} - \frac{2}{3}\frac{\partial v_j}{\partial x_j}\delta_{il}\right) \qquad (1.41)$$

The first tensor describes a pure compression, as we will show in just a second, the second one is purely antisymmetric, while the third one is traceless and symmetric.[2] Since in equation

---

[2] In the theory of groups, the term $\partial v_i/\partial x_l$ does not belong to an irreducible representation of the group of rotations, in contrast to the three terms in which it has been decomposed.

1.38 R_{il} is a tensor, so must be V_{il}; it must therefore be built by linear combination of A, B, and C, which are pure tensors, perhaps with distinct coefficients.

Before writing V_{il}, it helps to notice that the second term, B, that is, the antisymmetric one, cannot contribute to viscosity. The easiest way to show this is to consider the rotation of a rigid body, where $\vec{v} = \vec{\omega} \wedge \vec{r}$. In this case, there is no relative speed between two nearby points: their separation never changes, and therefore there can be no viscosity. It is easy to check that A and C vanish as they must, whereas the antisymmetric term is $\epsilon_{ilj}\omega_j \neq 0$, where ϵ_{ijl} is the usual completely antisymmetric tensor. It follows that the antisymmetric term cannot appear in the viscous stress tensor, because it would produce a viscous force even in a situation where no force of this kind can exist.

The most general viscous tensor is therefore given by

$$V_{il} = \rho\eta\nabla \cdot \vec{v}\, \delta_{il} + \nu\rho \left(\frac{\partial v_i}{\partial x_l} + \frac{\partial v_l}{\partial x_i} - \frac{2}{3}\frac{\partial v_j}{\partial x_j}\delta_{il} \right) \qquad (1.42)$$

where the coefficients η and ν are called, respectively, *kinematic coefficient of bulk viscosity* and *kinematic coefficient of shear viscosity*.[3] We shall show shortly that $\nu > 0$, and we state without proof (which would however be identical) that $\eta > 0$ also holds.

This is the origin of the names *bulk* and *shear*. We can easily see that the first term does not vanish when the fluid changes volume, whereas the second term vanishes because its trace is zero. In order to do this, let us consider a segment ending in x and $x+dx$; after an infinitesimal time dt, the ends of the segment will be in $x + v_x(x)dt$ and $x + dx + v_x(x + dx)dt \approx x + dx + [v_x(x) + dv_x/dx\ dx\ dt]$. This means that the length of the segment, originally dx, has become $dx(1 + dv_x/dx\ dt)$. Therefore, an infinitesimal volume becomes

$$\delta V \rightarrow \delta V (1 + \nabla \cdot \vec{v}dt) \qquad (1.43)$$

---

[3] Sometimes, $\eta' \equiv \rho\eta$ and $\nu' \equiv \rho\nu$ are called dynamic coefficients.

which shows that the fractional variation of the volume per time unit is given by

$$\frac{1}{\delta V}\frac{d\delta V}{dt} = \nabla \cdot \vec{v} \qquad (1.44)$$

Therefore, there is a volume change in the fluid only if $\nabla \cdot \vec{v} \neq 0$. Thus the first term in equation 1.42 can be present only if the fluid varies in volume, while the second term describes transformations of the fluid that leave its volume unchanged.

In many astrophysical situations, and certainly for all those considered in this book, we have $\eta \ll \nu$. Besides, the only case where we apply this equation is a rotating disk. We can easily see that in the case of pure rotation, there is no change in volume, so that $\nabla \cdot \vec{v} = 0$, and therefore bulk viscosity vanishes in any case.

We can therefore safely say that

$$V_{il} = \nu\rho\left(\frac{\partial v_i}{\partial x_l} + \frac{\partial v_l}{\partial x_i} - \frac{2}{3}\frac{\partial v_j}{\partial x_j}\delta_{il}\right) \qquad (1.45)$$

When this viscous stress tensor is inserted into equation 1.38, we have the equation of motion that replaces Euler's for the ideal fluid.

We must also change the energy conservation equation. It should not come as a surprise that viscosity heats a fluid— just think of what happens when we vigorously rub our hands. Therefore, viscosity increases the fluid entropy. We consider the derivative

$$\frac{D}{Dt}\left(\frac{v^2}{2} + \epsilon\right) \qquad (1.46)$$

where ϵ is the internal energy per unit mass. From the first principle of thermodynamics, we know that

$$d\epsilon = Tds - \frac{p}{M}dV = Tds - \frac{p}{\rho}\frac{dV}{V} = Tds - \frac{p}{\rho}\nabla \cdot \vec{v}dt \quad (1.47)$$

Using this equation and equation 1.38, we find that

$$\frac{D}{Dt}\left(\frac{v^2}{2}+\epsilon\right) = -\frac{\vec{v}}{\rho}\cdot\nabla p + \frac{\vec{v}}{\rho}\cdot(\nabla\cdot V) - \frac{p}{\rho}\nabla\cdot\vec{v} + T\frac{ds}{dt} = \tag{1.48}$$

$$-\frac{1}{\rho}\nabla\cdot(p\vec{v} - V\cdot\vec{v}) + T\frac{ds}{dt} - \frac{1}{\rho}V_{ij}v_{i,j} \tag{1.49}$$

Here I defined $\nabla\cdot V \equiv \partial V_{ij}/\partial x_j$ and $v_{i,j} \equiv \partial v_i/\partial x_j$. In order to see the meaning of the first term on the left-hand side, let us integrate on a given mass of fluid. We find

$$\int \frac{1}{\rho}\nabla\cdot(p\vec{v} - V\cdot\vec{v})\rho dV = \int(p\vec{v} - V\cdot\vec{v})\cdot d\vec{A} = \int \vec{v}\cdot\vec{f}dA \tag{1.50}$$

where the vector $\vec{f} = p\hat{n} - V\cdot\hat{n}$ is the external force exerted on the surface, which has as its normal $\hat{n}$. As a consequence, this integral is the work done by the forces (i.e., pressure and viscosity) on the mass element. On the other hand, we know that the quantity $v^2/2 + \epsilon$ can change only because of the forces' work, so that the last two terms of equation 1.48 must cancel each other:

$$\rho T\frac{ds}{dt} = v_{i,j}V_{ij} = \frac{1}{2}(v_{i,j}+v_{j,i})V_{ij} = \frac{1}{2}\left(\frac{1}{\nu\rho}V_{ij} + \frac{2}{3}\delta_{ij}v_{k,k}\right)V_{ij} \tag{1.51}$$

Here, the first equality is due to the fact that V is symmetric. Finally, remembering that V also has a null trace, we find the energy equation we were looking for:

$$\rho T\frac{ds}{dt} = \frac{1}{2\nu\rho}V_{ij}V_{ij} \tag{1.52}$$

which is obviously positive, provided $\nu > 0$.

1.7   Small Perturbations

Hydrodynamical equations possess a huge number of solutions, as well as techniques to obtain exact—or approximate

—solutions. Obviously we cannot illustrate the vast amount of knowledge that has accumulated in this field. We refer interested readers to the beautiful book by Landau and Lifshitz (*Fluid Mechanics* [1987b]). Much more modestly, we shall try here to describe four classes of solutions that have an immediate astrophysical relevance. First, we shall re-derive the properties of sound waves.

Let us consider,—for convenience,—a stable fluid at rest. In this case, $\vec{v} = 0$, $\rho = \rho_0$, and $p = p_0$. All the first derivatives of these quantities, with respect to time and space, vanish, so that the equations of hydrodynamics are trivially satisfied. Let us now consider small deviations from this solution, let us assume, namely, $\vec{v} = \delta\vec{v}, \rho = \rho_0 + \delta\rho, p = p_0 + \delta p$, where deviations from the zeroth-order solution are not assumed constant in space or in time. If we introduce these values in equations 1.6 and 1.10, we obtain the following equations for perturbed quantities (while the quantities in the zeroth-order solution are constant in space and time):

$$\frac{\partial}{\partial t}\delta\rho + \rho_0 \nabla \cdot \delta\vec{v} = -\nabla \cdot (\delta\rho\delta\vec{v})$$

$$\frac{\partial}{\partial t}\delta\vec{v} = -\delta\vec{v} \cdot \nabla\delta\vec{v} - \frac{1}{\rho_0 + \delta\rho}\nabla\delta p \qquad (1.53)$$

As we can see, there are three non linear terms in the unknown quantities $(\delta\rho, \delta p, \delta\vec{v})$; they make the solution of the problem very difficult. That is why we consider only infinitesimal perturbations, neglecting quadratic or higher-order terms in the unknown quantities. We thus obtain the following system:

$$\frac{\partial}{\partial t}\delta\rho + \rho_0 \nabla \cdot \delta\vec{v} = 0$$

$$\frac{\partial}{\partial t}\delta\vec{v} = -- -\frac{1}{\rho_0}\nabla\delta p \qquad (1.54)$$

In order to proceed further, we must specify the thermodynamical properties of our perturbations, since, as a general rule, the equation of state (which we can still write in the form $p = p(\rho, s)$) specifies the variation of pressure δp as a function of *two* thermodynamical quantities, one of which, δs, is not contained in the system above. It follows that we have to add a further equation in δs. Let us therefore suppose that small perturbations are adiabatic, $\delta s = 0$ (which automatically satisfies the energy conservation equation, 1.13). We shall soon discuss the validity of this assumption. In the meantime, since we suppose $s = $ constant, from the equation of state we find

$$\delta p = \left(\frac{\partial p}{\partial \rho}\right)_s \delta\rho \equiv c_s^2 \delta\rho \qquad (1.55)$$

The quantity c_s is called sound speed; we shall soon explain why. We remind the reader that for an ideal fluid

$$c_s^2 \equiv \left(\frac{\partial p}{\partial \rho}\right)_s = \frac{\gamma p}{\rho} \qquad (1.56)$$

where γ is the ratio between specific heats, and $\gamma = 5/3$ in many (though not all) cases of astrophysical relevance. We shall call polytropic each equation of state where γ is a constant.

We now introduce this relation between δp and $\delta\rho$ in the system above, then we take the derivative $\partial/\partial t$ of the first, take the divergence ($\nabla\cdot$) of the second, and eliminate the term $\partial/\partial t\nabla \cdot \delta\vec{v}$ between the two. We obtain

$$\frac{\partial^2}{\partial t^2}\delta\rho - c_s^2\nabla^2\delta\rho = 0 \qquad (1.57)$$

where $\nabla^2 = \partial^2/\partial x^2 + \partial^2/\partial y^2 + \partial/\partial z^2$ is the Laplacian. This equation is nothing but the well-known wave equation, with a speed of propagation c_s, whence the name sound speed. The generic solution of this equation is

$$\frac{\delta\rho}{\rho_0} = A\,e^{\imath(\vec{k}\cdot\vec{x}\pm\omega t)} \qquad (1.58)$$

subject to the restriction $\omega^2 = k^2 c_s^2$, and with A, the arbitrary wave amplitude, which is subject to the linear condition

$A \ll 1$, under which the equation was derived. All the solutions of this equation, once the boundary conditions are
fixed, can be expressed as linear superpositions of these fundamental solutions.

1.8   Discontinuity

One of the peculiarities of hydrodynamics is that it allows
discontinuous solutions, namely, on certain special surfaces—called surfaces of discontinuity—*all* physical quantities are discontinuous. From the mathematical viewpoint,
these solutions are just step functions: the left limit of quantity X is different from the right limit. On the other hand,
from the physical viewpoint, this discontinuity is not infinitely steep, as in the mathematical limit, but is thin
compared with all other physical dimensions; consequently,
it is reasonable to make the mathematical approximation of
infinitely steep discontinuities, apart from a very small number of problems.

We shall soon see that there are two different kinds of discontinuities. The first one is called tangential discontinuity,
which is present when two separate fluids lie one beside the
other, and the surface between them is not crossed by a flux
of matter. This kind of discontinuity is almost uninteresting:
it is unstable (in other words, any small perturbation of the
surface of separation leads to a complete mixture of the two
different fluids); as a consequence, it is short-lived. On the
other hand, the second kind of discontinuity (which is called
a *shock wave*, or *shock*) is a surface of separation between
two fluids, but there is a flux of mass, momentum, and energy through the surface. Shock waves have an extraordinary
importance in high-energy astrophysics because of their ubiquity and because matter, immediately after the shock wave
has passed, emits much more than before and can therefore
be detected by our instruments much more easily than matter
into which the shock wave has not passed yet.

Even if it may seem, at first, that discontinuous solutions may only take place for exceptional boundary conditions, the very opposite is true: shock waves are naturally produced within a wide range of phenomena. They are practically inevitable when the perturbations to which a hydrodynamic system is exposed are not infinitesimal. In other words, when the perturbations to which a system is exposed are small, sound waves are generated. If, on the other hand, perturbations are finite (i.e., not infinitesimal), shocks form. This is why we now proceed as follows. For the moment, let us assume that there is a mathematical discontinuity, and we shall see which properties these discontinuities must possess in order to be compatible with the equations of hydrodynamics. Later, we shall see how they are generated, in a specific model.

1.8.1 Surfaces of Discontinuity

For the sake of convenience, we shall place ourselves in a reference frame moving with the surface of discontinuity; we shall see later on that these surfaces cannot remain still, but there will always be a reference system moving with the surface, at least instantaneously. Let us also consider, for the sake of simplicity, a situation of plane symmetry (all quantities only depend on one coordinate, x, which is perpendicular to the discontinuity surface) and stationary, so that all quantities do not explicitly depend on time. Hydrodynamic equations in the form of equations 1.6, 1.28, and 1.36, can also be written as follows:

$$\frac{dJ}{dx} = 0 \tag{1.59}$$

where J is any kind of flux (mass, energy, or momentum). Let us now consider an interval in x with an infinitesimal length 2η, around the discontinuity surface, and integrate the above equation over this interval:

$$\int_{-\eta}^{+\eta} \frac{dJ}{dx} dx = J_2 - J_1 = 0 \tag{1.60}$$

where J_2 and J_1 are the expressions of fluxes, in terms of variables immediately after and before, respectively, the discontinuity surface. It is convenient to use the notation $[X] \equiv X_2 - X_1$.

As a consequence, hydrodynamic equations require continuous fluxes. In other words, physical quantities can be discontinuous, provided that fluxes are continuous: mass, energy, and momentum cannot be created inside the surface of discontinuity.

More explicitly, from the continuity of the mass flux we have

$$[\rho v_x] = 0 \tag{1.61}$$

whereas from the continuity of the energy flux we get

$$\left[\rho v_x \left(w + \frac{1}{2} v^2 \right)\right] = 0 \tag{1.62}$$

and from the continuity of the momentum flux

$$[p + \rho v_x^2] = 0 \tag{1.63}$$

for the component of the momentum perpendicular to the surface, and

$$[\rho v_x v_y] = 0 \quad [\rho v_x v_z] = 0 \tag{1.64}$$

for the two components of the momentum flux parallel to the surface.

From these equations we can see that there are two kinds of discontinuity. In the first one, mass does not cross the surface: the mass flux vanishes. This requires $\rho_1 v_1 = \rho_2 v_2 = 0$. Since neither ρ_1 nor ρ_2 can vanish, it follows that $v_1 = v_2 = 0$. In this case, the conditions of equations 1.62 and 1.64 are automatically satisfied, and the condition of equation 1.63 requires $[p] = 0$. Therefore,

$$v_{1x} = v_{2x} = 0 \quad [p] = 0 \tag{1.65}$$

whereas all the other quantities, v_y, v_z, ρ and any thermodynamic quantity apart from p, can be discontinuous. This kind of discontinuity is called tangential. A contact discontinuity is a subclass of the tangential discontinuities, where velocities are continuous but density is not. It is possible to show (Landau and Lifshitz 1987b) that this kind of discontinuity is absolutely unstable for every equation of state; if we perturb the surface separating the two fluids by an infinitesimal amount, and this perturbation has a sinusoidal behavior with a wavelength λ, the perturbation is unstable for any wavelength. The term *absolutely*, in this context, is the opposite of *convectively*: in the absolutely unstable perturbations, the amplitude of small perturbations, at a fixed point grows, whereas in the convectively unstable ones, small perturbations, growing in amplitude, are carried away by the flow of matter. In this latter case, at a given point, there may well be a transient of growing amplitude, but this is carried away by the motion of the fluid, so that the situation returns to its unperturbed state, at any fixed point. The result of this instability is a wide zone of transition, where the two fluids are mixed by turbulence. We shall discuss this type of instability in chapter 2 (sect. 2.5), so as to discuss, as well, the effect of the magnetic field. For sufficiently large magnetic fields, we can demonstrate that these discontinuities can be stabilized, but, in astrophysics, it is extremely difficult to come across magnetic fields this strong, so that we can reasonably say that contact and tangential discontinuities are always unstable in the actual situations we shall find.

In the second kind of discontinuity, namely shock waves, the mass flux does not vanish, therefore v_{1x} and v_{2x} cannot be null. From equation 1.64, it follows that tangential velocities must be continuous. Using the continuity of ρv_x and of tangential velocities, we can rewrite equations 1.61, 1.62, and 1.63 as

$$[\rho v_x] = 0 \tag{1.66}$$

$$\left[\frac{1}{2}v_x^2 + w\right] = 0 \tag{1.67}$$

$$[p + \rho v_x^2] = 0 \tag{1.68}$$

These conditions of continuity constitute what hydrodynamics imposes on shock waves; they are called Rankine-Hugoniot (RH) conditions.

1.8.2 Shock Waves

Since tangential velocity is continuous through a shock, it is a constant in the entire space. We can therefore choose a reference frame where it vanishes, and so we find ourselves in a reference system where the shock is at rest. Ahead of it (in the so-called *upstream* section) is a fluid, of density ρ_1 and pressure p_1, moving toward the shock with velocity v_1, directed along the shock normal. Alternatively, we may see the shock moving with speed V_s through a fluid that is not moving. Obviously, $V_s = v_1$, but the analysis of the RH conditions is simpler in the reference frame in which the shock is at rest.

The RH conditions allow us to determine the hydrodynamic and thermodynamic properties of the fluid after the shock, when we know the conditions ahead of the shock. This requires knowledge of the equation of state for the fluids in question, because the RH conditions contain p, ρ, and w. Here we shall discuss only the case of an ideal fluid, which is sufficient for astrophysical aims. An ideal fluid, with polytropic index γ, has $w = \gamma p/[(\gamma - 1)\rho] = c_s^2/(\gamma - 1)$. We now define the Mach number of the shock as the ratio $M_1 \equiv v_1/c_{s1}$ between the velocity at which the shock propagates in the unperturbed medium, and the sound speed in the same medium. We shall therefore find

$$\frac{\rho_2}{\rho_1} = \frac{v_1}{v_2} = \frac{(\gamma + 1)M_1^2}{(\gamma - 1)M_1^2 + 2}$$

$$\frac{p_2}{p_1} = \frac{2\gamma M_1^2}{\gamma + 1} - \frac{\gamma - 1}{\gamma + 1}$$

$$\frac{T_2}{T_1} = \frac{\left[2\gamma M_1^2 - (\gamma - 1)\right]\left[(\gamma - 1)M_1^2 + 2\right]}{(\gamma + 1)^2 M_1^2} \tag{1.69}$$

It is also useful to have the Mach number for the matter after the shock:

$$M_2^2 = \frac{2 + (\gamma - 1)M_1^2}{2\gamma M_1^2 - \gamma + 1} \tag{1.70}$$

These relations have very simple limits when $M_1 \to +\infty$, which is the so-called strong shock limit:

$$\frac{\rho_2}{\rho_1} = \frac{v_1}{v_2} = \frac{\gamma + 1}{\gamma - 1}$$

$$\frac{p_2}{p_1} = \frac{2\gamma M_1^2}{\gamma + 1}; \quad p_2 = \frac{2\rho_1 v_1^2}{\gamma + 1}$$

$$\frac{T_2}{T_1} = \frac{\gamma - 1}{\gamma + 1}\frac{p_2}{p_1} = \frac{2\gamma(\gamma - 1)}{(\gamma + 1)^2}M_1^2; \quad T_2 = 2\frac{\gamma - 1}{(\gamma + 1)^2}mv_1^2$$

$$M_2^2 = \frac{\gamma - 1}{2\gamma} \tag{1.71}$$

where m is the average mass of particles crossing the shock.

We can examine the stability of shocks with respect to the same kind of perturbations that make tangential discontinuities unstable, namely, ripples on the shock surface. This kind of instability, if present, is called corrugational. Such an analysis (D'yakov 1954) shows that plane shocks are unstable only for certain special kinds of equations of state, which, however, do not seem to be realized in nature. As a consequence plane shocks are essentially stable.

1.8.3   Physical Interpretation of Shock Waves

From the equations above, we can see that strong shocks compress moderately the undisturbed gas (a maximum factor of 4 for $\gamma = 5/3$), but they can heat it to a high temperature, since $T_2/T_1 \propto M_1^2$. Obviously, the source of energy for this heating must be the bulk kinetic energy of the incoming fluid. In fact, the speed of the fluid after the shock is smaller than the one before the shock, by the same factor by which density increases. Therefore, the shock transforms bulk

kinetic energy into internal energy. However, something else
is happening: we must take entropy into account.

From equation 1.69, we see that when $M_1 = 1$, $\rho_1 = \rho_2$,
$v_1 = v_2$, and $p_1 = p_2$; in this case, matter undergoes no
transformation. However, let us consider what happens if
$M_1 < 1$. In this case, we would get from the above equations $v_1 < v_2$ and $T_1 > T_2$; in other words, we would have
transformed a part of the fluid's internal energy into kinetic
energy, with no other effect. Now, as we know from elementary courses, one of the formulations of the second principle of thermodynamics is that it is impossible to realize a
transformation whose only consequence is the total transformation of heat into work. Well, if it were possible to have
$M_1 < 1$, we would have done exactly this: a transformation
whose only consequence is the total transformation of heat
($T_1 > T_2$) into work ($v_2 > v_1$, and this larger kinetic energy
could be immediately transformed into work). It follows that
the second principle of thermodynamics prevents the existence of subsonic shocks: shock waves are always supersonic.[4]

Entropy, just like any thermodynamic quantity, is discontinuous across shocks, $s_2 \neq s_1$. However, thanks to the second
principle of thermodynamics, we know that we cannot realize $s_1 > s_2$; it follows that, at the shock, we must have a
production of entropy:

$$s_2 > s_1 \tag{1.72}$$

For ideal fluids, we can easily demonstrate that this implies
$M_1 > 1$ (see problem 2). In fact, shock waves constitute the
most effective form of dissipation in an ideal fluid. This jump
of entropy at the shock also explains why we have used, for
the equation of energy conservation, equation 1.28, rather
than equation 1.13: entropy is conserved before and after the
shock, not at the shock.

Since $M_1 > 1$, as we have just shown, it is also easy to
see that we always have $\rho_2 > \rho_1$, $v_1 > v_2$, $p_2 > p_1$, $T_2 > T_1$,

---

[4]Such considerations can be stated in general, not just for polytropic
fluids, see for example Landau and Lifshitz (1987b).

and $M_2 < 1$: the shock appears supersonic with respect to the unshocked fluid, but it appears subsonic with respect to the shocked fluid.

1.8.4 Collisional and Noncollisional Shocks

We have just seen that shock waves transform part of the kinetic energy of the incoming fluid into (internal) thermal energy of the outgoing fluid, with creation of entropy. The generation of entropy is due to collisions between atoms or molecules of the fluid in question. It follows that the thickness of the shock wave must be large enough to allow each atom, or molecule, to undergo collisions with the other particles of the fluids, so as to transform at least a part of its ordered kinetic energy into internal kinetic (disordered) energy. Therefore, the thickness of the shock will be given, in order of magnitude, by the mean free path λ of a particle, since the speed of atoms (or molecules) is changed by 90° on a scale length of that order of magnitude. Using the well-known relation

$$\lambda n \sigma = 1 \tag{1.73}$$

where n is the space density of targets, and σ the cross section of the process that scatters particles, we can see that in the atmosphere, where $n \approx 10^{20}$ cm^{-3} and $\sigma \approx \pi a_B^2$, with $a_B \approx 10^{-8}$ cm Bohr's radius of an atom, $\lambda \approx 10^{-4}$ cm. Obviously, this value is very small, and therefore the approximation of an infinitely thin shock surface is certainly good. This kind of shock, where direct collisions between atoms transform bulk kinetic energy into disordered kinetic energy, is called collisional. All the shocks we study in laboratories or in the atmosphere are collisional shocks.

When we try to apply the above-mentioned relation to astrophysical situations, we come across a paradox: *in primis*, most atoms are ionized, so that they are no longer the size of Bohr's atom, but have nuclear dimensions, which are much smaller ($\approx 10^{-13}$ cm). But, even more important, the typical densities are much smaller, typically $n \approx 1$ cm^{-3} instead of $n \approx 10^{20}$ cm^{-3}. This difference alone would make the shock's thickness macroscopic ($\approx 10^{16}$ cm), and nonnegligible. However, astrophysical matter is mostly ionized, and electrons and atomic nuclei are subject to accelerations and deflections due to electric and magnetic fields. We assume, therefore, that a mixture of disordered—and partly transient—electric and magnetic fields is responsible for making the nuclei's momenta isotropic. In this case, the shock's thickness must be comparable with the Larmor radius for a proton, because the proton, which carries most of the energy and momentum of incoming matter, is deflected by the typical magnetic field over a distance of this order of magnitude. We have, therefore,

$$\lambda \approx r_{\rm L} = \frac{mvc}{eB} \approx 10^{10} \; {\rm cm} \frac{v}{10^4 \; {\rm km \; s}^{-1}} \frac{10^{-6} \; G}{B} \qquad (1.74)$$

where we used standard values for the speed of a proton emitted by a supernova, and for a typical galactic magnetic field. The consequence is that this thickness is sufficient to make the idealization of an infinitely thin discontinuity acceptable. The shocks where the agents responsible for the isotropization of the bulk kinetic energy are electromagnetic fields are called noncollisional. The best known example is the shock between the solar wind and the Earth's magnetosphere, at about 10^5 km from the Earth.

We must also add that in noncollisional shocks, it is absolutely not obvious whether the temperature of electrons after the shock will be the same as that of the ions; in fact, as a general rule, we shall have $T_e < T_i$, and, sometimes, even $T_e \ll T_i$. The reason for this will be explained in chapter 2 when we discuss hydromagnetic shocks.

1.8.5   Formation of a Shock

The formation of a shock wave depends on the nonlinearity of hydrodynamical equations as well as, naturally, on the details of the situation we take into consideration. Of course, these complications preclude an analytic treatment, apart

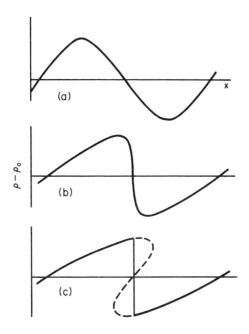

Figure 1.1. Formation of a shock (Landau and Lifshitz 1987b).

from a simple case: very special waves of noninfinitesimal amplitude, called *simple waves*. Landau and Lifshitz (1987b) discuss them, whereas we shall simply sum up their main features. Let us consider a perturbation with a finite (i.e., not infinitesimal) amplitude, within an otherwise homogeneous fluid (see fig. 1.1). It is possible to show for simple waves, too, that the speed of propagation is greater where density is greater. In particular, the points near the crest of the wave will move more quickly than those near its belly, and the difference between the two velocities is not infinitesimal, because the amplitude of the wave has been assumed finite. Therefore, the crest will reach the belly in a finite time, thus forming a surface of discontinuity. The only waves that manage to avoid this destiny are those whose density never decreases in the direction of motion; but, apart from this peculiar case, any perturbation with a finite amplitude evolves toward a discontinuity. Therefore, we do not necessarily need unusual initial conditions to give rise to shock waves: a strongly perturbed medium (i.e., a medium having perturbations of finite amplitude) will do.

1.9   Self-similar Solutions

Among the solutions to hydrodynamic problems, there is a whole class of analytical solutions that though at times idealized, can be extremely useful as the first approximation of more complex problems. These are called self-similar solutions.

The starting point is that hydrodynamics does not contain dimensional constants, unlike other physical theories such as gravity, which necessarily contains G, special relativity, which, in its turn, contains c, electromagnetism, which contains c and e, and quantum mechanics, which contains $\hbar$. It follows that the solution to hydrodynamic problems, which must have dimensions of velocity, density, and temperature, must be built from the dimensional constants contained in the boundary conditions and, obviously, from $\vec{r}$ and t. In most conditions, this is largely sufficient to identify the quantities with the right dimensions to be included in the solution to the problem. Occasionally, however, the problem in question may contain few dimensional constants, so that the form of the solution is strongly constrained, allowing us to obtain a solution with little effort.

For example, let us consider a strong explosion releasing an energy E, within such a small radius, that we shall assume a pointlike release of energy (i.e., everything takes place at $r = 0$), which is also instantaneous (i.e., the explosion takes place at time $t = 0$). The site of the explosion is surrounded by a uniform polytropic medium with density ρ_0, which is also very cold, so that we may assume $T_0 = 0$. A pointlike explosion in a uniform medium obviously has spherical symmetry. The explosion generates a shock wave that propagates in the medium. What is the time dependence of the shock radius R_s?

The answer must have the dimension of length, which we can build only from the dimensional quantities E, ρ_0, and t; indeed, the problem does not contain other dimensional quantities. We can easily see that the only combination with the right dimension is

$$R_s(t) = \beta \left(\frac{Et^2}{\rho_0} \right)^{1/5} \tag{1.75}$$

where β is a dimensionless quantity. Since it is impossible to build a dimensionless quantity from the variables and constants of the problem, β is a pure number, which we shall soon determine. The velocity of the shock is

$$V_s = \frac{dR_s}{dt} = \frac{2R_s}{5t} \tag{1.76}$$

This simple argument shows that it is already possible to determine the time evolution of the shock and its velocity, except for a numerical factor, without solving any equation. Before going on, please note that, assuming $T_0 = 0$, we have automatically assumed that the shock is strong. Therefore, we shall use equation 1.71. Note that in these equation the velocity v_2 is the velocity of postshock matter with respect to the shock, since the equations have been derived in the reference frame of the shock. On the other hand, we need the velocity of matter immediately behind the shock in the reference frame where the shock moves with velocity v_1 within a medium at rest; this is obviously given by v_1 minus the value given in equation 1.71. Thus we have

$$v_2 = \frac{2V_s}{\gamma + 1} \qquad \rho_2 = \frac{\gamma + 1}{\gamma - 1}\rho_0 \qquad p_2 = \frac{2\rho_0 V_s^2}{\gamma + 1} \tag{1.77}$$

We can now push the dimensional argument further and inquire, for example, about the value of ρ behind the shock, at distance r from the center at time t. The answer must have the dimension of density, and the only quantity with this dimension is ρ_0. Therefore, $\rho = \rho_0 R(r,t)$, where R is a dimensionless function of r and t. However, r and t do have dimensions and we must therefore find a dimensionless combination of r and t, as well as constants and boundary conditions (E, ρ_0). Since we have only four quantities at our disposal, with three independent dimensions (mass, length, time), obviously it is possible to build only one dimensionless quantity:

$$\xi \equiv \frac{r}{R_s} = \frac{r}{\beta} \left(\frac{\rho_0}{Et^2} \right)^{1/5} \tag{1.78}$$

Taking into account equation 1.77, it is convenient to use

$$\rho = \frac{\gamma + 1}{\gamma - 1} \rho_0 R(\xi) \tag{1.79}$$

where the initial constant has been chosen to have the boundary condition for the function R,

$$R(1) = 1 \tag{1.80}$$

In the same way, we find for the other quantities

$$v = \frac{2V_s}{\gamma + 1} V(\xi) \qquad p = \frac{2\rho_0 V_s^2}{\gamma + 1} P \tag{1.81}$$

and the boundary conditions for these equations are now

$$V(1) - 1 \quad P(1) - 1 \tag{1.82}$$

Why is what we have just done extremely convenient? Because now all hydrodynamic quantities are no longer a function of two independent variables, but of only one (ξ), so that a partial-derivative system has been reduced to an ordinary-derivative one, which is much easier to solve. It is worthwhile to illustrate this point in detail. We shall use equations 1.6,

1.10, and 1.13, which must be written in spherical symmetry:

$$\frac{\partial \rho}{\partial t} + \frac{\partial(\rho v)}{\partial r} + \frac{2\rho v}{r} = 0$$

$$\frac{\partial v}{\partial t} + v\frac{\partial v}{\partial r} = -\frac{1}{\rho}\frac{\partial p}{\partial r}$$

$$\left(\frac{\partial}{\partial t} + v\frac{\partial}{\partial r}\right)\ln\frac{p}{\rho^\gamma} = 0 \qquad (1.83)$$

With the definitions of equations 1.78, 1.79, and 1.81, the equations above become (using $\dot{X} \equiv dX/d\xi$)

$$-\xi\dot{R} + \frac{2}{\gamma+1}(\dot{R}V + R\dot{V}) + \frac{4}{\gamma+1}\frac{RV}{\xi} = 0$$

$$-2\xi\dot{V} - 3V + \frac{4\dot{V}V}{\gamma+1} = -\frac{\gamma-1}{\gamma+1}\frac{2\dot{P}}{R}$$

$$-3 - \frac{\dot{P}\xi}{P} + \frac{2}{\gamma+1}\frac{V\dot{P}}{P} + \gamma\xi\frac{\dot{R}}{R} - \frac{2\gamma}{\gamma+1}V\frac{\dot{R}}{R} = 0, \qquad (1.84)$$

which is an ordinary-derivative system with a single independent variable ξ.

One of the equations in this system can also be easily integrated (Landau and Lifshitz 1987b), but the analytical result is not particularly enlightening; the numerical solution can be easily obtained through standard methods, and is shown in figure 1.2 for $\gamma = 5/3$.

From this we see that most of the mass is contained in a thin region, $\delta R/R \approx 0.1$, immediately behind the shock, whereas the central region is full of hot, light gas (density tends to zero in the center, whereas pressure remains finite), which obviously contributes its pressure to push forward the spherical crown containing a large portion of the mass.

The reason why these solutions are called self-similar is quite obvious from figure 1.2. If we express all quantities in units of their value immediately after the shock, and the radius in units of the shock radius, the radial profile of these

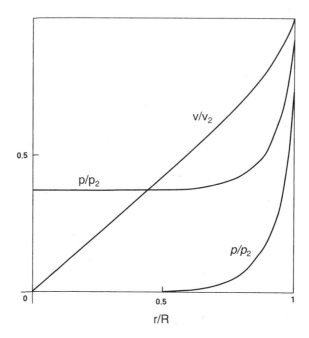

Figure 1.2. Numerical solution for the main hydrodynamic quantities in Sedov's solution for $\gamma = 5/3$.

quantities does not vary in time: the quantities always remain identical to themselves.

We still have to determine the constant β. The energy density is $\rho(v^2/2+\epsilon)$. Since, for an ideal fluid $\epsilon = p/[(\gamma-1)\rho]$, we find

$$E = \int_0^{R_s} \rho \left(\frac{1}{2}v^2 + \frac{p}{(\gamma-1)\rho} \right) 4\pi r^2 dr \qquad (1.85)$$

or, in dimensionless terms,

$$1 = \frac{\beta^5}{\gamma^2 - 1} \frac{16\pi}{25} \int_0^1 \xi^2 d\xi [R(\xi)V^2(\xi) + P(\xi)] \qquad (1.86)$$

The total energy is obviously a constant, because we have neglected ab initio any dissipative process; we find numerically that $\beta \approx (2.02)^{1/5}$ when $\gamma = 5/3$. The solution for a strong explosion within a homogeneous medium is due to Sedov (1946) and Taylor (1950).

The method of the self-similar solution easily lends itself to generalizations. In Sedov's solution, we assume energy conservation, thus neglecting energy losses due to radiation. In astrophysics problems, this is a good approximation only for early conditions. After some time, radiation losses reduce thermal energy, and we can idealize the situation as a gas with a temperature $T = 0$ rapidly expanding in a cold medium. In this case, the total energy is not conserved, but the linear momentum P must be. We can easily see that the time evolution of the shock radius must have the following form:

$$R_{\mathrm{s}} = \beta \left(\frac{Pt}{\rho_0} \right)^{1/4}$$

This solution is called a *snowplough*, since it can be shown that mass accumulates immediately behind the shock, even more than in the Sedov-Taylor case. Two more cases where self-similar solutions provide an immediate answer will be discussed in the problems.

1.9.1 Self-similar Solutions of the Second Kind

In the previous section, we showed how to obtain an exact solution for hydrodynamics equations, when the problem is so idealized that there is a lack of dimensional constants. If this were the entire scope of self-similar solutions, it would be very restricted indeed. But it is possible to derive important self-similar solutions in some cases where the problem is not lacking in dimensional constants. This is the so-called *second kind* of self-similar solution (Barenblatt 1996).

In order to keep the discussion concrete, let us consider a specific example of a self-similar flow of the second kind. Take a one-dimensional, plane problem, where an explosion of energy E occurs at time $t_0 < 0$ at $x = 0$, in a cold ($T = 0$)

medium, with density stratification

$$\rho = \frac{\rho_0}{2} \frac{1}{\cosh(kx_0) + \cosh(kx)} \qquad (1.87)$$

This problem has many dimensional parameters $(E, k, x_0,$ $t_0, \rho_0)$, so that there is no need, at first sight, to invoke a self-similar solution. And, at least immediately after being set off, the explosion will surely *not* be self-similar.

Yet one may wonder what occurs at very late times, when the shock has progressed far enough that the density may be approximated as

$$\rho = \rho_0 \exp(-kx) \qquad (1.88)$$

We may restrict our attention to the positive side, $x > 0$, because the problem is obviously symmetric in $|x|$. In other words, we are considering times much later than t_0 and distances much larger than x_0, so that the solution may be expected to have settled onto an evolutionary path that is independent of t_0 and x_0. We call this an *asymptotic* solution. We may now think (naively!) that there are still three dimensional quantities in the problem, E, k, and ρ_0, which would still spoil a self-similar form for the asymptotic solution, but this is incorrect. In fact, if an asymptotic solution exists at this stage, it will be the one that has *always* propagated down the density gradient given by the previous equation. But this has infinite mass, so that the asymptotic solution must necessarily have infinite energy. So, the explosion energy may not be used in deriving the asymptotic solution.

Let us be clear about this: the physical (as opposed to the asymptotic) solution has finite energy and finite mass, but the asymptotic solution to which it will tend must have infinite energy because it has crossed an infinite amount of mass. At this point, the equation of motion for the position of the shock front, X_s, can be built only by means of the quantities k and t:

$$\frac{dX_s}{dt} = \frac{\alpha}{kt} \qquad (1.89)$$

where α is a pure number that we now have to specify. This equation has the obvious solution

$$kX_s = -\alpha \ln |t| \qquad (1.90)$$

where the constant of integration has been chosen in such a way as to allow the shock to reach $+\infty$ at $t = 0$. It follows that the solution is valid for $t < 0$, and that the asymptotic limit is reached for $t \to 0^-$. In this way, we also obtain that the density of matter immediately *before* the shock is

$$\rho_b = \rho_0 |t|^\alpha \qquad (1.91)$$

We expect once again to be able to write all quantities in suitable dimensional form. For instance, for the density we have

$$\rho = \rho_b R(\xi) \qquad (1.92)$$

where ξ is some adimensional function of position. It is natural to use

$$\xi \equiv k(x - X_s) \qquad (1.93)$$

rather than the expression x/X_s, which is more suitable to a spherically symmetric problem: the quantity $k(x - X_s)$ is translationally invariant, while x/X_s is not. In much the same way, we obtain

$$v = \frac{\alpha}{kt} V(\xi), \quad p = \frac{\alpha^2}{k^2 t^2} \rho_b P(\xi) \qquad (1.94)$$

The adimensional quantities, V, P, and R are, of course, subject to the usual Rankine-Hugoniot jump conditions for strong shocks, which give

$$R(0) = \frac{\gamma+1}{\gamma-1}, \quad V(0) = P(0) = \frac{2}{\gamma+1} \qquad (1.95)$$

The equations for the postshock hydrodynamics, with these substitutions, become

$$\dot{V} + (V-1)\frac{\dot{R}}{R} = -1 \tag{1.96}$$

$$(V-1)\dot{V} + \frac{\dot{P}}{R} = \frac{V}{\alpha} \tag{1.97}$$

$$(V-1)\left(\frac{\dot{P}}{P} - \gamma\frac{\dot{R}}{R}\right) = \frac{2}{\alpha} + \gamma - 1 \tag{1.98}$$

Here, $\dot{Y} \equiv dY/d\xi$. Once again, we have greatly simplified the system of partial differential equations, transforming it into a system of ordinary differential equations.

We must now discuss what fixes α. If we solve the above system for $\dot{V}$, we find

$$\dot{V} = \frac{1 - 2/\alpha + V(V-1)R/(\alpha P)}{(V-1)^2 R/P - \gamma} \tag{1.99}$$

We see that the denominator of the above equation is *not* positive definite. Numerical integration reveals that it *does* change sign as we move from $\xi = 0$ (the shock front) toward $\xi \to -\infty$, while the numerator does not simultaneously vanish. This is physically unacceptable, because there is no obvious reason why the physical quantities in the downstream region should not be finite and well behaved. Thus we are forced to vary α until we obtain the simultaneous vanishing of the numerator and denominator of the previous equations, which keeps V and its derivatives finite. From a numerical computation (Chevalier 1990) it is found that $\alpha = 5.669$ for $\gamma = 4/3$, and $\alpha = 4.892$ for $\gamma = 5/3$. Thanks to these exponents, we see that the shock front, which moves according to $kX_s = -\alpha \ln|t|$, as $t \to 0^-$, accelerates so quickly that it reaches infinity in a finite time, implying that this problem really ought to be dealt with using special relativistic hydrodynamics (see the problems).

The physical meaning of the vanishing of the denominator is easy to find[5]: it corresponds to the point where the outflow velocity v equals the sound speed, which is called the *sonic point*.

What have we done so far? We have the solution to an approximate problem, where equation 1.88 substitutes for equation 1.87 everywhere. What is our hope? That the exact solutions for the distribution of equation 1.87 will approach with time the solution with density given by equation 1.88. So we see that the first difference between solutions of the first and second kind is that, while the former are exact solutions of the given problem, the latter are solutions of an approximate problems and thus may be used only as asymptotic solutions as time passes. The second difference lies in the fact that the critical exponent (2/5 in the Sedov-Taylor solution, α in this problem) is easily derived in the former from dimensional arguments, while in the latter it is used to guarantee some regularity of condition of the global solution.

Lastly, this self-similar solution will be of use to us if we have independent physical arguments that lead us to suspect that many, possibly all, exact solutions will converge to the self-similar one in due time. In the concrete example studied here, such argument is made possible by the existence of the sonic point. In fact, in this case we know that matter that has just crossed the shock is not in causal contact with the regions of material that were shocked a long time ago, exactly because this matter is moving supersonically, with respect to those regions, toward $-\infty$. Thus, the properties of just-shocked matter cannot be influenced by the boundary conditions of the problem, which is why we expect all exact solutions to converge asymptotically to the self-similar solution just found. Thus, the fact that this shock accelerates with time plays a crucial role in making the self-similar solution possible and useful. Furthermore, this also shows that the self-similar solution just found is an asymptotic solution

---

[5] So easy that it is left as an exercise to the reader.

for the propagation of a shock in *any* density distribution that
has as the asymptotic form of equation 1.88, not just equation
1.87. In particular, this solution is used as the asymptotic
form for solutions through exponential atmospheres (where
$\rho = \rho_0$ before for $x \leq x_0$) and for breakout of shocks in
stellar atmospheres.

More discussion and examples of self-similar flows of the
second kind can be found in Barenblatt's book (1996); it
should also be noted that the solution to the strong explosion with relativistic hydrodynamics, to be discussed shortly, is of the second kind, while the Sedov solution for the same
problem, in the Newtonian limit, is of the first kind.

1.10   Relativistic Hydrodynamics

There are classes of astrophysical sources that surely display relativistic motions. The most extreme among these are
gamma-ray bursts (which reach factors of Lorentz $\Gamma \approx 100$),
and BL Lacs, for which $\Gamma \approx 10$. Other classes of relativistic
sources are mini-quasars and SS 433. Other sources may also
give rise to relativistically expanding matter, such as certain special type Ic supernovae (SN1998bw showed spectral
lines indicating a speed $\approx 6 \times 10^4$ km s^{-1} about one month
after the explosion) and soft gamma-ray repeaters. This
shows that we must discuss hydrodynamics in relativistic conditions. It is also worth noting that this generalization is necessary also when *internal* velocities are close to the speed of
light, and that, therefore, the following discussion may prove
important for other classes of sources as well.

Just like any other field theory, relativistic hydrodynamics
will have an energy-momentum tensor T^{ik}, whose form can
be determined from the simple theorem according to which,
if two tensors of the same rank are identical in an inertial
reference system, they are always identical. Let us then consider the tensor T in the reference frame in which the fluid
is instantaneously at rest. We must have $T^{00} = e$, where e is

the density per unit volume of internal energy, whereas the components $T^{0i}/c = T^{i0}/c = 0$, since these are momentum densities, vanishing for a fluid at rest. As for the components $T^{\alpha\beta}$ ($\alpha, \beta = 1 - 3$), we remember that $T^{\alpha\beta}dA_\beta$ gives the momentum flux per unit time through the infinitesimal element of surface directed along dA_β; the momentum flux per unit time is simply the force per surface unit. In a fluid at rest, $T^{\alpha\beta}dA_\beta = pdA^\alpha$, from which it follows immediately that $T^{\alpha\beta} = p\delta^{\alpha\beta}$. Let us also use $\eta^{\alpha\beta} = (1, -1, -1, -1)$, which is Minkowski's metric tensor. Therefore, in the reference frame where the fluid is at rest,

$$T = \begin{pmatrix} e & 0 & 0 & 0 \\ 0 & p & 0 & 0 \\ 0 & 0 & p & 0 \\ 0 & 0 & 0 & p \end{pmatrix} \tag{1.100}$$

When the fluid is no longer at rest, we shall have to use its four-speed $u^j = dx^j/ds = (\gamma, \gamma\vec{v}/c)$, where $\vec{v}$ is the Newtonian velocity, and $\gamma = 1/\sqrt{1 - v^2/c^2}$ is the Lorentz factor. The four-speed of an object at rest is obviously $(1, 0, 0, 0)$. We can easily see now that the tensor

$$T^{ik} = (e + p)u^i u^k - p\eta^{ik} \tag{1.101}$$

reduces to the form 1.100 for a fluid at rest. Therefore, this is the expression of the energy-momentum tensor in relativistic hydrodynamics.

The equations of hydrodynamics, of course, are given by the covariant generalization of the Newtonian conservation equations,

$$\frac{\partial T^{ik}}{\partial x^k} = \frac{\partial T^{i0}}{\partial(ct)} + \frac{\partial T^{i\alpha}}{\partial x^\alpha} = 0 \tag{1.102}$$

From this equation, we can easily see that the energy flux is $cT^{0\alpha}$, while the momentum component flux is $T^{\alpha\beta}$ because the momentum density is $T^{\alpha 0}/c$.

The equation of mass conservation has no meaning in special relativity and is certainly not contained in equation 1.102.

However,the baryon and lepton numbers of particles are conserved. Obviously, once we have defined as n the number of particles per unit volume in the reference system where the fluid is at rest, nu^i is a flux of particles, whose conservation is expressed by

$$\frac{\partial(nu^i)}{\partial x^i} = 0 \qquad (1.103)$$

Together, equations 1.102 and 1.103 imply an adiabatic flow, just like in Newtonian hydrodynamics. In order to see it, let us calculate

$$u^i \frac{\partial T^k_{i}}{\partial x^k} = u^i u_i \frac{\partial((e+p)u^k)}{\partial x^k} + (e+p)u^k u^i \frac{\partial u_i}{\partial x^k} - u^i \frac{\partial p}{\partial x^i} = 0 \qquad (1.104)$$

where we have used the equations of motion (eq. 1.102) and the definition of T (eq. 1.101). If we now recall that $u^i u_i = +1$, and therefore $u^i \partial u_i / \partial x^k = 0$, we shall find

$$\frac{\partial[(e+p)u^k]}{\partial x^k} - u^k \frac{\partial p}{\partial x^k} = 0 \qquad (1.105)$$

Now we can also use

$$\frac{\partial(e+p)u^k}{\partial x^k} = \frac{\partial(nu^k)[(e+p)/n]}{\partial x^k}$$

$$= \frac{\partial(nu^k)}{\partial x^k} \frac{e+p}{n} + nu^k \frac{\partial[(e+p)/n]}{\partial x^k}$$

$$= nu^k \frac{\partial[(e+p)/n]}{\partial x^k} \qquad (1.106)$$

where we have used equation 1.103 in the preceding equation to obtain

$$nu^k \left(\frac{\partial[(e+p)/n]}{\partial x^k} - \frac{1}{n}\frac{\partial p}{\partial x^k} \right) = 0 \qquad (1.107)$$

The second law of thermodynamics for reversible transformations tells us that $d(e+p)/n = Tds + dp/n$, where s is the entropy per particle—or specific entropy, if you like. We now rewrite this equation as

$$u^k \frac{\partial s}{\partial x^k} = 0 \qquad (1.108)$$

The operator $u^k(\partial/\partial x^k)$ is obviously the relativistic generalization of the convective derivative. This equation tells us that specific entropy does not change, for a given mass element, just like the vanishing of the convective derivative of the specific entropy in Newtonian hydrodynamics (eq. 1.13) tells us that the entropy of a given mass element remains constant.

1.10.1 Shock Waves in Relativistic Hydrodynamics

Just like in Newtonian hydrodynamics, relativistic shock waves are possible, provided that the conservation of the fluxes of the particles' number (which takes the place of the mass flux), of momentum, and of energy hold. Once again, we can choose the reference frame in which the shock is instantaneously at rest, and matter arrives with speed directed along the x axis, which is also the normal to the shock surface. Therefore we have

$$[T^{xx} = (e+p)(u^x)^2 + p] = 0; \; [T^{0x} = (e+p)u^0 u^x] = 0; \; [nu^x] = 0$$
$$(1.109)$$

These equations can be rewritten more explicitly as

$$(e_1 + p_1)v_1^2\gamma_1^2 + p_1 c^2 = (e_2 + p_2)v_2^2\gamma_2^2 + p_2 c^2$$
$$(e_1 + p_1)v_1\gamma_1^2 = (e_2 + p_2)v_2\gamma_2^2$$
$$n_1 v_1 \gamma_1 = n_2 v_2 \gamma_2 \qquad (1.110)$$

where indexes 1 and 2 stand, respectively, for the quantities before and after the shock. These equations, which are the relativistic equivalent of Rankine-Hugoniot's conditions, are called Taub's conditions.

These equations can once again be solved with respect to the velocity of the shock and give, therefore, the relation between the corresponding quantities on different sides of the shock, provided the equation of state has also been specified. For the velocities, we obtain the following relations:

$$\frac{v_1}{c} = \sqrt{\frac{(p_2 - p_1)(e_2 + p_1)}{(e_2 - e_1)(e_1 + p_2)}} \quad \frac{v_2}{c} = \sqrt{\frac{(p_2 - p_1)(e_1 + p_2)}{(e_2 - e_1)(e_2 + p_1)}}$$

$$\frac{v_r}{c} = \sqrt{\frac{(p_2 - p_1)(e_2 - e_1)}{(e_1 + p_2)(e_2 + p_1)}} \tag{1.111}$$

where v_r is the relative velocity between the two fluids, namely, the shocked and the unshocked one.

However, the general solution is not very interesting, since a very relativistic shock is perforce a strong shock: in astrophysical situations, we always deal with sound speed $c_s \ll c$. Besides, matter behind the shock has a huge internal energy for shocks with $\gamma_1 \gg 1$, so that we can use the ultra-relativistic equation of state: $p = e/3$. Under these hypotheses, in the limit $\gamma_1 \gg 1$, the preceding equations simplify considerably. We use $p_1 = 0$ and $e_1 = n_1 mc^2$ (in the case of strong shocks, internal motions are of negligible importance in the material before the shock), and $p_2 = e_2/3$. We thus obtain

$$e_2 = (2\gamma_1^2 - 1)e_1 \quad \gamma_2^2 = \frac{9}{4}\frac{e_2 - e_1}{2e_2 - 3e_1} \quad \gamma_r^2 = \frac{1}{2}\gamma_1^2 + \frac{1}{2}$$

$$\tag{1.112}$$

The limit $\gamma_1 \gg 1$ of these equations is especially useful:

$$e_2 = 2\gamma_1^2 e_1 \quad \gamma_r^2 = \frac{1}{2}\gamma_1^2 \quad \gamma_2^2 = \frac{9}{8} \quad n_2 = \sqrt{8}\gamma_1 n_1 \quad (1.113)$$

From this we can see that the velocity of the fluid after the shock, with respect to the shock itself, tends to a constant,

$c/3$, which, even in the hyperrelativistic limit, makes the shock subsonic with respect to postshock matter. Notice, in fact, that the sound speed for the hyperrelativistic equation of state $p = e/3$ is given by $c/\sqrt{3}$. Therefore $M_2 = v_2/c_{s2} = 1/\sqrt{3} < 1$.

We also find the famous result according to which the energy density per unit of volume is proportional to the square of the Lorentz factor of the shock. This occurs because, on the one hand, the density of matter increases by one factor of Lorentz ($n_2 \propto \gamma_1$), and, on the other hand, because the gas is also heated by one Lorentz factor.

1.10.2 The Strong Explosion

Let us consider once again an instantaneous, pointlike explosion that expands in a cold surrounding medium (therefore $p = 0$), with a constant rest mass density, $e = \rho_0 c^2$. In relativistic hydrodynamics, too, the strong explosion is described by a self-similar solution (of the second kind).

Let us consider an equation describing the four-speed of the shock, dx_s^μ/ds. We note that this quantity must be a four-vector, but the available quantities, E, ρ_0, and c, are not; the only four-vector available is x_s^μ, the instantaneous position of the shock. Besides, the four-speed is dimensionless, and the only (relativistic scalar) length available is s, so that we find

$$\frac{dx_s^\mu}{ds} = \alpha \frac{x_s^\mu}{s} \qquad (1.114)$$

The quantity α can only be a pure number, because, among the quantities at hand (E, ρ_0, c, s), only c and s are relativistic invariants; from them we can therefore form no dimensionless quantity. However, we can use the fact that we know the value of α in the Newtonian limit. Indeed, when all velocities are small with respect to c, $s \approx ct$, $dx_s^\mu/ds \approx 1/c \; dR_s/dt$; and thus, for $v \ll c$, the above equation simplifies to

$$\frac{dR_s}{dt} = \alpha \frac{R_s}{t} \qquad (1.115)$$

Finally, if we compare with equation 1.76, we see that the solution corresponding to the strong, instantaneous, and point-like explosion requires

$$\alpha = \frac{2}{5} \tag{1.116}$$

We can find at once the dependence of the shock speed on time. By integrating equation 1.114 with the help of equation 1.116, we find

$$x_s^\mu = As^{2/5} \qquad \frac{dx_s^\mu}{ds} = \frac{2}{5}As^{-3/5} \tag{1.117}$$

If we use proper time instead of the observer's time, the solution for the motion of the shock wave is identical to that of the Newtonian case. If we now recall that $dx^\mu/ds = V_s\Gamma/c$, where Γ is the Lorentz factor of the shock, we find

$$s = \left(\frac{5V_s\Gamma}{2cA}\right)^{-5/3} \tag{1.118}$$

whence

$$ds = \frac{cdt}{\Gamma} = dV_s\frac{d}{dV_s}\left(\frac{5V_s\Gamma}{2cA}\right)^{-5/3} \tag{1.119}$$

which can be integrated immediately. We shall now focus on the hyperrelativistic limit, where $c - V_s \ll c$, and $\Gamma \gg 1$. In this case, the integration of the preceding equation gives

$$\Gamma^2 t^3 = \text{constant} \tag{1.120}$$

which is the equation replacing equation 1.76 in the hyperrelativistic case. Now we can easily find the position of the shock using $1/\Gamma^2 \equiv 1 - (V_s/c)^2$. The position of the shock is given by

$$R_s(t) = \int V_s(t')dt' \tag{1.121}$$

Once again, if we keep only the most significant terms in $1/\Gamma$, we find $V_s/c \approx 1 - 1/(2\Gamma^2)$, when $\Gamma \gg 1$; and therefore

$$R_s(t) \approx \int c\left(1 - \frac{1}{2\Gamma^2(t')}\right)dt' = ct\left(1 - \frac{1}{8\Gamma^2} + \mathcal{O}(1/\Gamma^4)\right)$$

$$\tag{1.122}$$

where we have used equation 1.120 in order to calculate the integral. Once again, we have determined the time dependence of the shock's radius without having to solve the full hydrodynamics equations. Furthermore, in the hyperrelativistic limit, we can find

$$s \propto \Gamma^{-5/3} \tag{1.123}$$

When looking at the conditions at the shock, or Taub's equation (1.113), we can define new functions

$$e(r,t) = 2\Gamma^2 e_1 r_1(\xi) \quad \gamma^2(r,t) = \frac{1}{2}\Gamma^2 r_2(\xi) \quad n(r,t) = \sqrt{8}\Gamma n_1 r_3(\xi) \tag{1.124}$$

that allow us to fix convenient boundary equations (i.e., at the shock) for all unknown functions:

$$r_1(0) = r_2(0) = r_3(0) = 1 \tag{1.125}$$

What is the self-similarity variable ξ? We notice that the above equations are written in the reference frame where the shock is at the origin of the coordinates and is instantaneously at rest. Therefore, the quantities r_1, r_2, and r_3 are defined in the shock's reference frame and can depend neither on any external dimensional quantity, such as radius or duration of the explosion (which we have supposed instantaneous and pointlike), nor on the shock's distance from the origin of the coordinates, because the shock *is* at the origin of the coordinates. Therefore, they can depend only on the radial distance from the shock, X', as well as on time, s, since, in the shock's reference frame, we can use only the shock's own time. Thus we must have

$$\xi \equiv \frac{X'}{s} \tag{1.126}$$

However, it is more convenient to express this quantity in terms of quantities defined in the observer's reference frame. If we call r the distance of the point where we want to calculate the quantities from the explosion site, in the reference system where the explosion site is at rest, we obviously find

$$X' = \Gamma(R_s - r) \tag{1.127}$$

and, in the same way,

$$s \propto x_\mu s^{3/5} \propto \frac{R_s}{\Gamma} \qquad (1.128)$$

where we used equation 1.123. We find, in the end,

$$\xi \equiv \frac{R_s - r}{R_s}\Gamma^2 \qquad (1.129)$$

This is the self-similarity variable.

It is then possible to find an exact analytical solution to hydrodynamic equation 1.102 for the problem of the strong explosion in a hyper-relativistic regime (Blandford and Mc-Kee 1976). Defining

$$\chi \equiv 1 + 8\xi \qquad (1.130)$$

we find

$$r_1(\chi) = \chi^{-17/12} \quad r_2(\chi) = \chi^{-1} \quad r_3(\chi) = \chi^{-7/4} \quad (1.131)$$

Now we can show that the total energy is conserved. From equation 1.101, we can see that the total energy density behind the shock is given by

$$\epsilon = \left(\frac{4}{3}\gamma^2 - 1\right)e \approx \frac{4}{3}\gamma^2 e \qquad (1.132)$$

where we have used the hyper-relativistic equation of state $(e = 3p)$ and $\gamma \gg 1$. The total energy is

$$E = \int_0^{R_s} \frac{16\pi}{3}e\gamma^2 r^2 dr \qquad (1.133)$$

Inserting equations 1.124, 1.130, and 1.131, we find

$$E = \frac{8\pi}{17}\rho_0 c^5 t^3 \Gamma^2 \qquad (1.134)$$

which gives a constant, provided $t^3\Gamma^2$ is a constant of motion, as already obtained (eq. 1.120).

1.11   The De Laval Nozzle

One of the classic problems of modern astrophysics is how to explain the formation of the jets of matter emitted by many different kinds of sources. Some of these jets are certainly relativistic, with Lorentz factors of the order of $\Gamma \approx 5$ and beyond, corresponding to a speed $v \gtrsim 0.98c$. Since the highest known sound speed is that of the hyper-relativistic fluid, with an equation of state $p = \rho c^2/3$ corresponding to $c_s = c/\sqrt{3}$, these jets are largely supersonic.

Though, according to common opinion, magneto-hydro-dynamic mechanisms play a fundamental role (see chapter 8), it is still worth noting that there is a purely hydrodynamic mechanism of collimation and acceleration. Since this mechanism can play a role in any case, even in the presence of magneto-hydrodynamic effects, it seems worth discussing.

The problem of the continuous acceleration of matter beyond sound speed is different from the creation of a shock wave, since it is based on a counterintuitive hydrodynamic property, which is worth explaining before discussing nozzles.

Let us consider a stationary and irrotational flow of matter. In this case, Bernoulli's theorem holds (eq. 1.24):

$$w + \frac{1}{2}v^2 = w_0 \qquad (1.135)$$

where the constant is the same in the entire space, since we have assumed the flow to be irrotational. The constant on the left-hand side has been computed where the velocity vanishes. Of course, the higher the velocity, the smaller the enthalpy w. The maximum value is reached for an expansion in a vacuum, $v_{max} = \sqrt{2w_0}$. Since the flow is adiabatic, $dw = dp/\rho$, and therefore dp and dw have the same sign: v increases as p decreases. The funny and unusual thing is that this is not true for the mass flux $j = \rho v$. Indeed, from Euler's equation for stationary flows (eq. 1.10) we find that $vdv = -dp/\rho = -c_s^2 d\rho/\rho$, so that

$$\frac{d\rho}{dv} = -\frac{\rho v}{c_s^2} \qquad (1.136)$$

and for the mass flux we find:

$$\frac{d(\rho v)}{dv} = \rho \left(1 - \frac{v^2}{c_s^2}\right) . \qquad (1.137)$$

The mass flux increases with speed as long as the flow is subsonic but decreases when the flow becomes supersonic. Therefore, j will have a maximum value, which can be determined as follows for a polytropic gas. We have

$$w = \frac{c_s^2}{\gamma - 1} \qquad (1.138)$$

and, besides, the maximum of j is reached for $v = c_s$, sometimes also called critical point; therefore, using Bernoulli's theorem (eq. 1.135) at the critical point

$$w_c + \frac{1}{2}v_c^2 = \frac{c_s^2}{\gamma - 1} + \frac{1}{2}c_s^2 = \frac{c_{s0}^2}{\gamma - 1} \qquad (1.139)$$

where c_{s0} is the sound speed when $w = w_0$ and $v = 0$, and the subscript c indicates the value at the critical point. We find that, at the critical point,

$$c_{cs} = c_{s0}\sqrt{\frac{2}{\gamma + 1}} \qquad (1.140)$$

which shows that here the outflow speed is still smaller than the sound speed in the initial fluid. Furthermore

$$\rho_c = \rho_0 \left(\frac{2}{\gamma + 1}\right)^{1/(\gamma-1)} \qquad p_c = p_0 \left(\frac{2}{\gamma + 1}\right)^{\gamma/(\gamma-1)}$$
$$(1.141)$$

which shows that density and pressure, at the critical point, are lower than their initial values. Putting together these equations, we find that, still at the critical point,

$$j_{max} = \rho_0 c_{s0} \left(\frac{2}{\gamma + 1}\right)^{\frac{\gamma+1}{2(\gamma-1)}} . \qquad (1.142)$$

This is the maximum value of the mass flux.

It is also possible to derive the run of all quantities, as a function of speed, along a flux line. In fact, we know that the flow is adiabatic,

$$P = p_0 \left(\frac{\rho}{\rho_0} \right)^\gamma \qquad \rho = \rho_0 \left(\frac{T}{T_0} \right)^{1/(\gamma-1)} \tag{1.143}$$

and, by replacing these equations in Bernoulli's theorem, equation 1.135, we find

$$X \equiv 1 - \frac{(\gamma-1)v^2}{2c_{s0}^2}, \ \ T = T_0 X, \ \ \rho = \rho_0 X^{1/(\gamma-1)}, \ \ p = p_0 X^{\gamma/(\gamma-1)} \tag{1.144}$$

Eliminating the speed, we find a formula that will be useful in the following:

$$j = \rho v = \left(\frac{p}{p_0} \right)^{1/\gamma} \left[\frac{2\gamma p_0 \rho_0}{\gamma - 1} \left(1 - \left(\frac{p}{p_0} \right)^{(\gamma-1)/\gamma} \right) \right]^{1/2} \tag{1.145}$$

This function vanishes for $p = p_0$ and for $p = 0$ and reaches, therefore, its peak between these two values.

After this premise, let us consider the following hydrodynamic mechanism, which is used in order to accelerate a gas to high speed: a vessel with a gentle neck that tapers to a minimum section S_{min}; we shall also assume that the pressure inside the container, p_0, is larger than the pressure outside, p_e. Of course, we (naively!) hope that when the difference between the external pressure and the internal pressure pushes the fluid through the bottleneck, since the flux $Q = \rho v S$ must remain constant, the fluid may reach an arbitrarily large speed.

We shall assume that the speed along the pipe is everywhere perpendicular to the section of the pipe itself, which is more or less true if the pipe narrows very gently. In this case, the pipe's flux will be given by $Q = \rho v S = jS$. Since we already know that $j \leq j_{max}$, we also know that the pipe has a maximum capacity, Q_{max}, which is determined as follows. When $Q = Q_{max}$, $j = j_{max}$, and this value of j can be reached only in the narrowest point of the pipe, where $S = S_{\min}$. If this were not so, and if $j = j_{\max}$ when $S = S_1 > S_{\min}$, since $Q = jS$ is a constant, we should have $j > j_{\max}$ per $S_{\min} < S < S_1$, which is obviously impossible, as we said beforehand. Therefore, the maximum Q is obtained when $j = j_{\max}$ at the end of the pipe, where $S = S_{\min}$; and therefore

$$Q_{\max} = j_{\max} S_{\min} \qquad (1.146)$$

At this point, the pressure in the pipe is as large as the critical pressure, p_c, which we have just calculated.

Now, let us imagine that we can change the external pressure p_e, starting from the value p_0. When $p_e = p_0$ there will be no outflow, and as long as $p_0 - p_e \ll p_e$, the outflow will be weak. However, as p_e decreases, the flow must increase, until it reaches the peak value we have just determined, $Q_{\max}$, which corresponds to a value of $p_e = p_c$. If we further lower the external pressure, we see that the further pressure fall cannot occur in the pipe because it would produce $j > j_{\max}$, which is impossible. The fall of pressure along the pipe is limited: the minimum pressure we can realize at the end of the pipe is p_c; if $p_e < p_c$, the rest of the pressure fall will be realized outside of the pipe, without giving rise to further acceleration. We can therefore see that it is useless to lower the external pressure further: the maximum speed we can reach is the sound speed, and there can be no further acceleration to supersonic regimes.

The problem is that we have considered a nozzle that narrows continuously, whereas we should consider a de Laval nozzle (which takes its name from the Swedish physicist who discussed it first in the nineteenth century), like the one shown in the lower part of the figure 1.3, which initially narrows and then widens out again.

In this nozzle, after its narrowing, it is possible to maintain the capacity $Q = jS$ constant even if j decreases as speed increases for $v > c_s$, according to equation 1.137. Besides, since ρ decreases as speed increases (eq. 1.136), and

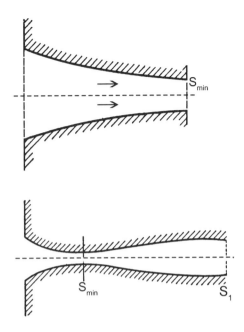

Figure 1.3. A naive nozzle (*upper*) and a de Laval nozzle
(*lower*).

$dp = d\rho/c_{\mathrm{s}}^2$, we see that, as speed increases, pressure decreases.

We can put this argument in a formula as follows. According to Euler's equation (eq. 1.10),

$$v\frac{dv}{dx} = -\frac{1}{\rho}\frac{dp}{dx} = -\frac{c_{\mathrm{s}}^2}{\rho}\frac{d\rho}{dx} \tag{1.147}$$

where we have used the relation $dp = c_{\mathrm{s}}^2 d\rho$, which holds for
adiabatic flows, like the one we are considering. The equation
for the mass flux gives

$$\rho v S = \text{constant} \tag{1.148}$$

Taking the derivative of this and dividing by $\rho v S$ we obtain

$$\frac{1}{\rho}\frac{d\rho}{dx} + \frac{1}{v}\frac{dv}{dx} + \frac{1}{S}\frac{dS}{dx} = 0 \qquad (1.149)$$

and, using equation 1.147, we find

$$(1 - M^2)\frac{1}{v}\frac{dv}{dx} + \frac{1}{S}\frac{dS}{dx} = 0 \qquad (1.150)$$

Here, $M = v/c_s$ is Mach's outflow number. In the nozzle's narrowest point, $dS/dx = 0$, which implies that either $1 - M^2 = 0$ or $dv/dx = 0$. The latter is a disastrous hypothesis for us, since it means that the flow stops its acceleration. On the other hand, if $1 - M^2$ changes sign when dS/dx also changes sign, dv/dx can maintain its sign, which is obviously positive. However, we have already seen for the bottleneck that $M = 1$ in its narrowest point, when the difference of pressure between inside and outside is large enough. Therefore, $M = 1$ at the bottleneck, and the flow has $dv/dx > 0$ all along the nozzle: the fluid keeps accelerating, even after passing the sonic point. Thus a supersonic outflow is generated.

How long will the acceleration last? If the nozzle has an end section $S_1 > S_{\min}$, the mass flux at the outlet will be $j_1 = j_{\max} S_{\min}/S_1 < j_{\max}$; from equation 1.145 we find that this corresponds to a certain pressure p_1. If the external pressure $p_e < p_1$, the whole pipe will be used for the acceleration, and the rest of the pressure fall will take place outside the pipe, without further acceleration. On the other hand, if $p_e > p_1$, the counterpressure exerted by the external material generates a shock wave that slows the outflow (see Landau and Lifshitz 1987b).

Therefore, if we place a negligible external pressure outside a de Laval nozzle, (almost) the whole pressure fall from p_0 to p_e can be used to accelerate the jet; or, in other words, (almost) the whole pressure fall takes place inside the pipe. There follows that it is possible to accelerate a fluid to a supersonic speed with purely hydrodynamic mechanisms, provided we have a suitable constraint namely, a nozzle with a bottleneck, in the form of a de Laval nozzle, rather than a simple bottleneck, which can only reach the sound speed.

1.12   Problems

1. Derive equations 1.28, assuming we know equations 1.6, 1.10, and 1.13. (Hint: Start by looking for an expression for $(\partial/\partial t)(\rho v^2/2 + \rho\epsilon)$, and remember to use the second principle of thermodynamics in the form $d\epsilon = Tds + pd\rho/\rho^2$, as well as eq. 1.17).

2. The specific entropy for an ideal gas is $s = c_V \ln(p/\rho^\gamma)$. Show that the entropy of the gas after the shock, s_2, is higher than the one before the shock, s_1, only if $M_1 > 1$.

3. Suppose an instantaneous and pointlike explosion takes place in a medium where density varies according to

$$\rho = \frac{A}{r^\alpha}$$

with $\alpha < 3$ (why?). Determine the time evolution of the shock radius.

4. In Sedov's solution, we assume the conservation of energy, not of the linear momentum. Why? (Hint: Is the total momentum conserved in Sedov's solution? How is it possible?).

5. Let us assume that a continuous source with a luminosity L lit up a long time ago. Determine the dependence of the shock radius on time, at late times.

6. How can you change the solution for the strong explosion, in relativistic hydrodynamics, when you assume that the energy release is not instantaneous but takes place at a constant rate L? And if mass density in the surrounding medium varies according to $\rho = A/r^\alpha$?

7. Given a de Laval nozzle with an external section $S_1 > S_{min}$ and a negligible external pressure ($p_e \ll p_1$), determine Mach's number for the outflow. (Hint: Remember that the motion is adiabatic, and therefore $p \propto \rho^\gamma$ and $\rho \propto T^{1/(\gamma-1)}$.)

8. It is possible to find self-similar solutions where the shock wave accelerates with time, within relativistic hydrodynamics. How can this be done? (Careful: Self-similar analysis alone will not do. For the solution, see Perna and Vietri (2002) and references therein.)

Chapter 2

Magnetohydrodynamics and Magnetic Fields

In the preceding chapter, we introduced the equations of hydrodynamics, Newtonian and relativistic, neglecting all electromagnetic phenomena related to the fluids in question. However, in high-energy astrophysics, the fluids' temperatures are very high, and most atoms are completely ionized. This is especially true for hydrogen and helium, the most abundant constituents of cosmic matter, because their ionization potentials are rather low. As a consequence, astrophysical fluids are mostly ionized. In this situation, electric fields are irrelevant, thanks to matter's charge neutrality and to the abundance of free charges: Together, these two facts guarantee the short out of any electric field. However, if the fluid is immersed in a magnetic field, its motion relative to B generates an electric field, and this in turn generates currents. These currents are affected by magnetic fields and generate new magnetic fields, thus creating a complex and interesting physical situation. Because of their low density, astrophysical fluids have a low resistivity, unlike laboratory fluids; that is why here I discuss the properties of ideal magnetohydrodynamics. After introducing the fundamental equations and the properties of waves, I shall discuss the flux-freezing theorem, buoyancy and reconnection, the existence of two–temperature fluids, and shock waves. We shall then consider the origin of astrophysical magnetic fields. After a brief review of the contemporary notions about the main magnetic fields in the stars, in the galaxy, and in other galaxies and clusters of galaxies, we shall consider two kinds of batteries, namely, the phenomena generating the initial magnetic field. Finally, I shall briefly discuss astrophysical dynamos.

2.1 Equations of Motion

Magnetohydrodynamics describes the behavior of fluids, which are at least partly ionized, in the presence of electromagnetic fields; this description holds in the limit of mean free paths short with respect to all macroscopic lengths of the problem. By mean free path we mean not only those between collisions among particles of various species, but also the Larmor radii for particles in the magnetic field, be they protons, nuclei, or electrons. In some exceptional cases, the equations of magnetohydrodynamics even apply to fluids where mean free paths are large (see, e.g., Binney and Tremaine 1987), but we shall not discuss any of those cases here. Moreover, we shall assume that all velocities are small in comparison with the speed of light.

The equation expressing mass conservation continues to hold without any change:

$$\frac{\partial \rho}{\partial t} + \nabla \cdot (\rho \vec{v}) = 0 \qquad (2.1)$$

On the other hand, the equation expressing the conservation of momentum needs a correction. The typical astrophysical fluids have zero charge density. In these conditions, the net force exerted by electric fields vanishes, unlike the one exerted by magnetic fields. We know from elementary courses that a current $\vec{j}$ immersed in a magnetic field is subject to a force per unit volume (which is actually the Lorentz force per unit volume, in the absence of a net charge density) $\vec{j} \wedge \vec{B}/c$. This external force must be included in Euler's equation:

$$\frac{\partial \vec{v}}{\partial t} + \vec{v} \cdot \nabla \vec{v} = -\frac{\nabla p}{\rho} + \frac{\vec{j} \wedge \vec{B}}{\rho c} \tag{2.2}$$

The equation of energy conservation must be modified too, since, as is well known, the presence of currents implies a resistance, and resistance implies dissipation and therefore heating. From elementary courses we know that the heating rate (namely, the heat released per unit time and volume) by a current flowing inside a medium of given conductivity σ is given by j^2/σ. So long as we neglect dissipative processes, we have, per unit mass,

$$T\frac{Ds}{Dt} = 0 \tag{2.3}$$

In the presence of resistive dissipation, as mentioned above, there is a heating term, given by j^2/σ per unit volume and time; therefore, the equation of energy conservation, in the presence of this kind of dissipation (i.e., resistive) is given by

$$\rho T\frac{Ds}{Dt} = \frac{j^2}{\sigma} \tag{2.4}$$

Why should there be a current in our fluid? Because there might be an electric field inducing a current. In elementary courses, we saw Ohm's law, according to which, in a medium at rest, an electric field $\vec{E}'$ generates a current $\vec{j} = \sigma \vec{E}'$, where σ is the conductivity of the medium. In our case, the fluid is moving in the reference system of the laboratory (or, for the astronomers, of the observer!), while electric and magnetic fields are measured in the reference frame of the laboratory. It follows that the electric field $\vec{E}'$ perceived by the fluid in its reference frame is different from the laboratory one. Let us recall the Lorentz transformations for the electric field:

$$\vec{E}' = \gamma \left(\vec{E} + \frac{\vec{v} \wedge \vec{B}}{c} \right) - \frac{\gamma^2}{\gamma+1} \frac{\vec{v}(\vec{v} \cdot \vec{B})}{c^2} \tag{2.5}$$

where $\vec{v}$ and γ are the velocity of the fluid in the observer's reference frame and its Lorentz factor. Let us assume that the fluid is Newtonian, $v \ll c$. Therefore, in the preceding equation, we shall only keep terms of smallest order in v/c:

$$\vec{E'} = \vec{E} + \frac{\vec{v} \wedge \vec{B}}{c} \tag{2.6}$$

is the electric field that the fluid perceives in its reference system. As we shall soon explain, the two terms $\vec{E}$ and $\vec{v} \wedge \vec{B}/c$ are of the same order in v/c.

At this point, the form of Ohm's law is

$$\vec{j} = \sigma \left(\vec{E} + \frac{\vec{v} \wedge \vec{B}}{c} \right) \tag{2.7}$$

In principle, the conductivity σ is a tensor, and it is easy to guess that in the presence of strong magnetic fields, conductivity (just like all scattering coefficients, such as heat transport or viscosity) is different, according to whether we consider the parallel or perpendicular direction to the magnetic field. However, the coefficients of conduction along these two directions differ only by a factor $3\pi/32 = 0.295$ (Spitzer 1962), and we shall therefore leave out a distinction between the two directions, treating σ as a scalar. Naturally, $\vec{j}$ is the spatial component of a four–vector, which is thus transformed when we leave the fluid's reference frame to return to the laboratory frame. However, the Lorentz transformation of $\vec{j}$ mixes it with the charge density, which we assumed as zero in our fluids. Thus $\vec{j'} = \gamma \vec{j} \approx \vec{j}$, because we assumed that fluids are Newtonian, $v \ll c$. Therefore, under these hypotheses, $\vec{j}$ is not changed by the Lorentz transformations.

Let us now discuss the form of Maxwell's equations. Obviously, for the electric field we have

$$\nabla \cdot \vec{E} = 0 \qquad \nabla \wedge \vec{E} = -\frac{1}{c} \frac{\partial \vec{B}}{\partial t} \tag{2.8}$$

In Gauss's equation, the right-hand side vanishes because we assumed the fluid is electrically neutral; therefore, the density of local charge is zero. It follows that the only reason for the existence of an electric field, in magnetohydrodynamics, is induction. It is then easy to see from the second part of equation 2.8 that in order of magnitude, $E \approx vB/c$, which explains why the two terms in equation 2.6 are of the same order of magnitude.

Finally, the equations for $\vec{B}$ are

$$\nabla \cdot \vec{B} = 0, \qquad \nabla \wedge \vec{B} = \frac{4\pi}{c}\vec{j} \qquad (2.9)$$

These are the well-known Maxwell's equations, apart from the fact that we neglected the displacement current, $1/c \, \partial\vec{E}/\partial t$. Since the electric field, in magnetohydrodynamics, is exclusively due to induction, from the equation for the curl of E we find that $E \approx LB/cT$, where L and T are typical lengths and time scales on which the electromagnetic field varies, and L/T obviously is a typical velocity. Thence the displacement current, in order of magnitude, equals $\approx LB/c^2T^2$. This must be compared with the curl of B, which, in order of magnitude, is B/L. The ratio between the two terms equals $L^2/(cT)^2 \approx (v/c)^2$. As a consequence, the displacement current is smaller than the curl of B by a factor $(v/c)^2$, which, under our hypotheses, is $\ll 1$. Therefore, the displacement current can be neglected.

In the preceding equations, the current density $\vec{j}$ and the electric field $\vec{E}$ can be completely eliminated, thanks to the following manipulation. $\vec{j}$ can be replaced in equations 2.2 and 2.4 by $c/4\pi \, \nabla \wedge \vec{B}$, in accordance with the second part of equation 2.9. Moreover, we shall rewrite Ohm's law (eq. 2.7) as

$$\frac{\vec{j}}{\sigma} - \frac{\vec{v} \wedge \vec{B}}{c} = \vec{E} \qquad (2.10)$$

Also, here $\vec{j}$ can be replaced by $c/4\pi \, \nabla \wedge \vec{B}$. At this point, we can take the curl of both members and remove the curl of E, thanks to the second part of equation 2.8, thus obtaining (after using $\nabla \cdot \vec{B} = 0$)

$$\frac{\partial \vec{B}}{\partial t} = \nabla \wedge (\vec{v} \wedge \vec{B}) + \frac{c^2}{4\pi\sigma}\nabla^2\vec{B} \qquad (2.11)$$

which is very useful since, at this point, the system of equations of Magnetohydrodynamics only contains B and no longer j or E.

2.1.1   The Limit of Ideal Magnetohydrodynamics

The equations derived in the preceding paragraph are further simplified in the limit of ideal magnetohydrodynamics, namely, when $\sigma \rightarrow \infty$. In this case, in fact, the equation of energy conservation (eq. 2.4) goes essentially back to its form of pure entropy conservation:

$$\frac{Ds}{Dt} = 0 \qquad (2.12)$$

which simply expresses the fact that, when $\sigma \rightarrow \infty$ and dissipation can therefore be entirely neglected, the fluid conserves once again the entropy of each mass element.

Euler's equation maintains its form,

$$\frac{\partial \vec{v}}{\partial t} + \vec{v} \cdot \nabla \vec{v} = -\frac{\nabla p}{\rho} + \frac{(\nabla \wedge \vec{B}) \wedge \vec{B}}{4\pi\rho} \qquad (2.13)$$

just like mass conservation, equation 2.1. On the other hand, equation 2.11 simplifies:

$$\frac{\partial \vec{B}}{\partial t} = \nabla \wedge (\vec{v} \wedge \vec{B}) \qquad (2.14)$$

The three equations above and equation 2.1 are the fundamental equations of ideal magnetohydrodynamics.

In order to realize what happens when the limit of ideal magnetohydrodynamics does not hold, let us consider equation 2.11, leaving out the curl instead of the Laplacian. We have

$$\frac{\partial \vec{B}}{\partial t} = \frac{c^2}{4\pi\sigma}\nabla^2\vec{B} \qquad (2.15)$$

which can be immediately recognized as the equation of heat transfer. The solutions of this equation are, by definition, the emblem of dissipation: even though, at first, there is a certain quantity of heat concentrated in a small region of space, after a while this quantity of heat spreads to the whole space, becoming unobservably small. The magnetic field, subject only to the Laplace term, obeys the same law, on the dissipation time scale,

$$T_d = \frac{4\pi\sigma L^2}{c^2} \qquad (2.16)$$

The coefficient $c^2/(4\pi\sigma)$ is called *magnetic diffusivity* and is obviously smaller the closer we are to the limit of ideal magnetohydrodynamics.

In order to understand whether the approximation of ideal magnetohydrodynamics is suitable to astrophysical situations, let us calculate the time T_d for some concrete situations. Let us take for σ the conductivity of a totally ionized gas of pure hydrogen (Spitzer 1962):

$$\sigma = 6.98 \times 10^7 \frac{T^{3/2}}{\ln\Lambda} s^{-1} \qquad (2.17)$$

where T is the gas temperature, expressed in degrees kelvin, and $\ln\Lambda \approx 30$ is an approximate factor called the Coulomb logarithm. For a stellar interior, for which $T \approx 10^7 K$, and $L \approx 10^{11}$ cm, we find $T_d \approx 3 \times 10^{11}$ yr, larger than the age of the universe, $T_H \approx 10^{10}$ yr. For a gas cloud, the galaxy in its entirety, or a cluster of galaxies, which have huge dimensions, $T_d \gg T_H$, *a fortiori*. Even for smaller objects, this time is long. In the innermost regions of the accretion disk around a black hole, which have dimensions corresponding to ≈ 100 Schwarzschild radii (corresponding to $\approx 10^{15}$ cm), we find that $T_d \approx 10^{20}$ yr. In other words, the astrophysical fluids are so huge that the dissipation time for the magnetic field is almost always longer than the age of the universe: the magnetic field is (almost, see section 2.7) never dissipated, in astrophysical conditions.

Another way to decide whether this is a good approximation is to compare the two terms on the right-hand side of equation 2.11. The ratio between the one containing the curl and the other containing the Laplace term is

$$\mathcal{R}_{\mathrm{m}} = \frac{4\pi\sigma v L}{c^2} \qquad (2.18)$$

where, once again, v and L are speed and length characteristic of the problem under examination. The quantity $\mathcal{R}_{\mathrm{m}}$ is called the magnetic Reynolds number. When $\mathcal{R}_{\mathrm{m}} \gg 1$, the Laplace term in equation 2.11 can be neglected, so that we are at the limit of ideal magnetohydrodynamics. It is easy to see that in all the cases mentioned above, $\mathcal{R}_{\mathrm{m}} \gg 1$.

2.1.2   Equations of Motion in a Conservative Form

We saw in chapter 1 that the fundamental equations of hydrodynamics can be written in an explicitly conservative form. This can also be done in magnetohydrodynamics.

Euler's equation without magnetic terms has been written as

$$\frac{\partial(\rho v_i)}{\partial t} = -\frac{\partial}{\partial x_k} R_{ik} \qquad (2.19)$$

In order to rewrite the new equation 2.13 in the same form, we use the identity

$$(\nabla \wedge \vec{B}) \wedge \vec{B} = (\vec{B} \cdot \nabla)\vec{B} - \frac{1}{2}\nabla B^2 \qquad (2.20)$$

In order to prove it, we must take into account the fact that $\nabla \cdot B = 0$. We can show that equation 2.19 still holds with the definition

$$T_{ik} = p\delta_{ik} + \rho v_i v_k - \frac{1}{4\pi}\left(B_i B_k - \frac{1}{2}B^2\delta_{ik}\right) \qquad (2.21)$$

which we rewrite as:

$$T_{ik} = R_{ik} + M_{ik} \qquad (2.22)$$

where R_{ik} is the Reynolds stress tensor, which we came across before, and M_{ik} is called *Maxwell stress tensor*:

$$M_{ik} = -\frac{1}{4\pi}\left(B_i B_k - \frac{1}{2}\delta_{ik}B^2\right) \qquad (2.23)$$

It describes the flux per unit time, of the ith component of momentum, through a surface whose normal is directed along the kth coordinate. Therefore, we see that even a magnetic field can carry momentum. The question whether in concrete astrophysical situations, such as pulsar winds and black holes' jets, the hydrodynamic term $(p\delta_{ik} + \rho v_i v_k)$ or the magnetic term dominates is still open.

It will not be surprising to discover, at this point, that the magnetic field can also carry energy. In a sense, this is already well known from elementary courses, where we learn that the Poynting vector, $c\vec{E}\wedge\vec{B}/4\pi$, is responsible for carrying energy, and that the magnetic field has a density of energy equal to $B^2/8\pi$. This is still true in magnetohydrodynamics, and the only question is how to get rid of the electric field appearing in the Poynting vector. Let us consider Ohm's law (eq. 2.7) in the limit of ideal magnetohydrodynamics, $\sigma \to \infty$. In this case, we must obviously have

$$\vec{E} = -\frac{\vec{v}\wedge\vec{B}}{c} \qquad (2.24)$$

In fact, if this equality did not hold in the ideal limit, we would have infinite currents and thus infinite forces, which is obviously unphysical. This relation between E and B is very important and will often be used in the coming chapters. At this point, the Poynting vector can be rewritten. The easiest way is to rewrite the energy conservation in an explicitly conservative form, and check it post facto, but we shall leave this as an exercise for the reader:

$$\frac{\partial}{\partial t}\left(\frac{\rho v^2}{2} + \rho\epsilon + \frac{B^2}{8\pi}\right) = -\nabla\cdot\left(\rho\vec{v}\left(\frac{v^2}{2} + w\right) + \frac{1}{4\pi}\vec{B}\wedge\left(\vec{v}\wedge\vec{B}\right)\right)$$

$$(2.25)$$

2.2 The Force Exerted by the Magnetic Field

We have just seen (eq. 2.13) that the magnetic field exerts a force density on the fluid, which is given by

$$\vec{f} = \frac{1}{4\pi}(\nabla \wedge \vec{B}) \wedge \vec{B} \qquad (2.26)$$

which we want to rewrite in order to make it more transparent. To this aim, we shall use once again the mathematical relation

$$(\nabla \wedge \vec{B}) \wedge \vec{B} = (\vec{B} \cdot \nabla)\vec{B} - \frac{1}{2}\nabla B^2 \qquad (2.27)$$

In order to prove it, we must use the fact that $\nabla \cdot \vec{B} = 0$. Thus we find

$$\vec{f} = -\frac{1}{8\pi}\nabla B^2 + \frac{1}{4\pi}(\vec{B} \cdot \nabla)\vec{B} \qquad (2.28)$$

The last term can be rewritten. We note, first of all, that the operator $\vec{B} \cdot \nabla$ represents B times the derivative along a magnetic line:

$$(\vec{B} \cdot \nabla)\vec{B} = B\frac{\partial}{\partial s}\vec{B} \qquad (2.29)$$

where s is a coordinate along the magnetic field line. As a general rule, $\vec{B}$ varies in modulus and in direction:

$$\vec{B} = B\hat{b} \qquad (2.30)$$

where $\hat{b}$ is a unit vector along $\vec{B}$. We have

$$\frac{1}{4\pi}(\vec{B} \cdot \nabla)\vec{B} = \frac{1}{4\pi}B\frac{\partial}{\partial s}B\hat{b} = \frac{B^2}{4\pi}\frac{\partial\hat{b}}{\partial s} + \frac{\hat{b}}{8\pi}\frac{\partial B^2}{\partial s} \qquad (2.31)$$

The first term on the right-hand side is the change of direction of the field when we move along a line. Referring to figure 2.1, we can define

$$\frac{\partial\hat{b}}{\partial s} = \frac{\hat{n}}{R} \qquad (2.32)$$

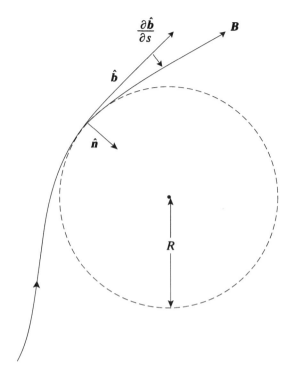

Figure 2.1. The origin of magnetic tension. From Dendy 1994.

where $\hat{n}$ is a unit vector toward the center of the tangent circle,[1] which has a radius R.

On the other hand, the second term on the right has the same form as the gradient of magnetic pressure, $\nabla B^2/8\pi$, but opposite sign; therefore, it cancels the component of the gradient along the magnetic field. Defining an operator

$$\nabla_\perp \equiv \nabla - \hat{b}\frac{\partial}{\partial s} \qquad (2.33)$$

we finally find that

$$\vec{f} = -\frac{1}{8\pi}\nabla_\perp B^2 + \frac{B^2}{8\pi R}\hat{n} \qquad (2.34)$$

---

[1] The correct term is *osculating* from the Latin *osculare*, meaning "to kiss."

which is the expression we were looking for. Now we see that the magnetic field exerts two forces on the fluid, which are perpendicular to the lines of force. This is entirely reasonable, since these are obviously Lorentz forces, $\propto \vec{v} \wedge \vec{B}$, which are always perpendicular to the field. The first force is the gradient of the magnetic pressure, perpendicular to the field; where the field is more intense, and the lines of force denser, the force is stronger. The second term is directed toward the center of the osculating circle; therefore, it tends to shorten the magnetic field line. This term is wholly similar to the force exerted by a violin string when plucked: it exerts a force that tends to bring the string back to its minimum length. The same happens in the magnetic field, whence the name *magnetic tension*.

In order to understand this term better, see problem 2.

2.3   Magnetic Flux Freezing

We have examined the behavior of the solutions of equation 2.11 in the case where the dominating term is the dissipative one. Since we have demonstrated that the dissipative term is always negligible in astrophysical conditions, we must examine the behavior of the solutions of equation 2.11 when the dissipative term is absent; in other words, let us examine the behavior of the solutions of equation 2.14. Alfvèn (1942) suggested a very useful interpretation of this equation, which we present in Dendy's (1994) simple form.

Let us consider, at time t_0, a closed line, painted in our fluid, and trace any open surface having as border this closed line (see fig. 2.2). We compute the flux of the magnetic field through this open surface,

$$\Phi_0 = \int \vec{B} \cdot d\vec{A} \qquad (2.35)$$

Now we follow for a time t, not necessarily infinitesimal, the motion of fluid elements that constitute the closed line. At the new instant $t_1 = t_0 + t$, these fluid elements form a new

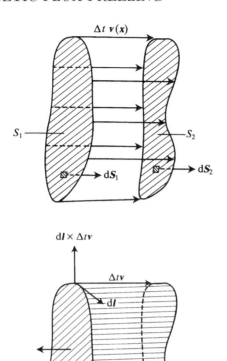

Figure 2.2. The theorem of flux conservation. From Dendy 1994.

closed line, different from the first one. If we now calculate the flux of the magnetic field Φ_1 through any open surface insisting on the new closed curve G_1, we find that

$$\Phi_0 = \Phi_1 \tag{2.36}$$

In other words, the flux of the magnetic field through a closed line, which moves together with the fluid's particles, remains constant. This property is called *freezing of the magnetic flux*.

It represents the opposite of the dissipation term, leading to a decrease in the flux thus computed.

We now move on to the proof of this simple theorem. The astrophysical consequences of flux freezing will be discussed later in this chapter. Readers who are not interested in the mathematical details can move directly to the next section.

Let us consider a series of adjacent fluid elements, constituting a finite surface (S_0) at the time t_0, and the same elements at the time $t_1 = t_0 + dt$, which will evidently constitute a new surface S_1. What we want to do is to establish a connection between the magnetic flux

$$\Phi_0 = \int_{S_0} \vec{B}(\vec{x}, t_0) \cdot d\vec{S}_0 \tag{2.37}$$

and the same quantity, at the moment t_1, on the surface S_1:

$$\Phi_1 = \int_{S_1} \vec{B}(\vec{x} + dt\vec{v}, t_0 + dt) \cdot d\vec{S}_1 \tag{2.38}$$

In equation 2.38, we can make a Taylor expansion up to terms of the first order in the (small) time interval $dt = t_1 - t_0$:

$$\Phi_1 \approx \int_{S_1} \vec{B}(\vec{x} + dt\vec{v}, t_0) \cdot d\vec{S}_1 + dt \int_{S_1} \frac{\partial}{\partial t}[\vec{B}(\vec{x} + dt\vec{v}, t_0)] \cdot d\vec{S}_1 \tag{2.39}$$

In the second term in the left-hand side of the above equation, the integral is extended to the surface S_1, which differs from S_0 for first–order terms in dt. In its turn, this term is of the first order in dt, and therefore, if the integral is extended to the surface S_0 instead of S_1, an error is made that, in the expansion of Φ_1, gives only a quadratic contribution in dt and is therefore negligible. We have, therefore,

$$\Phi_1 \approx \int_{S_1} \vec{B}(\vec{x} + dt\vec{v}, t_0) \cdot d\vec{S}_1 + dt \int_{S_0} \frac{\partial}{\partial t}[\vec{B}(\vec{x} + dt\vec{v}, t_0)] \cdot d\vec{S}_0 \tag{2.40}$$

The advantage of this approximate form is that it relates the magnetic flux through two distinct surfaces, *at the same time.*

We can thus use Maxwell's first equation, $\nabla \cdot \vec{B} = 0$, in the following way. Let us consider a closed, simultaneous surface, constituted by S_0, S_1, as well as the small strip joining them, S_2 (see fig. 2.2). Since this is a closed surface, we know that, thanks to Gauss's theorem,

$$\oint \vec{B} \cdot d\vec{S} = \int \nabla \cdot \vec{B} dV = 0 \qquad (2.41)$$

where we must use the *external* normal, in the surface integral; instead, in equation 2.37, we have used the *internal* normal. The above equation can be written as

$$\Phi_0 = \int_{S_1} \vec{B}(\vec{x} + dt\vec{v}, t) \cdot d\vec{S}_1 + \int_{S_2} \vec{B} \cdot d\vec{S}_2 \qquad (2.42)$$

Inserting equation 2.40 here, we find

$$\Phi_1 = \Phi_0 + dt \int_{S_0} \frac{\partial \vec{B}}{\partial t} \cdot d\vec{S}_0 - \int_{S_2} \vec{B} \cdot d\vec{S}_2 \qquad (2.43)$$

We can easily see from figure 2.2 that the element of surface $d\vec{S}_2$ on the strip separating the surfaces S_0 and S_1 is given by $d\vec{S}_2 = d\vec{x} \wedge dt\vec{v}$, where $d\vec{x}$ is the element of length along the edge of S_0. We thus find

$$\int_{S_2} \vec{B} \cdot d\vec{S}_2 = \int_{S_2} \vec{B} \cdot (d\vec{x} \wedge dt\vec{v}) = dt \oint (\vec{v} \wedge \vec{B}) \cdot d\vec{x} \quad (2.44)$$

where the contour integral is computed on the edge of S_0, and I used the identity $\vec{a} \cdot (\vec{b} \wedge \vec{c}) = \vec{b} \cdot (\vec{c} \wedge \vec{a})$. Finally, thanks to the curl theorem, this can be rewritten as

$$dt \oint (\vec{v} \wedge \vec{B}) \cdot d\vec{x} = dt \int_{S_0} \nabla \wedge (\vec{v} \wedge \vec{B}) \cdot d\vec{S}_0 \qquad (2.45)$$

If we include this equation in equation 2.43, we find

$$\Phi_1 = \Phi_0 + dt \int \left(\frac{\partial \vec{B}}{\partial t} - \nabla \wedge (\vec{v} \wedge \vec{B}) \right) \cdot d\vec{S}_0 \qquad (2.46)$$

We can now see from equation 2.14 that the term to be integrated in the right–hand side vanishes identically. As a consequence, the two fluxes are equal, which is Alfvèn's interpretation of equation 2.14.

2.4   Small Perturbations in a Homogeneous Medium

I shall discuss in detail the case of the small perturbations of a magnetohydrodynamic equilibrium, not only in order to study how the introduction of the magnetic field enriches the existing types of waves with respect to the simple hydrodynamic case, but also in order to explain how to study the stability of arbitrary astrophysical configurations. The method can be summed up in four fundamental steps:

1. Choose a configuration of equilibrium.

2. Introduce small deviations from the solution of equilibrium, and linearize the equations with respect to these small deviations.

3. Guess (or infer) the space-time dependence of the small perturbations, and infer the conditions in which there are no identically vanishing solutions.

4. Identify the properties of the different modes.

Proceeding according to this program, let us choose a configuration of equilibrium. We shall consider here a homogeneous, infinite system at rest. In general, however, any solution can be studied with this technique, even inhomogeneous and time-dependent ones (but it is harder). Let us therefore take a solution

$$p = p_0, \quad \rho = \rho_0, \quad \vec{v} = 0, \quad \vec{B} = B_0 \hat{z} \qquad (2.47)$$

where the quantities p_0, ρ_0, and B_0 are constant in space and time. We can check that this is really a solution of equilibrium by introducing it in the fundamental equations of magnetohydrodynamics, equations 2.1, 2.12, 2.13, and 2.14, and noticing that they are (obviously) satisfied.

Let us now consider a situation where the physical quantities differ only slightly from the equilibrium solution, namely,

$$p = p_0 + \delta p, \quad \rho = \rho_0 + \delta\rho, \quad \vec{v} = \delta\vec{v}, \quad \vec{B} = B_0 \hat{z} + \delta\vec{B} \quad (2.48)$$

Since we are looking for the perturbations corresponding to acoustic waves, we assume that perturbations are isentropic. In problem 3 we ask students to determine what happens to nonisentropic perturbations. For isentropic perturbations,

$$\delta p = c_{\mathrm{s}}^2 \delta \rho \tag{2.49}$$

which we shall assume and use from now on.

The equation of mass conservation, equation 2.1, taking into account the homogeneity of the zero-order solution (eq. 2.47) gives

$$\frac{\partial \delta \rho}{\partial t} + \rho_0 \nabla \cdot \delta \vec{v} + \delta \rho \nabla \cdot \delta \vec{v} + (\delta \vec{v} \cdot \nabla) \delta \rho = 0 \tag{2.50}$$

In this equation, we can see that the last two terms on the left-hand side are quadratic in small perturbations, whereas the other two terms are linear. Therefore, as long as the relative amplitude (namely, $\delta \rho / \rho$) of perturbations is $\ll 1$, the quadratic terms are much smaller than the linear ones and can be neglected. Linearizing the equations, step number 2 of the above-mentioned program, consists in this: keeping the terms linear in small perturbations, and dropping the higher order ones. We can therefore reduce the whole to

$$\frac{\partial}{\partial t} \frac{\delta \rho}{\rho_0} = -\nabla \cdot \delta \vec{v} \tag{2.51}$$

We can now proceed in the same way in perturbing equation 2.13 and equation 2.14, whereas equation 2.12 is automatically satisfied. We can thus obtain the perturbed form of equation 2.13:

$$\frac{\partial \delta \vec{v}}{\partial t} = -c_{\mathrm{s}}^2 \nabla \frac{\delta \rho}{\rho_0} + \frac{(\nabla \wedge \delta \vec{B}) \wedge \vec{B}_0}{4 \pi \rho_0} \tag{2.52}$$

Let us now perturb equation 2.14:

$$\frac{\partial}{\partial t} \delta \vec{B} = \nabla \wedge (\delta \vec{v} \wedge \vec{B}_0) \tag{2.53}$$

We notice, however, that $\vec{B}_0$ has vanishing space derivatives, so we can further simplify it thanks to the identity (where we have also used $\nabla \cdot \vec{B} = 0$)

$$\nabla \wedge (\vec{v} \wedge \vec{B}) = (\vec{B} \cdot \nabla)\vec{v} - (\vec{v} \cdot \nabla)\vec{B} - \vec{B}\nabla \cdot \vec{v} \qquad (2.54)$$

Finally, we obtain

$$\frac{\partial}{\partial t}\delta\vec{B} = B_0 \frac{\partial}{\partial z}\delta\vec{v} - B_0 \hat{z}\nabla \cdot \delta\vec{v} \qquad (2.55)$$

We have thus completed step 2 of the above program.

Now we must find a solution for equations 2.51, 2.52, and 2.55. It may seem difficult, but we get considerable help by analogy with the waves' equations. We know that, when we have systems of equations with partial derivatives with *constant coefficients*, solutions can be expanded in a Fourier series, and a solution can be found for each Fourier amplitude. Namely, let us suppose

$$\frac{\delta\rho}{\rho_0} = re^{\iota\phi}, \qquad \delta\vec{v} = \vec{V}e^{\iota\phi}, \qquad \frac{\delta\vec{B}}{B_0} = \vec{b}e^{\iota\phi}, \qquad \phi \equiv \vec{k}\cdot\vec{r} - \omega t \qquad (2.56)$$

The quantities $\vec{k}$ and ω are, for the moment, totally arbitrary; obviously, they simply identify the Fourier coefficient in the analysis of solutions. Moreover, $\vec{k}$ defines the direction of motion of the wave front.

Introducing these equations into equations 2.51, 2.52, and 2.55, we shall find

$$\omega r = \vec{k}\cdot\vec{V}$$

$$\omega\vec{V} = c_{\mathrm{s}}^2 r\vec{k} - \frac{(\vec{k}\wedge\vec{b})\wedge\hat{z}B_0^2}{4\pi\rho_0}$$

$$\omega\vec{b} = -k_z\vec{V} + \hat{z}\vec{k}\cdot\vec{V} \qquad (2.57)$$

This is a simple linear system, which is purely homogeneous, in the unknown quantities r, $\vec{V}$, and $\vec{b}$. We therefore know that it admits only the trivial solution $r = \vec{V} = \vec{B} = 0$, unless the determinant of the system vanishes. Therefore, the condition for the existence of solutions that are not identically zero is the vanishing of the determinant of this system.

Before going on, let us pause for a few comments. Imposing the vanishing of the above-mentioned determinant, we shall find the only condition that must be satisfied by the amplitudes of Fourier's expansion. The general solution will thus be given by a linear combination, with arbitrary coefficients, of the solutions corresponding to each term of the Fourier series. Does this method *always* give us the way to find the solution? The answer is no. The method of the Fourier series with respect to any variable w (no matter whether it is t or one of the components of $\vec{r}$) works only when the zero-order problem is homogeneous in the variable w in question. In fact, the coefficients appearing in the linearized equations are, as a general rule, functions of the zero-order solution. If this is homogeneous in w (namely, if w never appears), when you analyze it in a Fourier series, the dependence on w disappears, the derivatives to w being replaced by a multiplication by $\imath k_w$. This is also true if the zero-order problem is homogeneous in w, but not in other variables. On the other hand, if the zero-order problem is not homogeneous in w, the dependence of all physical quantities can be broken up into factors as $f(w)$, but the concrete form of $f(w)$ must be determined each time.[2]

It proves convenient to break up the wave vector into its components parallel $k_\parallel$ and perpendicular $k_\perp$ to the undisturbed magnetic field, $B_0 \hat{z}$:

$$\vec{k} = k_\parallel \hat{z} + k_\perp \hat{y} \qquad (2.58)$$

We can define the Alfvèn velocity v_A as follows:

$$v_A^2 \equiv \frac{B_0^2}{4\pi\rho_0} \qquad (2.59)$$

---

[2] It is perhaps worth recalling at this point, that a semihomogeneous problem (namely, a problem that is homogeneous for $z > 0$ and for $z < 0$, but not identical under the transformation $z \to -z$, as, e.g., a heavy liquid lying on a light one) is not a homogeneous problem in z, and therefore the solution of small perturbations is not $e^{\imath k_z z}$

With a little patience, the determinant of equation 2.57 can be brought to the form

$$
\det \begin{pmatrix}
\omega^2 - v_A^2 k_\parallel^2 & 0 & 0 \\
0 & \omega^2 - v_A^2 k_\parallel^2 - \left(c_s^2 + v_A^2\right) k_\perp^2 & -c_s^2 k_\parallel k_\perp \\
0 & -c_s^2 k_\parallel k_\perp & \omega^2 - c_s^2 k_\parallel^2
\end{pmatrix} = 0
$$

$$(2.60)$$

or, written in full,

$$
\left(\omega^2 - v_A^2 k_\parallel^2\right)\left(\omega^4 - \omega^2 \left(c_s^2 + v_A^2\right) k^2 + c_s^2 v_A^2 k^2 k_\parallel^2\right) = 0
$$

$$(2.61)$$

We can see, therefore, that the condition for a solution to the equations for small perturbations is a relationship between the frequency of the wave ω and the wave-number k. This relationship is called *dispersion relation*. This is due to the fact that $v_f = \omega/k$, thus determined, is not the true velocity of the wave, but only its phase velocity. Its true velocity is the group velocity, defined as

$$
v_g \equiv \frac{\partial \omega}{\partial k}
$$

$$(2.62)$$

We know that there is a dispersion phenomenon (namely, the velocity of the wave depends on its wavelength) each time that phase and group velocity differ, thence the name of equation 2.61.

Occasionally, it may happen that for a given k, ω has an imaginary part; recalling that all physical quantities depend on time like $\propto e^{-i\omega t}$, this means that the phenomenon in question is either strongly damped or strongly (exponentially) amplified. If even just one of the solutions is amplified, we can speak of *instability* of the zero-order solutions: the small perturbations, even if at first they have a small amplitude, tend to grow without bounds. The amplitude at which these instabilities stop growing cannot, in general, be determined from the linear analysis. We must perform a more complex analysis in which we also consider nonlinear terms in the development of perturbations.

However, in this case, there is no instability. It is easy to see, in fact, that there are three kinds of different solutions of equation 2.61. The first has

$$\omega^2 = v_A^2 k_\parallel^2 \qquad (2.63)$$

It shows no dispersion and is called Alfvèn's wave. It has the peculiarity of being exclusively transmitted along the direction of the magnetic field, as shown by the presence of only the parallel component $k_\parallel$ in the space dependence, $e^{i\vec{k}\cdot\vec{x}} = e^{ik_\parallel z}$.

The second and third wave correspond to the other two solutions of equation 2.61:

$$\omega^2 = \frac{k^2}{2}\left(c_s^2 + v_A^2 \pm \sqrt{\left(c_s^2 - v_A^2\right)^2 + 4c_s^2 v_A^2 k_\perp^2/k^2} \right) \qquad (2.64)$$

They are called magnetosonic waves, fast or slow, according to the sign in front of the root.

The fact that the dispersion is even in k, namely, that it exclusively depends on k^2 and not on k, means that the two directions of propagation are equivalent (see eq. 2.56). However, the existence of three absolutely distinct solutions for ω^2 means that there are three completely independent modes of oscillation. In principle, these modes could be identified by finding linearly independent and orthogonal solutions of equation 2.57, but we shall choose a more physical and intuitive route. First of all, let us consider equation 2.52 and take, for the moment, only its component along the z axis, namely the direction of the unperturbed field. Thus we find

$$\omega \delta v_z = k_z \frac{\delta p}{\rho_0} \qquad (2.65)$$

which tells us that each perturbation of the velocity along the axis of the magnetic field is exclusively due to the gradient of pressure and is not connected to perturbations of the magnetic field.

Let us consider the component of equation 2.52 along the y axis:

$$\omega \rho_0 \delta v_y = k_\perp \delta p + \frac{B_0}{4\pi}(k_\perp \delta B_z - k_\parallel \delta B_y) \qquad (2.66)$$

The equation $\nabla \cdot \vec{B} = 0$ becomes

$$k_\parallel \delta B_z = -k_\perp \delta B_y \qquad (2.67)$$

which can be introduced in the preceding equation to give

$$\omega \rho_0 k_\perp \delta v_y = k_\perp^2 \delta p + k^2 \frac{B_0 \delta B_z}{4\pi} = k_\perp^2 \delta p + k^2 \frac{\delta B_z^2}{8\pi}. \qquad (2.68)$$

Combining equations 2.65 and 2.68, we find

$$\vec{k} \cdot \delta \vec{v} = \frac{k^2}{\omega \rho_0} \delta \left(p + \frac{B^2}{8\pi} \right) \qquad (2.69)$$

This equation obviously indicates that these waves have a restoring force, the *total* pressure, namely, the sum of the fluid's pressure and of magnetic pressure. That is why they are called magnetosonic. They can be either fast or slow, because the initial values of δp and $\delta B^2/8\pi$ may have either the same or the opposite sign. Where they have the same sign, the restoring force is stronger and the waves are faster. Where they have opposite signs, the two restoring forces tend to cancel each other, and the wave will also propagate more slowly.

In order to examine the third type of wave, let us consider once again the third component of equation 2.52, the one along the x direction which is perpendicular to $\vec{B}_0$ and $\vec{k}$. We find

$$\omega \rho_0 \delta v_x = -\delta B_x \frac{k_\parallel B_0}{4\pi} \qquad (2.70)$$

From this equation, we can see that the pressure plays no role in Alfvèn's wave: the restoring force is entirely due to the magnetic tension. Also, the wave is transverse, as we noticed beforehand. Finally from equation 2.59 we see that the preceding equation can be rewritten as

$$\frac{v_x}{v_A} = -\frac{\delta B_x}{B_0} \qquad (2.71)$$

2.5 Stability of Tangential Discontinuities

Let us now apply this technique to the study of tangential discontinuities. We consider two incompressible fluids to keep the analysis moderately simple; the conclusions that we will reach here will remain qualitatively correct for compressible fluids as well, even though we will not prove this here. Let the plane xy be the undisturbed surface of separation of the fluids; in other words, one fluid is found at $z < 0$, the other at $z > 0$. The zero-order solution is a motion of the two fluids along the y axis with different velocities: v_1 and v_2. Let us also assume different values of pressure and of the magnetic field on the two sides of the separating surface, $z = 0$ (but of course identical values of the total pressure, $p + B^2/8\pi$). Lastly, we take the magnetic field to be everywhere parallel to the surface separating the two fluids, $z = 0$. Since all derivatives vanish, this is obviously a configuration of equilibrium.

We now consider a situation close to equilibrium, with slightly perturbed physical quantities:

$$\vec{v} = \vec{v}_{1,2} + \delta\vec{v}, \quad p = p_{1,2} + \delta p, \quad \vec{B} = \vec{B}_{1,2} + \delta\vec{B} \quad (2.72)$$

Here the δX terms are small perturbations, and the subscripts 1 and 2 denote the regions with $z < 0$ or $z > 0$, respectively.

We define a new velocity

$$\vec{u} \equiv \frac{\vec{B}}{\sqrt{4\pi\rho}} \qquad (2.73)$$

Since the fluids are incompressible and the zero-order solutions homogeneous, the equation for mass conservation, equation 2.1 and Maxwell's equation (eq. 2.9), to first order, give

$$\nabla \cdot \delta\vec{v} = 0, \quad \nabla \cdot \delta\vec{u} = 0 \tag{2.74}$$

Equation 2.14, still to first order, gives

$$\frac{\partial}{\partial t}\delta\vec{u} = (\vec{u} \cdot \nabla)\delta\vec{v} - (\vec{v} \cdot \nabla)\delta\vec{u} \tag{2.75}$$

whereas Euler's equation (eq. 2.13) yields

$$\frac{\partial}{\partial t}\delta\vec{v} + (\vec{v} \cdot \nabla)\delta\vec{v} = -\frac{1}{\rho}\nabla\left(\delta p + \rho\vec{u} \cdot \delta\vec{u}\right) + \vec{u} \cdot \nabla\delta\vec{u} \tag{2.76}$$

Taking the divergence of this equation, and using equation 2.74, we obtain

$$\nabla^2\left(\delta p + \rho\vec{u} \cdot \delta\vec{u}\right) = 0 \tag{2.77}$$

Let us now assume, for small perturbations, a dependence on a position like

$$\delta X \propto \exp(\imath(\vec{k} \cdot \vec{r} - \omega t) + \kappa z) \tag{2.78}$$

where the vector $\vec{k}$ is in plane xy. When we replace this form for small perturbations in the above equation, we find $k^2 = \kappa^2$, so that in order to be sure that small perturbations vanish at infinity, we must take

$$\kappa = |k| \text{ when } z < 0$$
$$\kappa = -|k| \text{ when } z > 0 \tag{2.79}$$

We can now eliminate δv_z between the two eqs. 2.75 and 2.76 to obtain

$$\delta p + \rho\vec{u} \cdot \delta\vec{u} = -\delta u_z\frac{\imath\rho}{\kappa\vec{k} \cdot \vec{u}}((\omega - \vec{k} \cdot \vec{v})^2 - (\vec{k} \cdot \vec{u})^2) \tag{2.80}$$

This equation holds separately on both sides of the surface of separation $z = 0$, and therefore represents two equations.

So far, we have not specified what happens to the surface of separation. We may assume that the surface moves away from the unperturbed position, $z = 0$, by a small amount, ζ, depending on position as

$$\zeta \propto \exp(\imath(\vec{k} \cdot \vec{r} - \omega t)) \qquad (2.81)$$

The condition that the mass flux through this surface vanishes is automatically satisfied (why?), but we must still make sure that the total pressure $p + B^2/8\pi$ is continuous across the discontinuity, and that the component of the perturbed magnetic field normal to the perturbed surface also vanishes. The latter condition simply comes from the definition of surface of separation. These three conditions give

$$(\delta p + \rho \vec{u} \cdot \delta \vec{u})_1 = (\delta p + \rho \vec{u} \cdot \delta \vec{u})_2 \qquad (2.82)$$

$$\delta u_{1z} - \vec{u}_1 \cdot \zeta = 0 \qquad (2.83)$$

$$\delta u_{1z} - \vec{u}_1 \cdot \zeta = 0 \qquad (2.84)$$

Using the first of these three equations and equation 2.80, we find the system

$$-\left(\delta u_z \frac{\imath \rho}{\kappa \vec{k} \cdot \vec{u}}((\omega - \vec{k} \cdot \vec{v})^2 - (\vec{k} \cdot \vec{u})^2)\right)_1$$

$$= -\left(\delta u_z \frac{\imath \rho}{\kappa \vec{k} \cdot \vec{u}}((\omega - \vec{k} \cdot \vec{v})^2 - (\vec{k} \cdot \vec{u})^2)\right)_2 \qquad (2.85)$$

$$\delta u_{1z} - \vec{u}_1 \cdot \zeta = 0 \qquad (2.86)$$

$$\delta u_{1z} - \vec{u}_1 \cdot \zeta = 0 \qquad (2.87)$$

This is a system of three equations in three unknown quantities, ζ, u_{1z}, and u_{2z}, which has only a trivial null solution, unless the determinant vanishes. This condition can be written as

$$(\omega - \vec{k} \cdot \vec{v}_1)^2 + (\omega - \vec{k} \cdot \vec{v}_2)^2 = (\vec{k} \cdot \vec{u}_1)^2 + (\vec{k} \cdot \vec{u}_2)^2 \qquad (2.88)$$

which is, obviously, the dispersion relation we were looking for.

Now we can easily see that the system has no complex roots if

$$2(\vec{k} \cdot \vec{u}_1)^2 + 2(\vec{k} \cdot \vec{u}_2)^2 - (\vec{k} \cdot (\vec{v}_1 - \vec{v}_2))^2 > 0 \quad (2.89)$$

which can be rewritten as

$$k_i k_j (2u_{1i}u_{1j} + 2u_{2i}u_{2j} - v_i v_j) > 0 . \quad (2.90)$$

Here $\vec{v} \equiv \vec{v}_1 - \vec{v}_2$ is the relative velocity between the two fluids. We know from elementary algebra that this form is positive-definite if the determinant and the trace of the matrix in parentheses are positive. Thus we have, as conditions of stability, that

$$B_1^2 + B_2^2 > 2\pi\rho v^2, \; (\vec{B}_1 \wedge \vec{B}_2)^2 > 2\pi\rho \left((\vec{B}_1 \wedge \vec{v}_1)^2 + (\vec{B}_2 \wedge \vec{v}_2)^2 \right)$$
$$(2.91)$$

If the density is also discontinuous, the formulae above are still correct, with the replacement

$$\rho \to \frac{2\rho_1\rho_2}{\rho_1 + \rho_2} \quad (2.92)$$

In the hydrodynamic limit, when $\vec{B} \equiv 0$, it is impossible to satisfy the first part of equation 2.91, and the discontinuity is absolutely unstable. We speak in this case of *Kelvin-Helmholtz instability*, and we find that it grows at the rate:

$$\omega = \frac{\vec{k} \cdot (\vec{v}_1 + \vec{v}_2) \pm 2\imath (\vec{k} \cdot v_1 - \vec{k} \cdot \vec{v}_2)}{4} \quad (2.93)$$

The presence of an imaginary part in ω immediately implies an instability. In fact, if we remember that we have taken all perturbations $\propto \exp(-\imath\omega t)$, we see that, as time goes by, the amplitude of small perturbations grows without bounds. This is the first case of instability we have found. It can be noticed that, in this case, there is instability for each value of k, whereas it may happen, in general, that instability exists for just a finite interval in k.

On the other hand, when there is a magnetic field, this can stabilize the surface of separation. However, the required magnetic fields are rather large.

2.6 Two-Temperature Fluids

One of the most characteristic phenomena of plasmas is that, in astrophysical conditions, a fluid constituted only by electrons may be at a temperature T_e different from the fluid constituted by ions T_i; normally, the inequality $T_e \lesssim T_i$ holds. In order to understand the reason for this difficulty in reaching the thermal equilibrium of an astrophysical plasma, we shall leave out radiative processes. Obviously these, always favoring the emission from electrons, tend to exacerbate the relation $T_e < T_i$, instead of softening it.

The fundamental idea is very simple: the thermal equilibrium, namely, the fact that electrons and ions have a Maxwell-Boltzmann velocity distribution at the same temperature, is reached by means of collisions, which redistribute energy between particles and make the velocity distribution isotropic. However, since the proton is much more massive than the electron, the collision between the two is quasi elastic: the electron bounces back with almost its initial energy, transferring very little momentum to the proton. If the mass ratio were $m_p/m_e \rightarrow \infty$, the momentum transfer would vanish exactly. In an *impulsive* approximation; in other words, when the electron is so fast that its trajectory is very close to a straight line, the fraction of kinetic energy of the electron transferred to the proton is $\delta E_K/E_K \approx (E_p/E_K)^2$, where $E_p = e^2/b$ is the potential energy at the point of closest approach, b. However, the impulsive approximation is valid only if $E_p \ll E_K$, whence we can see that the energy transfer between electrons and protons is rather inefficient. (Why are we not considering *slow* collisions?)

This topic can be made rigorous through relaxation times (Spitzer 1962). Let us now introduce these different time scales, and then consider what happens to a group of electrons and protons with initially arbitrary distributions of velocity and energy.

Let us consider a particle with a mass m, a velocity v, and a kinetic energy $E = mv^2/2$, which moves in a medium constituted by n_f particles per unit of volume, with a mass m_f and a temperature T. The time of deflection, that is, the time in which the particle m loses memory of its initial direction of motion, is

$$t_d = \frac{v^3}{A_d[\Phi(E/kT) - G(E/kT)]} \qquad (2.94)$$

where A_d is the scattering constant:

$$A_d \equiv \frac{8\pi e^4 n_f \ln \Lambda}{m^2} \qquad (2.95)$$

and $\ln \Lambda$, the *Coulomb logarithm*, is a factor that depends only weakly on all quantities, $\ln \Lambda \approx 6 - 30$, and describes the relative importance of weak collisions (i.e., the most frequent ones!) with respect to strong ones. The functions Φ and G are defined as

$$\Phi(x) \equiv \frac{2}{\sqrt{\pi}} \int_0^x e^{-y^2} \, dy , \qquad G(x) \equiv \frac{\Phi(x) - x\Phi'(x)}{2x^2} \qquad (2.96)$$

Obviously, $\Phi(x)$ is the usual error function, which tends to 0 for $x \to 0$, and $\Phi \to 1$ when $x \to \infty$; $G(x)$ tends to 0 for $x \to 0$ and for $x \to \infty$.

On the other hand, the time scale on which energy is redistributed is

$$t_E = \frac{v^3}{A_d G(E/kT)} \qquad (2.97)$$

which differs significantly from t_d: in fact, $t_d/t_E = 4G(E/kT)/(\Phi(E/kT) - G(E/kT)) \to 0$ when $E/kT \to \infty$. In other words, a fast particle is easily deflected, but it is more difficult to steal its energy.

In the same way, we can obtain the time scale on which identical particles with *comparable* velocities will redistribute their energies and make their velocities isotropic. In fact, it is possible to use equations 2.94 and 2.97 to this aim, by using

$E/kT = 3/2$. In this case, we see that $t_{\rm d}/t_{\rm E} \approx 1.1$, and it is therefore possible to define a time of self-collision

$$t_{\rm c} = \frac{m^{1/2}(3kT)^{3/2}}{5.7\pi n\, e^4 \ln \Lambda} \tag{2.98}$$

This time scale is sufficient to redistribute energies and to make velocities isotropic, because $t_{\rm c} \approx t_{\rm E} \approx t_{\rm d}$.

Finally, when we have two kinds of particles with a Maxwell-Boltzmann velocity distribution but with different temperatures, the time scale on which the thermal equilibrium is reached is

$$\frac{dT}{dt} = \frac{T_{\rm f} - T}{t_{\rm eq}} \tag{2.99}$$

where the time of equipartition $t_{\rm eq}$ is given by

$$t_{\rm eq} = \frac{3m\, m_{\rm f}}{8\sqrt{2\pi} n_{\rm f} e^4 \ln \Lambda} \left(\frac{kT}{m} + \frac{kT_{\rm f}}{m_{\rm f}} \right)^{3/2} \tag{2.100}$$

Let us now consider a group of electrons and protons with initially arbitrary distributions of velocities, but with energies of the same order of magnitude. At first, the collisions of electrons with protons lead to an *isotropy* of velocities, without, however, leading to a significant exchange of energies, because, as we have already seen, $t_{\rm d} \ll t_{\rm E}$ when the velocities of electrons are higher than the average velocities of the protons. Thus, it is easy to see that $t_{\rm c} \ll t_{\rm E}$ for electrons, and therefore self-collisions bring electrons towards a Maxwell-Boltzmann distribution. The same is true for protons, but the self-collision time $t_{\rm c} \propto \sqrt{m}$, so that protons become isotropic and reach the Maxwell-Boltzmann distribution on a longer time scale than electrons, by a factor $\sqrt{m_{\rm p}/m_{\rm e}} \approx 43$. Finally, let us compare $t_{\rm eq}$ with $t_{\rm c}$ for the electrons: we find $t_{\rm c}/t_{\rm eq} \approx m_{\rm e}/m_{\rm p}$. If we now compare $t_{\rm c}$ for the protons with $t_{\rm eq}$, we find $t_{\rm c}/t_{\rm eq} \approx \sqrt{m_{\rm e}/m_{\rm p}}$. Therefore, the time in which equipartition is reached (i.e., the same temperature) is the longest of all.

Therefore, if we consider plasmas on time scales $\gg t_{\text{eq}}$, they will be characterized by a single temperature for electrons and protons. On the other hand, if we consider the fluid on shorter time scales, or, alternatively, radiative times are $\lesssim t_{\text{eq}}$, since electrons lose energy through radiation more easily than protons because of their small mass, the temperature difference between the two fluids is likely to increase, not to decrease. The main (but not the only) case in which this property finds its application is in advection-dominated accretion flows, which we shall discuss in chapter 7. Another very important case in which differences of temperature might be generated is after shocks, as we shall discuss at the end of section 2.8.

Finally, we must emphasize that t_{eq} represents an upper limit to the time scale on which complete thermal equilibrium is established. A series of noncollisional phenomena, called *collective* processes, contribute to supplement the effectiveness of collisions; they generally involve interactions between waves and particles (Begelman and Chiueh 1988). The discussion of these phenomena, which are, in any case, extremely uncertain, lies outside the scope of this book.

2.7   Magnetic Buoyancy and Reconnection

The approximation of ideal magnetohydrodynamics is often excellent, but there are still two important phenomena that dissipate the magnetic field. The first one, the so-called Parker instability (1955), is important in the sense that it is a mechanism for the expulsion of the magnetic flux, which is totally independent of dissipation. On the other hand, the second one, magnetic reconnection, depends on conductive dissipation, in certain special situations, which we shall describe.

2.7.1   Magnetic Buoyancy

Many astrophysical phenomena are intrinsically turbulent; a classical example is the convective heat transfer in stars. In these conditions, it may happen that the average element of fluid is in dynamic equilibrium, with a gradient of pressure that balances the force of gravity; however, occasionally bubbles of fluid with a magnetic field larger than average may form. There is a dynamic difference between these two types of fluid elements; as we shall see, the bubbles with high magnetic content are lighter than the average fluid, tend to float, and are likely to be expelled.

Let us consider an atmosphere stratified along the direction z. An atmosphere is a region where gravitational attraction is mainly due to an external element. In a star, the atmosphere includes the outermost strata, which contain an infinitesimal fraction of the total mass. On the other hand, in an accretion disk, the gravitational attraction is due to an external object (the compact object, neutron star or black hole, around which the disk orbits). The average element of fluid must be in dynamic equilibrium.[3] In other words, $\vec{v} = 0$. This is possible because in Euler's equation, with the contribution of the gravitational potential ϕ,

$$\frac{\partial \vec{v}}{\partial t} + (\vec{v} \cdot \nabla)\vec{v} = -\frac{\nabla p}{\rho} - \nabla\phi \qquad (2.101)$$

the right-hand side vanishes; specializing to a plane atmosphere, we must have

$$\frac{dp}{dz} = -\rho g \qquad g \equiv \frac{d\phi}{dz} \qquad (2.102)$$

For the element of fluid containing a significant contribution from a magnetic field to its pressure, on the other hand, we must apply Euler's equation (eq. 2.13), together with the gravitational term

$$\frac{Dv_z}{Dt} = -\frac{1}{\rho_m}\frac{dp_m}{dz} + \frac{((\nabla \wedge \vec{B}) \wedge \vec{B})_z}{4\pi\rho_m} - g \qquad (2.103)$$

---

[3] We neglect here convective motions, which push fluid elements alternatively up and down along the radial direction, in order to keep the analysis simpler.

where the subscript m indicates that the values of the magnetic bubble might be different from those of the nonmagnetized fluid. We have already seen that the term producing the magnetic force can be rewritten as

$$(\nabla \wedge \vec{B}) \wedge \vec{B} = -\frac{1}{2}\nabla B^2 + (\vec{B} \cdot \nabla)\vec{B} \qquad (2.104)$$

In our system, we can assume that the second term on the right vanishes. This happens, for example, when the magnetic field is mostly along the direction x or y, whereas the main variation is along the normal direction, that is, z. This simplifies our treatment compared to the general case, which, however, reaches the same conclusions (Parker 1955). We can therefore simply write the following form of Euler's equation:

$$\rho_m \frac{Dv_z}{Dt} = -\frac{d}{dz}\left(p_m + \frac{B^2}{8\pi}\right) - \rho_m g \qquad (2.105)$$

Initially, the magnetic bubble will be at a distance $z_0 < 0$ from the star surface, but convective motions will tend to push it toward the surface. When this happens, we can suppose that the bubble is in local pressure equilibrium at each moment (otherwise, it would be compressed, or it would quickly expand in order to reach equilibrium), and that it does not exchange heat with the external medium, because motions are fast compared to any process of entropy exchange. In this situation, the bubble density cannot remain constant. If the bubble were made only of fluid, like the surrounding medium, its thermodynamic evolution would be dictated by the equation $p \propto \rho^\gamma$, with $4/3 < \gamma \leq 5/3$ appropriate for an ideal fluid. However, since there is a magnetic field, we need to take into account the evolution of the magnetic pressure. From the flux-freezing theorem, we know that $BR^2 =$ constant, since the bubble surface is obviously $4\pi R^2$. Therefore, the magnetic pressure scales as $R^{-4} \propto V^{-4/3} \propto \rho^{4/3}$. As a consequence, the magnetic field behaves like a polytropic fluid, with the index $\gamma = 4/3$.

The sum of gas and magnetic field can thus be approximated by a fluid with $p \propto \rho^{\gamma_m}$, with $\gamma_m < \gamma$. It follows that, if the pressure is reduced by δp on an infinitesimal path, in the nonmagnetized medium the density is reduced by

$$\frac{\delta \rho}{\rho} = \frac{\delta p}{\gamma p} \tag{2.106}$$

whereas the same reduction in the pressure of the surrounding medium is realized in the bubble with a drop of density

$$\frac{\delta \rho_{\rm m}}{\rho_{\rm m}} = \frac{\delta p}{\gamma_{\rm m} p} \tag{2.107}$$

Since $\gamma_{\rm m} < \gamma$, we find that

$$\frac{\delta \rho_{\rm m}}{\rho_{\rm m}} = \frac{\gamma}{\gamma_{\rm m}} \frac{\delta \rho}{\rho} > \frac{\delta \rho}{\rho} \tag{2.108}$$

The reduction of density, while the magnetic bubble rises to the surface, is larger than that of the surrounding medium. The bubble is therefore lighter and is pushed toward the surface by Archimedes's principle; this property is called magnetic *buoyancy*. In order to see it in detail, we must remember that $p_{\rm m} + B^2/8\pi$, the pressure in the bubble, must equal p, the pressure outside the bubble, at *any* height z. Therefore

$$\frac{dp}{dz} = \frac{d}{dz}\left(p_{\rm m} + \frac{B^2}{8\pi}\right) = -\rho g \tag{2.109}$$

where I used equation 2.102. Now we only need to introduce this into equation 2.105:

$$\rho_{\rm m}\frac{Dv_{\rm z}}{Dt} = (\rho - \rho_{\rm m})g \tag{2.110}$$

Since, as we have just shown, $\rho > \rho_{\rm m}$ at any height, it follows that the magnetized bubble is subject to a net force toward the surface: the principle of Archimedes. From this equation it is easy to see that the velocity acquired by the magnetic bubble is of the order $\eta v_{\rm A}$, where the factor η is not well specified by this analytical discussion. From numerical simulations, we find $\eta \approx 0.1$.

What happens to the bubble as it reaches the surface of the star or of the disk? The bubble, which is in the void, now expands freely and spreads the magnetized matter in the surrounding medium. In this way, the star can rid itself of some magnetic field.

2.7.2 Reconnection

Another mechanism that leads to the dissipation of the magnetic field takes place when two regions with a magnetic field in opposite directions are pushed one against the other. In this case, the fluid separating the two regions is compressed along one direction and escapes along another; there is therefore a transition between a magnetic field $+B$ and $-B$ on a shorter and shorter distance. In that case, it is no longer possible to neglect the dissipative term j^2/σ, even for astrophysical values of conductivity. Besides, in the presence of strong dissipative heating, magnetohydrodynamic instabilities may cause the formation of turbulence, which increases the resistivity up to values several orders of magnitude larger than the usual ones (we speak then of *anomalous* resistivity). Because of dissipation, the magnetic field is destroyed, and its whole topology is altered.

We shall discuss here at an elementary level the velocity with which reconnection can take place. Let us imagine two clouds, pushed one against the other along the z direction, which tally along a distance L (fig. 2.3). Each cloud has a magnetic field with the same modulus $B\hat{x}$ but with opposite sign, directed along the x axis. By assuming stationary conditions, at what speed can the cloud come closer?

In Euler's equation (eq. 2.19), we once again use the identity equation 2.104. Assuming stationary conditions, we have

$$(\vec{v} \cdot \nabla)\vec{v} = -\frac{1}{\rho}\nabla p + \frac{(\nabla \wedge \vec{B}) \wedge \vec{B}}{4\pi\rho} \qquad (2.111)$$

We expect that the fluid moves mainly along the x direction. In this case, the component of Euler's equation along the z

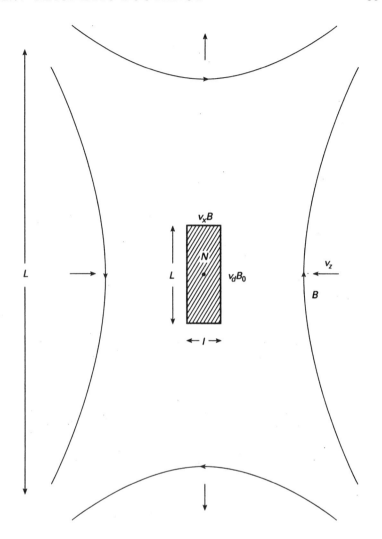

Figure 2.3. Geometry of the reconnection area, which takes place in the hatched zone. Soward and Priest 1977.

direction is

$$\frac{d}{dz}\left(p + \frac{B^2}{8\pi}\right) = 0 \qquad (2.112)$$

where I used the identity equation 2.104 in order to rewrite the magnetic term in Euler's equation. This shows that the quantity $p + B^2/8\pi$ is constant along z. Near the plane $z = 0$, where the reconnection takes place, the magnetic field vanishes, so that the pressure reaches its peak for $z = 0$. This pushes the fluid out of the region. In fact, the component of Euler's equation along the x axis is

$$\rho v \frac{dv}{dx} = -\frac{dp}{dx} \tag{2.113}$$

where the magnetic term disappears because $\vec{B} = B\hat{x}$. In order to simplify computations, let us assume that the fluid is incompressible, $\rho = $ constant. Therefore

$$\frac{\rho v^2}{2} + p = \text{constant} \tag{2.114}$$

along the x axis. The pressure difference between the central area, where reconnection takes place, and the outside, toward which the fluid escapes, is obviously $\triangle p = B^2/8\pi$. Therefore, the outflow velocity is given by

$$\frac{v^2}{2} = \frac{B^2}{8\pi\rho} \to v = v_{\mathrm{A}} \tag{2.115}$$

This outflow along the axis x must be balanced, in stationary conditions, by an equal inflow along the z axis. Having assumed that the fluid is incompressible, the equality of mass fluxes gives

$$v_{\mathrm{d}} L = v_{\mathrm{A}} l \tag{2.116}$$

where L is the dimension along the x axis of the reconnection area, and $l \ll L$ the one along the z axis, and v_{d} is the velocity of the influx, as well as of the destruction of the magnetic field.

In order to determine l, let us notice that, in stationary conditions, the rate at which the density of magnetic energy is carried to the destruction area (the small zone with $-l \le z \le l$) must equal the dissipation rate. The first is obviously given by $R_1 = v_{\mathrm{d}} B^2/8\pi l$, the second one is given by $R_2 = j^2/\sigma$.

We can roughly evaluate j by Maxwell's equation (eq. 2.9), thus obtaining $j = cB/4\pi l$. Setting $R_1 = R_2$, we find

$$l = \frac{c^2}{2\pi\sigma v_d} \tag{2.117}$$

We can now rewrite the solutions we have just found for v_d and l using the magnetic Reynolds number, equation 2.18:

$$v_d = \frac{2v_A}{\mathcal{R}_m^{1/2}} \qquad l = \frac{2L}{\mathcal{R}_m^{1/2}} \qquad \mathcal{R}_m \equiv \frac{4\pi\sigma L v_a}{c^2} \tag{2.118}$$

This result is only partly encouraging: it is true that the velocity $v_d \propto v_A$, but the coefficient of proportionality, $\mathcal{R}_m^{-1/2} \approx 10^{-7}$, is still very small, for typical values of $\mathcal{R}_m \approx 10^{14}$. The best way out of this impasse (Petschek 1964) is to realize that reconnection need not necessarily take place in a zone of transverse dimension L. In fact, it may consist of one point only: referring to Figure 2.3, the four arched lines should simply be deformed to create a cross, with only one connecting point. Various phenomena help this configuration to maintain a high destruction rate for the magnetic field. On one hand, the lines of force of the newly reconnected magnetic field are endowed with tension and help to push the fluid out of the reconnection region. On the other hand, the pressure vacuum caused by the quick outflow of matter swallows the still magnetized fluid into the reconnection region. A detailed analysis of this configuration (called type X reconnection) lies outside the scope of this book; we shall simply state without proof that the reconnection speed is much larger than in the above-mentioned case. We find indeed (Soward and Priest 1977)

$$v_d = \frac{\pi v_A}{4(\ln \mathcal{R}_m + 0.74)} \approx (0.01 - 0.1)v_A \tag{2.119}$$

which is the typical value used in all modern numerical estimates. It also offers the advantage of a weak (logarithmic) dependence on both the configuration parameters and the conductivity.

Where does the magnetic energy thus dissipated end up? Obviously, it ends up in the currents and, through resistivity, in the internal energy of the gas carrying the currents, namely, *in primis*, electrons, which, after a while, will share this internal energy with protons.

2.8 Shock Waves

The description of discontinuities in magnetohydrodynamics is more complex than in hydrodynamics. However, the following circumstance in astrophysics allows us a remarkable simplification. Except for the regions surrounding pulsars, astrophysical fluids are never dominated by the magnetic field, namely,

$$\frac{B^2}{8\pi p} = \frac{\gamma v_A^2}{2c_s^2} \lesssim 0.1 - 0.3 \qquad (2.120)$$

It follows that the dynamic importance of magnetic fields can always be approximated as small. This circumstance also allows us to neglect the fact that shock waves in magnetohydrodynamics must always be super-Alfvénic, that is, $V_s > v_A$, where V_s is the velocity of the shock wave. In fact, since $v_A = \sqrt{B^2/(4\pi\rho)}$ and $B^2/8\pi \lesssim \rho c_s^2$, the condition $V_s > v_A$ is equivalent to $V_s > c_s$, a condition that we already know as absolutely necessary.

In magnetohydrodynamics, there are two distinct kinds of discontinuity without mass crossing of the discontinuity surface, whereas in hydrodynamics there is only one. The new discontinuity, sometimes called *rotational* discontinuity, is stable to corrugating perturbations. The other kind of discontinuity, namely, that generalizing tangential discontinuities, can be stabilized with a rather intense magnetic field, as we discussed in section 2.5. These two circumstances together might make us think that in magnetohydrodynamics there can be discontinuities other than shock waves, which are stable as well, but this is not the case.

First, the magnetic field necessary to stabilize tangential discontinuities is larger than the one of equation 2.120. Second, it is difficult to imagine a concrete physical situation where only a rotational discontinuity is generated. It follows that all discontinuities contain a tangential component, which is not stabilized, thus creating a wide turbulent interface between the two fluids. It follows in turn that essentially all discontinuities without flux crossing (in astrophysics) are unstable in magnetohydrodynamics and in hydrodynamics.

As far as shock waves are concerned, the conditions on the discontinuity surface are always the same: continuity of the fluxes of mass, momentum, and energy. We use equations 2.1, 2.19, and 2.25 and indicate with X_n and X_t the components of a vector X normal and tangential to the shock surface, respectively. We find, for the continuity of the mass and momentum fluxes (respectively) perpendicular and parallel to the shock, and of energy:

$$[\rho v_n] = 0; \quad \left[p + \rho v_n^2 + \frac{1}{8\pi}\left(B_t^2 - B_n^2\right)\right] = 0; \quad \left[\rho v_n v_t - \frac{1}{4\pi}B_n B_t\right] = 0$$

$$\left[\rho v_n \left(\frac{v^2}{2} + w\right) + \frac{1}{4\pi}\left(v_n B^2 - B_n(\vec{v}\cdot\vec{B})\right)\right] = 0 \quad (2.121)$$

We must add the usual electromagnetic conditions to these equations: the components B_n and E_t are continuous. Since we are in the limit of ideal magnetohydrodynamics, $\vec{E} = -\vec{v}\wedge\vec{B}/c$ (eq. 2.24), and we have additional conditions:

$$[B_n] = 0 \qquad [B_n v_t - B_t v_n] = 0 \qquad (2.122)$$

The analysis of these equations is simple in two limiting cases. In the first one, the magnetic field before the shock is normal to the shock surface, $B_t = 0$. This situation is called *parallel*, because the magnetic field is parallel to the normal of the shock. In this case, it is easy to show that $B_t = 0$ also after the shock. Since the normal magnetic field and the one tangential to the shock are continuous,

$$B_{n1} = B_{n2} \qquad B_{t1} = B_{t2} = 0 \qquad (2.123)$$

we can easily see that this case can be exactly reduced to the purely hydrodynamic case, as if the magnetic field did not exist. On the other hand, if we suppose $B_n = 0$, (perpendicular case), we can see from the third of the equations in equation 2.121 that v_t is continuous, so that we can once again enter a reference frame where $v_t = 0$. Moreover, we can see from the second of the equations in equation 2.122, and from the first of the equations in equation 2.121 that

$$\frac{B_{t1}}{B_{t2}} = \frac{\rho_1}{\rho_2} \tag{2.124}$$

The other equations in equation 2.121 reduce to

$$[\rho v] = 0 \quad \left[p + \frac{1}{2}v^2 + \frac{B_t^2}{8\pi} \right] = 0 \quad \left[\frac{1}{2}v^2 + w + \frac{B_t^2}{8\pi\rho} \right] = 0,$$

$$\tag{2.125}$$

which are almost identical to those of ordinary hydrodynamics, apart from the fact that we had to introduce the contributions of the magnetic field to pressure and density of energy.

We must now discuss the temperature of electrons after the shock, namely, whether we have $T_e = T_i$. Strictly speaking, this discussion also applies to purely hydrodynamic shocks, but we postponed it to discuss the physics of fluids with two temperatures (section 2.6). We saw in chapter 1 that in astrophysical shocks, what transforms ordered kinetic energy into internal kinetic energy is not collisions between particles, but variable electromagnetic fields. Since electrons have the same charge as ions, they are subject to the same forces, but also to much larger accelerations due to their small mass: $m_e/m_p \approx 1/1836$. In these conditions it is certainly possible that electrons may radiate copiously and may dissipate the internal kinetic energy generated in the shock wave. It follows that they come out of the shock with a temperature $T_e < T_i$, and even $T_e \ll T_i$. This means that the shock dissipates a fraction of its own energy, but, since before the shock, where the kinetic energy is mostly directed, electrons possess only a fraction $m_e/m_p \ll 1$ of the total energy, even if all of this energy were dissipated, there would be no important dynamic consequences on the Rankine-Hugoniot conditions. The ratio T_e/T_i depends in detail on the processes transforming directed energy into internal energy; at the moment, this quantity is not well determined, neither by theoretical arguments nor by numerical simulations. After the shock, we must wait at least a time t_{eq} (eq. 2.100) for the establishment of thermal equilibrium between electrons and ions.

But electrons need not reach the protons' temperature, even if we completely neglect radiative processes. In fact, there is no obvious reason why the shock thickness should be so large as to allow the relative thermalization of protons and electrons. If these processes are ineffective, the electrons will transform into internal energy the same fraction of their directed energy as protons (barring radiative processes, of course), which is, in any case, smaller by a factor of m_e/m_p than protons'. In conclusion, we can say that in absence of radiative losses within the shock thickness and of heat transfer from protons to electrons, the temperature of electrons behind the shock is a factor m_e/m_p lower than that of protons, which we calculate from the usual Rankine-Hugoniot (or Taub) equations. Successive processes of energy transfer from protons to electrons tend to increase T_e, provided they are faster than the radiative cooling that electrons undergo in any case.

2.9 Magnetic Fields in Astrophysics

Without claiming completeness, in this section we shall describe what we currently know about astrophysical magnetic fields.

2.9.1 Observations

Table 2.1 shows, in summary, the magnetic properties of all astrophysical sites in which we could make direct

Table 2.1. Values of the magnetic field in astrophysical objects.

Astrophysical site	$B/1$ G	Coherence	Method*
Intergalactic field	$<10^{-9}$		RFE
Clusters of galaxies	10^{-6}	100 kpc?	RFI
Quasars and radio galaxies	100	<1 pc	
Milky Way			
Regular component	3×10^{-6}	1 kpc	RFE, RFP, POP
Random component	2×10^{-6}	<55 pc	RFE, RFP, POP
Molecular clouds	10^{-4}	10 pc	Z21
Maser sources, dense clouds	10^{-2}	$<10^{16}$ cm?	ZOH
Accretion disks	10^{9}	Turbulent	ST
Gamma ray burst	10^{3}?		PSG

*RFE, Faraday rotation in external sources; RFI, intrinsic Faraday rotation; PR, radio polarization; RFP, Faraday rotation in pulsars; POP, optical polarization in the dust; Z21, Zeeman effect in the hydrogen line at 21 cm; ZOH, Zeeman effect in the OH molecule lines; ST, theoretical estimates based on the energy transport mechanism; STT, theoretical estimates based on the deceleration rate or on the position of Alfvèn radius; CX, cyclotron lines in the X band; PS, polarization in synchrotron emission; ZO, Zeeman effect in optical lines; ML, *in loco* measurement; PST, theoretical estimate. Adapted from Zel'dovich, Ruzmaikin, and Sokoloff 1990.

Astrophysical site	$B/1$ G	Coherence	Method
Pulsar			
Large distance field	10^8–10^{14}	Dipolar field	STT
Surface field	10^{11}–10^{13}	Multipolar	CX
Soft gamma ray repeaters	10^{15}	Dipolar field	STT
Magnetic white dwarfs	10^6–10^8	Dipolar field	PS
Ap Stars	10^4	Dipolar field	ZO
Sun			
Polar field	0.5	0.1–1 R.	ZO, RFI
Subphotospheric azimuthal field	10^3	Azimuthal field	
Corona	10^{-5}		
Planets			
Jupiter	4	Dipolar field	ML
Saturn	0.2	Dipolar field	ML
Earth	1	"	"
Mercury	3×10^{-3}	Dipolar field	ML
Mars	6×10^{-4}	Dipolar field	ML

97

measurements. This table adequately illustrates the perva-
siveness of magnetic fields: except for regions completely outside galaxies and clusters, we have either direct measurements
of the magnetic field, or really strong arguments indicating
the presence of important magnetic fields. The only exception, as we said before, regards the existence of an intergalactic magnetic field.

Another comment concerns the techniques of measurement, as well as theoretical estimates. With the exception of
the Zeeman effect, which should be well known from elementary courses, all arguments and methods will be described in
this book. The most important method is certainly Faraday
rotation, which consists of the fact (see appendix A) that
a polarized wave can be linearly decomposed into the sum
of two waves with opposite circular polarization, propagating within a magnetized medium with different velocities. In
appendix A, we show that the two waves, respectively, with
a right-hand and a left-hand polarization but identical frequency, have slightly different wave vectors:

$$k_\pm = \frac{\omega}{c}\sqrt{1 - \frac{\omega_p^2}{\omega(\omega \pm \omega_c)}} \qquad (2.126)$$

where ω_p and ω_c are, respectively, the plasma and cyclotron
frequencies,

$$\omega_p \equiv \sqrt{\frac{4\pi n\, e^2}{m_e}}, \qquad \omega_c \equiv \frac{eB_\parallel}{m_e c} \qquad (2.127)$$

and $B_\parallel$ is the component of the magnetic field along the direction of propagation. Since the space-time dependence of
the wave is $\propto e^{i(kx-\omega t)}$, the two waves, which cover the same
path, reach the observer with slightly different phases:

$$\triangle \phi = \int ds(k_+ - k_-) \approx \frac{2\pi e^3}{m_e^2 c^2 \omega^2} \int_0^D nB_\parallel ds \qquad (2.128)$$

In this equation, the product $nB_\parallel$ is integrated along the
entire line of sight, from the source to us. Therefore, when

observing, we shall see a wave given by the linear combination of the two waves with opposite circular polarization, but slightly different phases, which, when summed up, give origin to a wave with linear polarization, in which, however, the polarization plane is inclined by a $\triangle\phi$ angle with respect to the original one.

Let us now suppose that we observe a source showing a measurable amount of linear polarization. There are many sources of this kind in nature, including many types of active galactic nuclei, pulsars, and so on. In fact, it does not even matter whether we know either the type of source or its properties; the background source acts exclusively as a probe of the medium in which light propagates, to measure the magnetic field. If we made measurements at one frequency only, we could say nothing. However, since we can continuously vary the frequency of observation, $\omega/2\pi$, we can see that the polarization axis smoothly varies with ω, like $1/\omega^2$. This effect allows us to measure the integral of $nB_\parallel$ along the line of sight. As a general rule, for extragalactic measurements, this integral will receive the largest contribution from the path within galaxies, because only here is n appreciably different from zero. Besides, the contribution of the path within the Milky Way can be evaluated with a certain precision thanks to the observation of sources close on the plane of the sky, but always within our Galaxy. When this contribution is subtracted from the value obtained through observation, the rest is the external contribution.

The value of $\int nB_\parallel ds$ thus obtained obviously gives a *lower* limit to the true value of B. This is not because the integral measures only one component of B instead of the three independent ones that it has, since this distortion can be easily corrected in a statistical manner by multiplying the value obtained through observation for $\sqrt{3}$ (why?). The reason is that $B_\parallel$ can give both positive and negative contributions to the above-mentioned integral, which, statistically, tend to cancel. Moreover, when we try to derive a *typical* value of $B_\parallel$ from the estimate of the integral, there is always uncertainty regarding the length of the path to be used to reach an estimate of B. Despite these uncertainties, the measurements indicate values of B that are consistently different from zero.

A particularly significant value of B is derived for the magnetic field of the galaxy, in its ordered and in its turbulent components. This is due to the fact that the magnetic field has roughly the same energy density of the other two main elements of the interstellar medium in our Galaxy, namely, gas and nonthermal particles. The gas has a density of $n \approx 1$ cm^{-3} particles, and a temperature $T \approx 10^4$ K; nonthermal particles have an energy density of ≈ 1 eV cm^{-3}, whereas the magnetic field has an amplitude of about 3×10^{-6} G (Spitzer 1978). Such a magnetic field is called, for obvious reasons, an *equipartition* field. This property still has no detailed explanation, but we do know that many processes can exchange energy among these components. For example, if there are magnetic bubbles (i.e., zones in which the gas pressure is lower than the magnetic pressure) that have a lower (or higher) pressure than that of the gas, they will compress (or expand) until they reach pressure balance.

The magnetic field shows, at least in our galaxy, fluctuations on all scales smaller than 55 pc. On scales from some parsecs upward, it is generally thought that this is due to supernovae (see next paragraph), but on small scales this is certainly due to the presence of turbulence in the fluid in which the magnetic field is frozen. According to a well-known analysis (Kraichnan 1965), turbulence in the presence of a magnetic field should have velocity and magnetic field fluctuations v_λ and B_λ on a scale λ that are in this relationship:

$$\rho v_\lambda^2 \approx \frac{B_\lambda^2}{4\pi} \tag{2.129}$$

whereas the energy density per unit mass in fluctuations with wave-number k $E(k)$ scales as

$$E(k) = Ak^{-\alpha}, \qquad \alpha = 3/2 \tag{2.130}$$

What may appear surprising is that this kind of turbulence is directly observable, thanks to interstellar scintillation (Narayan 1992). Observations indicate a spectral index $\alpha \approx 5/3$, which is consistent with ordinary Kolmogorov-Okhubo turbulence. This is a surprising result, since the Kolmogorov-Okhubo theory applies to isotropic fluids without a magnetic field, which an astrophysical fluid should not be. Moreover, some measurements even report values of $\alpha = 2$ (Wilkinson, Narayan, and Spencer 1994), *a fortiori* in flagrant violation of Kraichnan's prediction.

2.9.2 Origin of Magnetic Fields

What is the origin of observed magnetic fields? First of all, we can notice that equation 2.11 does not contain a source term; in other words, if initially $B = 0$, it will remain zero forever. Therefore, at least the origin of the magnetic field must be searched for outside of magnetohydrodynamics. The effects that originate magnetic fields are called *batteries*, because, exactly like in a laboratory, nonelectromagnetic phenomena are called on to give origin to currents. It seems quite easy to invent plausible batteries, and we shall illustrate two models of this kind.

The next question is whether, given these initial fields, we can find plausible mechanisms for them to grow to observed values. Obviously, we can see that all the equations of magnetohydrodynamics are nonlinear and therefore contain at least the possibility to increase the amplitude of magnetic fields. The models that make this amplification are called *dynamos*.

Once suitably amplified, there is no doubt that at least a part of the magnetic fields that we observe are due to effects of contamination; for example, supernovae and, if they exist, superbubbles inject a huge amount of matter into the interstellar medium. This matter is magnetized and is certainly partly responsible for the variability of magnetic fields on small-scale distances in the galaxy. Indeed, the formation rate of supernovae in the galaxy is about one each 30 years, which implies that since the formation of the disk, there have been about 3×10^8 supernovae. The volume of the galaxy is that of a disk with a radius ≈ 10 kpc, and height ≈ 200 pc, corresponding to $V = 6 \times 10^{10}$ pc^3. Each supernova will have filled an average volume of 200 pc^3. The magnetic field observed in the Crab Nebula, at about 1 pc away from the pulsar left over from the explosion, is $\approx 10^{-4}$ G. Using flux conservation, when this matter expands to fill 200 pc^3, it will have a magnetic field of $\approx 5 \times 10^{-6}$ G, comparable to the disordered component of the galactic magnetic field.

In the same way, the loss of gas by galaxies in rich clusters, which explains the presence of hot material observed in the X band and of the iron line in the intergalactic medium, can explain the presence of strong magnetic fields at the center of the cluster. Quasars represent other sources of magnetized material, as well as radio galaxies. Therefore, we also expect that clusters containing objects of this kind possess significant magnetic fields.

Batteries

All the batteries proposed in the literature exploit the different e/m ratio of electrons and protons. The most important of these seem to be Biermann's battery and the cosmological battery.

The first battery proposed dates back to Biermann (1950) and works in the presence of rotation. Let us consider, for the sake of simplicity, a star composed exclusively of hydrogen, which rotates slowly. The partial pressures of electrons and protons must be identical, because the number of particles is the same, therefore,

$$p_e = p_p = \frac{1}{2}p \qquad n_e = \frac{1}{m_e + m_p}\rho_p = n_p \qquad (2.131)$$

In order to consider the equilibrium of protons and electrons separately, we must also consider two distinct Euler's equations:

$$\frac{\partial \vec{v}_p}{\partial t} + (\vec{v}_p \cdot \nabla)\vec{v}_p = -\frac{\nabla p_p}{\rho_p} - \nabla \phi + \omega^2 \vec{r}_\perp + \frac{\vec{f}}{\rho_p}$$

$$\frac{\partial \vec{v}_e}{\partial t} + (\vec{v}_e \cdot \nabla)\vec{v}_e = -\frac{\nabla p_e}{\rho_e} - \nabla \phi + \omega^2 \vec{r}_\perp - \frac{\vec{f}}{\rho_e} \quad (2.132)$$

Here $\vec{f}$ is any mutual force that protons and electrons exert on one another, $\vec{r}_\perp$ is the distance from the rotation axis, ω is the star's angular velocity of rotation, and the term $\omega^2 \vec{r}_\perp$ is the centrifugal force.

Let us start by considering a nonrotating star. In this case, we know that protons are in equilibrium, therefore $\vec{v}_p = 0$. We shall naively assume $\vec{f} = 0$, so that

$$\nabla p_p = -\rho_p \nabla \phi \quad (2.133)$$

We have just said that $p_p = p_e$, therefore, if we introduce this equation in Euler's equation for the electrons, we find that the right-hand side is

$$-\nabla p_e - \rho_e \nabla \phi = (\rho_p - \rho_e)\nabla \phi = (m_p - m_e)n_e \nabla \phi \approx \rho_p \nabla \phi \quad (2.134)$$

which shows that if $\vec{f} = 0$, the forces on the electrons are not balanced; the pressure gradient for the electrons exceeds by a factor $m_p/m_e \approx 1836$ the gravitational force. Of course, the electrons must also be stationary ($\vec{v}_e = 0$) in a star, so there will be an electric force $\vec{f}$ that balances the pressure gradient:

$$\nabla p_e = -\vec{f} \quad (2.135)$$

Physically, this is obviously an electrostatic force; because of the different positions of positive and negative ions, a light attractive force arises to balance the various gravitational forces. Another way to look at the same thing is this: since

$$\vec{f} = -en_e \vec{g} \quad (2.136)$$

we see that $\nabla \wedge \vec{g} = 0$ because we are in spherical symmetry. Thus we can write $\vec{g} = \nabla \phi_e$ and obtain

$$\vec{f} = -e n_e \nabla \phi_e \qquad (2.137)$$

When we introduce equation 2.135 into Euler's equation for protons, and use equation 2.131, we see that the condition for them to be stationary becomes

$$\nabla p = -\rho_p \nabla \phi \qquad (2.138)$$

where, however, we see the total, not the partial, pressure, as we expect.

Let us now consider the case of a rotating star, for which there is also a centrifugal term. Repeating the above procedure, and taking into account the presence of the electrostatic term (eq. 2.137), we know that in order for the protons to be in a centrifugal equilibrium, we must have, from the first Euler equation (eq. 2.132),

$$\nabla p_p = -\rho_p \nabla \phi + \rho_p \omega^2 \vec{r}_\perp + \vec{f} \qquad (2.139)$$

However, since $p_e = p_p$, the above expression also gives ∇p_e, which we can reintroduce in the second of Euler's equations (eq. 2.132), to obtain

$$\frac{\vec{f}_{tot}}{\rho_p} \equiv \nabla \phi - \omega^2 \vec{r}_\perp - \frac{2\vec{f}}{\rho_p} = \nabla \phi - \omega^2 \vec{r}_\perp + \frac{2e}{m_p} \nabla \phi_e \quad (2.140)$$

where I used equation 2.137. Now, can there be an electrostatic potential ϕ_e that makes $\vec{f}_{tot} = 0$? If so, we should also have $\nabla \wedge (\vec{f}_{tot}/\rho_p) = 0$. Therefore, since $\nabla \wedge \nabla X = 0$ for each scalar X, we should have

$$\nabla \wedge (\omega^2 \vec{r}_\perp) = 0 \qquad (2.141)$$

which is satisfied only if ω is exclusively a function of $r_\perp$, which never happens. In other words, since $\vec{f}$ is an electrostatic force, it is not possible to choose it so that $\vec{f}_{tot} = 0$, because $\nabla \wedge \vec{f}_{\text{tot}}/\rho_{\text{p}}$ must necessarily be $\neq 0$. Finally, the total force on the electrons, $\vec{f}_{\text{tot}}$, *does not* vanish.

What have we obtained? The total force on the protons is zero, the one on the electrons is not; there will be a current, in other words, a *motion of the electrons relative to protons*. This current, which is *not* due to electromagnetic forces, exactly as in the batteries we have at home, generates the initial magnetic fields. What intensity will these magnetic fields possess?

Like all vectors, $\vec{f}_{\text{tot}}/\rho_{\text{p}} \equiv \vec{X}$ can be decomposed into an irrotational part, $\vec{X}_{\text{i}}$, and a solenoidal part, $\vec{X}_{\text{s}}$, such as

$$\vec{X} = \vec{X}_{\text{i}} + \vec{X}_{\text{s}} \qquad \nabla \wedge \vec{X}_{\text{i}} = 0 \qquad \nabla \cdot \vec{X}_{\text{s}} = 0 \qquad (2.142)$$

The irrotational part can be balanced by an electrostatic potential and is therefore canceled: $\vec{X}_{\text{i}} = 0$. The solenoidal part remains, and we have, approximately, $X_{\text{s}} \approx \omega^2 L$, where L is a typical dimension of the problem. It generates a current:

$$\vec{j} = \sigma \left(\frac{m_{\text{p}}}{e} \vec{X}_{\text{s}} + \vec{E}_{\text{ind}} \right) \qquad (2.143)$$

In its turn, the current generates a magnetic field through

$$\nabla \wedge \vec{B} = \frac{4\pi}{c} \vec{j} \qquad (2.144)$$

and this generates the induction field $\vec{E}_{\text{ind}}$ through the equation

$$\nabla \wedge \vec{E}_{\text{ind}} = -\frac{1}{c} \frac{\partial \vec{B}}{\partial t} \qquad (2.145)$$

How long does this process last? At first, the magnetic field will be negligible, and therefore $\vec{j}$ will also be small; thus $\vec{E}_{\text{ind}} \approx -m_{\text{p}} \vec{X}_{\text{s}}/e$ is a constant. From the induction equation, we see therefore that $\vec{B}$ grows linearly with time. At least initially,

$$B \approx \frac{m_{\text{p}} c \omega^2 t}{e} \qquad (2.146)$$

However, after a time equal to the dissipation time, $T_d \approx 4\pi\sigma L^2/c^2$, ohmic dissipation becomes important, and the magnetic field stops growing and settles around the value

$$B \approx \frac{m_p c\omega^2 t}{e} T_d = \frac{4\pi\sigma m_p X_s L}{ec} \approx \frac{4\pi\sigma m_p \omega^2 L^2}{ec} \approx 500 \text{ G}$$

$$(2.147)$$

where I used values typical of the Sun: $\omega = 3 \times 10^{-6}$ s^{-1}, $L = R_\odot/2 \approx 3 \times 10^{10}$ cm, and $\sigma = 10^{17}$ s^{-1}. From this we can see that this battery can certainly give initial magnetic fields of nonnegligible amplitude, on the scale of a star.

However, this mechanism is at work in any situation in which an ionized gas is in centrifugal equilibrium. For example, we can apply it to the entire galaxy, but we must remember that the time of ohmic decay is longer than the Hubble age, and therefore the magnetic field that can be obtained in this way is limited by

$$B \approx \frac{m_p c\omega^2 t_H}{e} \qquad (2.148)$$

In a cosmological context, we can assume that this initial field can be produced before the gravitational collapse of the galaxy, because after that moment a galactic dynamo can start. Therefore, let us assume that the gravitational collapse takes place at $z \approx 3$, when the universe has an age that is only one-tenth of the present one. Also, we assume that the angular velocity of the galaxy varies with time, but we conserve the angular moment, $\omega \propto R^{-2}$, and scale down from the typical current value $\omega \approx 10^{-15}$ s^{-1}. We can then take as radius of peak expansion the triple of the present radius; it would be only double if the collapse were not dissipative, but we wish to include roughly the effect of dissipation. Thus we find

$$B_i \approx 10^{-20} \text{ G} \qquad (2.149)$$

in good agreement with numerical simulations (Kulsrud et al. 1997, $B_i \approx 10^{-21}$ G). The difference comes from the fact that not the entire vector $\omega^2 \vec{r}_\perp$ is rotational; the irrotational part must be subtracted, and this reduces the order of magnitude of this term to $<\omega^2 R$.

The cosmological battery is important because of the context in which it is originated. When the galaxies start to form, dark and baryonic matter acquire angular momentum through tidal interactions and start rotating. At that time, matter is decoupled from radiation, which therefore does not rotate, and forms an isotropic background. In this background, which is thermal and isotropic, an electron is subject to a force called *Compton drag* (see problem 5):

$$m_e \frac{d\vec{v}_e}{dt} = -\frac{4}{3} \epsilon_\gamma \sigma_T \frac{\vec{v}}{c} \qquad (2.150)$$

where ϵ_γ is the energy density of isotropic photons, and σ_T is the Thomson cross section. The force opposes the velocity of the electron, and thus represents a drag. The reason for this drag is that, in its reference frame, the electron scatters photons with a back-front symmetry, but because of its motion, more photons seem to arrive from the front than from the back (the same happens when you run in the rain and get soaked in front).

The same force is much weaker on protons because the Thomson cross section is $\propto m^{-2}$, therefore about 3×10^6 smaller. There is thus a force that preferentially slows down electrons and thus creates a current. After a while, an electric field will arise to prevent the electrons from lagging too much behind the protons, exactly as in Biermann's dynamo. Therefore, the equation of motion for the electrons becomes

$$m_e \frac{d\vec{v}_e}{dt} = -e\vec{E} - \frac{4}{3} \epsilon_\gamma \sigma_T \frac{\vec{v}}{c} \qquad (2.151)$$

We can now assume a stationary status, in which the total force on each electron vanishes:

$$-e\vec{E} - \frac{4}{3} \epsilon_\gamma \sigma_T \frac{\vec{v}}{c} = 0 \qquad (2.152)$$

Here we have neglected the component $\vec{v} \wedge \vec{B}$ of the Lorentz force, because we assume there is no initial magnetic field. Taking the curl of the previous equation, and using Maxwell's equation, we find

$$\frac{\partial \vec{B}}{\partial t} = \frac{4\sigma_{\mathrm{T}}\epsilon_\gamma}{3e}(\nabla \wedge \vec{v}) \qquad (2.153)$$

In a rotating galaxy, $\nabla \wedge \vec{v} = 2\vec{\omega}$, and the above equation can be integrated from the moment of formation of the galaxy (supposedly at $z = 10$) until now, thus obtaining

$$B_{\mathrm{i}} \approx 10^{-21}\ \mathrm{G} \qquad (2.154)$$

Dynamos

In order to illustrate the difficulties of building a plausible dynamo, let us notice first of all that our task is not only to increase the average amplitude of the magnetic field, but also to build an *ordered* magnetic field: in the case of the galaxy, to build a field that is coherent on a scale length comparable to the distance of the Sun from the galactic center, and for the interiors of stars, to privilege the dipole moment rather than higher order moments. Indeed we know that, in the case of the Sun and of the pulsars showing cyclotron lines, the surface field is not very different from the dipole one, which indicates that the components of the magnetic field with shorter coherence length are weaker than the most ordered field.

Furthermore, there are important antidynamo theorems (Cowling 1934) that demonstrate that it is impossible to build a poloidal magnetic field[4] with a velocity distribution axially symmetric around the z axis. It follows that dynamos are intrinsically three-dimensional objects.

Another difficulty is illustrated by the following erroneous argument. Since the electric field, in magnetohydrodynamics,

---

[4] We call *toroidal* the component along $\hat{e}_\phi$ of a vector, and the remaining component *poloidal*; this is contained in a plane passing through the axis z.

is due exclusively to induction (because of the large number of free electric charges), it will have an amplitude $\approx vB/c$, which induces a current $j \approx \sigma vB/c$. This current must generate the magnetic field B, which means that the curl of B, $\approx B/L$, is $j/c \approx vB/c^2$. Setting the two equal, we find that this requires a velocity $v \approx c^2/\sigma L$. Using once again the hydrogen conductivity (eq. 2.17), we find that in order to maintain the field of the Sun, the necessary velocity is $v \approx 10^{-6}$ cm. It may appear easy, but this is the relative velocity between the lines of the field and the material. In other words, the theory of the dynamo must explain the growth of the magnetic field in situations where the field is almost perfectly frozen in the fluid. When flux freezing holds, it is very easy to amplify the magnetic field; the element of fluid in which the field is frozen need only become smaller and smaller. However, when this happens, a magnetic field is created with a coherence length that is correspondingly small, instead of the ordered magnetic field we mentioned in the preceding paragraph.

The only situation in which we think that flux freezing is *on its own* the solution to the problem of the growth of the magnetic field is the generation of pulsars' magnetic fields. Pulsars are formed in the explosions of supernovae, in which the central parts of massive stars, with a radius $>10^{11}$ cm, collapse to form neutron stars with a radius $\approx 10^6$ cm. The reduction of the object's surface by a factor $>10^{10}$ is sufficient to generate magnetic fields $>10^{12}$G from stellar fields $>10^2$G. Furthermore, since the star already possesses an ordered dipole magnetic field, the field compressed by flux freezing will maintain the same ordered structure. However, it has been suggested that newly formed neutron stars may give rise to powerful dynamos, which can bring the magnetic field up to as much as 10^{17}G, the equipartition field in neutron stars (Thompson and Duncan 1994).

If we extend the preceding argument to stars, we get an incorrect result. Since the interstellar medium is endowed with a magnetic field $\approx 10^{-6}$G, a solar-mass star, which is formed from a molecular cloud with a density of 10^6 cm^{-3} particles, thus being subject to a contraction by a factor 10^6 in radius (to a solar radius $\approx 10^{11}$ cm), should have a magnetic field of $\approx 10^6$ G, which is certainly wrong. A part of this magnetic field must be lost during the very process of collapse, because, as we have seen, the magnetic field has pressure, which, if very large, can inhibit stellar collapse. If there were no flux loss, stars could not form. But it is even worse than that. Even after the star has formed, loss of magnetic field must continue due to a different reason: solar-mass stars undergo a phase, before hydrogen ignition, in which they are entirely convective (the so-called Hayashi stage). As we have seen in section 2.7.1, when the star is in a convective stage it tends to expel magnetic flux on a time scale that is estimated to be a few years (Parker 1979). Therefore, during the Hayashi stage, which lasts about 10^6 yr, the small-mass stars are left without a magnetic field by the Parker instability.

In fact, this is one of the arguments used to emphasize the necessity of the dynamo effect. The other argument is that the time for the decay of the magnetic field on Earth and on Jupiter is rather short: 10^4 yr for Earth and 10^7 yr for Jupiter. In order to establish this, you can use equation 2.16, the radius of the iron nucleus of Earth ($\approx R_E/2 \approx 3000$ km), and the conductivity of iron ($\sigma \approx 3 \times 10^{14}$ s^{-1}). For Jupiter, we can simply use a different radius, $R_J = 7 \times 10^9$ cm. On the other hand, there is evidence of the presence of a magnetic field on Earth with intensity (not orientation!) similar to the current one (determined from the magnetic properties of rocks) for a period of at least 10^9 yr. It follows that Earth and Jupiter (as well as the small-mass sunlike stars, for the reason explained in the preceding paragraph) must contain a dynamo.

How does a dynamo work? There are two fundamental ideas. The first one is that differential rotation transforms a magnetic field, which is at first assumed to be purely poloidal, into a toroidal one. This mechanism is caused by the freezing of the flux lines: when a fluid element of small size rotates, the differential rotation (namely, the fact that ω is not constant with the radius) makes the innermost parts rotate faster than the outer parts, so that it is stretched in the direction $\hat{e}_\phi$. After several rotations, the fluid element and the frozen magnetic field are tightly wrapped around the center. Since the intensity of the magnetic field is by definition proportional to the density of the field lines, the wrapping up implies an amplification of the field. This mechanism clearly takes place on the rotation time scale and is therefore very fast in stars. It is called *shearing* of the magnetic lines.

The second idea is that there is a mechanism transforming the toroidal field into a poloidal field, closing the amplification loop. We may begin with a purely toroidal field, but in a star which is both rotating and convectively unstable. We may then show (Parker 1979) that the Coriolis force generates loops of poloidal field. When the field is purely toroidal, a field line in a poloidal plane is represented by a single dot. But consider now a convective cell, where material rises close to the cell axis, fans out radially at the cell top, cascading back far from the axis toward the cell bottom, from where it is sucked in toward the axis again. As matter is drawn in toward the axis, it spins up because of angular momentum conservation, but most importantly, it is acted upon by the Coriolis force which sets it in motion about the cell axis. Thus, a single field line, though initially oriented in the toroidal direction, is plucked upward in the radial direction in its point closest to the cell axis, but it is also dragged by the rotating material in a spiral pattern around the same cell axis. When this deformation is plotted in a poloidal plane, the line, which must necessarily close on itself, appears like a closed loop of poloidal field: the fact that the line is not represented by a single dot in this plane already shows that the mechanism has generated a poloidal component from a purely toroidal configuration. But the real beauty of the mechanism is displayed when one considers what happens close to the top of the cell, for, naively, one might fear that the poloidal loop generated here might cancel that at the bottom. Instead, the reverse is true: the two reinforce each other. This happens because the Coriolis force changes sign for material fanning out from the axis, and because matter on the axis is lower (not higher) than that further out. Thus the projection of the same line onto the poloidal plane is a loop of the same orientation as that generated at the bottom of the cell. It is also easy to see what happens when we sum up all these poloidal loops, which is best accomplished by summing up the toroidal currents (all with the same sign!) that give rise to them. In this case, we get toroidal currents of the same sign distributed over the whole star, giving rise to a *large scale* poloidal field. This mechanism is called *Parker's mechanism*, for its inventor.

These two mechanisms (shearing and Parker's circulation) constitute the mechanisms for the growth of the magnetic field and identify the time scales on which the dynamo can work. However, a self-respecting theory must also predict at which level the growth of the magnetic field saturates. To this end, it is necessary to introduce some terms leading to the decay of the magnetic field. These terms must necessarily be nonlinear, otherwise the problem would simply be either unstable or dampened. The most commonly invoked mechanisms are ohmic dissipation (which is sometimes to be found in the form of magnetic reconnection) and Parker's instability. The first one heats up the medium in which the dynamo takes place, whereas the second one is purely noncollisional.

These arguments apply to a configuration without turbulence, in other words, one where some sort of averaging has been carried out, in space and in time. However, we saw before that antidynamo theorems seem to require turbulence, which must thus be properly included in a full discussion. This of course adds enormously to the technical difficulties of the problem.

The type of dynamo here briefly described is called $\alpha\omega$; the reader can find a simplified, illuminating example of it described by Tout and Pringle (1992), an exhaustive discussion by Cowling (1981), and a more recent assessment by Krause, Radler, and Rüdiger (1993), as well as Beck, Brandenburg, Moss, Shukurov and Sokoloff (1996).

Most authors accept that a dynamo mechanism may be at work inside stars. On the other hand, there are discussions about a galactic dynamo. It has been argued (Turner and Widrow 1988) that the magnetic fields generated by the cosmological battery are too small to be amplified by a galactic dynamo within the Hubble age, which led to various hypotheses according to which the initial magnetic field is not due to an astrophysical battery but is the product of one of the phase transitions that have characterized our universe at high temperatures. However, other authors (Kulsrud et al. 1997) say that the astrophysical mechanisms can certainly account for the observed magnetic fields, as well as for the large coherence lengths. The question remains unresolved at the writing of this book.

2.10 Problems

1. Consider a black hole with a mass M accidentally immersed in an interstellar medium composed of electric charges all of the same sign. Show that it can be charged with a maximum electric charge $Q_{\max}$, and determine it. Finally, compare it with the maximum electric charge that can be carried by a black hole, $Q_{\mathrm{BH}} = \sqrt{GM^2}$.

2. Consider a situation in which the lines of a magnetic field are very tightly wrapped on the surface of a cylinder. In what direction is the magnetic tension exerted?

3. Show that the entropy perturbations, in the problem studied in section 2.4, are decoupled from all the others and are independent of time (in some sense to be specified).

4. The solar wind has, in proximity of Earth, a density of $n_{\mathrm{SW}} = 4$ cm^{-3} protons, and a speed of $v_{\mathrm{SW}} = 4 \times 10^7$ cm s^{-1}. Earth has a dipole magnetic field with an intensity $B = 0.5$ G on the surface. Find out at what distance from Earth, in the Earth-Sun direction, a shock wave forms (the so-called *magnetopause*). This is what prevents the solar wind from hitting us.

5. Derive equation 2.150. It is better to use the reference frame of the electron and calculate the momentum of photons before and after they are scattered. In order to do this, use a thermal, isotropic distribution of photons, in the reference frame in which the electron has a velocity $\vec{v}_e$, and consider only the limit $v_e \ll c$. Moreover, take into account the fact that the differential cross section for the pure scattering of an unpolarized wave is $d\sigma_T/d\Omega = r_0^2/2(1 + \cos^2 \theta)$, where $r_0 = e^2/m_e c^2$ is the classic radius of the electron, and θ is the angle between the initial and the final direction of the wave. It is also possible to infer this formula in a dimensional manner, except for the factor $4/3$, of course. How?

Chapter 3

Radiative Processes

In this chapter, we shall present a brief overview of the main radiative processes relevant to high-energy astrophysics. The contents of this chapter are extremely classic: radiation transport, low-temperature thermal emission, bremsstrahlung, synchrotron, inverse Compton, pair production, and annihilation. Moreover, in the last section, I shall discuss the phenomena that dampen the energies of very high energy protons and photons in their cosmological propagation. For some of these processes, I shall not give formal derivations of the fundamental results; the reader can find them in the beautiful and very complete book by Rybicki and Lightman (1979).

3.1 Radiative Transport

A radiation field that can be specified through its *specific intensity* I_ν is defined as follows. We consider all the radiation to be directed in a small cone of angular opening $d\Omega$ around the reference direction, with frequency between ν and $\nu + d\nu$. I_ν is the quantity of energy that crosses a surface dA *normal* to the reference direction:

$$dE = I_\nu dA dt d\Omega d\nu \qquad (3.1)$$

115

The specific intensity is linked to the radiation flux F_ν and to the radiation pressure p_ν by the obvious relations

$$F_\nu = \int I_\nu \cos\theta d\Omega \qquad (3.2)$$

$$p_\nu = \frac{1}{c} \int I_\nu \cos^2\theta d\Omega \qquad (3.3)$$

A fluid element with volume dV can emit radiation and change the radiation field I_ν; in fact, it emits, in a time dt, an amount of energy

$$dE = j(\nu)dV\,d\nu d\Omega dt \qquad (3.4)$$

which is directed toward a small solid angle $d\Omega$ around a reference direction and consists of electromagnetic waves with a frequency in an infinitesimal interval $d\nu$ around the reference frequency ν. The quantity $j(\nu)$ is called the *coefficient of spontaneous emission*. Obviously, in crossing a volume $dV = dAds$, where ds is the radiation path along its direction of motion, the specific intensity must increase:

$$dI_\nu = j_\nu ds \qquad (3.5)$$

Furthermore, the fluid element can remove specific intensity from this beam of radiation, because of absorption. We simply define the *coefficient of absorption* α_ν as the fractional amount of specific intensity subtracted from the beam, per unit path length s. In other words,

$$dI_\nu = -\alpha_\nu I_\nu ds \qquad (3.6)$$

α_ν has units of L^{-1}, and it is obviously easy to see that, if the fluid is made of n particles per unit volume, each with cross section σ_ν, we have

$$\alpha_\nu = n\sigma_\nu \qquad (3.7)$$

It must be noted that we describe two different phenomena, namely, the absorption of photons and their scattering. The processes are different, because scattering conserves the number of photons, while absorption destroys them. This is the reason why the coefficient of emission j_ν depends on the specific intensity I_ν—because scattering subtracts photons from a certain direction and makes them reappear in another direction.

The equation of radiative transport is

$$\frac{dI_\nu}{ds} = j_\nu - \alpha_\nu I_\nu \tag{3.8}$$

which appears deceivingly simple. In fact, we can easily find a purely formal solution. Indeed, if we call

$$\tau_\nu \equiv \int \alpha_\nu ds \tag{3.9}$$

the *optical depth*, and

$$\frac{j_\nu}{\alpha_\nu} \equiv S_\nu \tag{3.10}$$

the *source function*, equation 3.8 has the solution

$$I_\nu(\tau_\nu) = I_\nu(0)e^{-\tau_\nu} + \int_0^{\tau_\nu} S_\nu(x)e^{x-\tau_\nu} dx \tag{3.11}$$

Why is this a purely formal solution? Because the coefficient of emission j_ν, and therefore also the source function S_ν, contain a contribution due to scattering of specific intensity I_ν from all the other directions toward the direction in question. Therefore, it is impossible to specify S_ν in absolute terms without reference to I_ν, unless the problem we are considering is entirely without scattering.

A particularly important case is when bodies with different properties are in thermal equilibrium with each other and with a radiation field. In this case, there must be spatial homogeneity ($dI_\nu/ds = 0$), and therefore $I_\nu = S_\nu$. Therefore the radiation field has a specific intensity that may depend only on T and ν. In fact, if this were not so, and I_ν depended also on the properties of the bodies in equilibrium with the radiation, it would be possible, after laying side by side two systems with *different* specific intensities but with the same temperature, to make energy flow from one to the other, with no other effect, thus violating the second principle of thermodynamics. It follows that, in thermal equilibrium, I_ν is a universal function. If we consider a *black body*, namely an object that absorbs all the light that falls over it, we see that this universal quantity must equal the emissivity of the blackbody. From elementary courses we know that

$$I_\nu \equiv B_\nu(T) = \frac{2h\nu^3}{c^2} \frac{1}{e^{h\nu/kT} - 1} \qquad (3.12)$$

It is easy to show, integrating on frequency, that

$$\epsilon = aT^4 \qquad a \equiv \frac{8\pi^5 k^4}{15c^3 h^3} \qquad p = \frac{1}{3}\epsilon \qquad (3.13)$$

where ϵ is the density of energy per unit of volume. Moreover, the flux of energy across a surface unit is given by

$$F = \sigma T^4 \qquad \sigma = \frac{ac}{4} = \frac{2\pi^5 k^4}{15c^2 h^3} \qquad (3.14)$$

In thermal equilibrium, we obviously have

$$j_\nu = \alpha_\nu B_\nu \qquad (3.15)$$

this relation will help us in the future, because once we have calculated the coefficient of emission for a thermal process, we can immediately infer its absorption coefficient.

Let us now go back to equation 3.11 and consider the idealized but important case in which the source function does not depend on space. This, for example, occurs in thermal equilibrium, where we also know that $S_\nu = B_\nu(T)$, and therefore,

$$I_\nu(\tau_\nu) = B_\nu(T) + e^{-\tau_\nu}(I_\nu(0) - B_\nu(T)) \qquad (3.16)$$

This equation tells us that when the optical depth is very large, $\tau_\nu \gg 1$, we can no longer see a background object (which would be the term $I_\nu(0)$), and we see only the thermal emission of the body located along the line of sight. On the other hand, if $\tau_\nu \ll 1$, we see that $I_\nu \approx I_\nu(0) + \tau_\nu(B_\nu(T) - I_\nu(0))$. According to which inequality is valid, $I_\nu > B_\nu(T)$ or $I_\nu < B_\nu(T)$, we see that the emission can be higher or lower than that of background sources, giving rise to absorption or emission lines, respectively. However, if $I_\nu(0) = 0$, we see that the total emission we perceive is proportional to τ_ν, and therefore to the total amount of material along the line of sight. This case ($\tau_\nu \ll 1$) is called optically thin, and its main feature is that we see (nearly) every photon produced.

We now discuss what happens when the body is not in thermal equilibrium. In this case, the role of $B_\nu(T)$ is played by the source function $S_\nu \equiv j_\nu/\alpha_\nu$, and we thus know that when $\tau_\nu \ll 1$, the total perceived intensity is $= \tau_\nu S_\nu = R\alpha_\nu S_\nu = Rj_\nu$, where R is the dimension of the object along the line of sight. In this case, the intensity of emission has the same spectrum as the emissivity. However, if $\tau_\nu \gg 1$, things change: the perceived intensity is S_ν (see eq. 3.16, where we have used S_ν instead of B_ν),

$$I_\nu \approx \frac{j_\nu}{\alpha_\nu} = S_\nu \qquad (3.17)$$

whose spectrum is not necessarily equal to that of j_ν, because the two terms differ by a factor α_ν. As a consequence, the spectrum changes character at the frequency where we have $\tau_\nu = 1$, which shows once more the importance of the optical depth.

Let us now consider a source with a dimension R. We shall say that it has an optical depth $\tau_\nu = \alpha_\nu R$ for a given process. It is useful to think of the scattering process as a *random walk*. We know then that the probability that the photon suffers N events of the kind in question is

$$P_N = \frac{\tau^N}{N!}e^{-\tau} \qquad (3.18)$$

namely, the process obeys Poisson statistics. Let us consider for example simple scattering. In this case, the relevant cross section is Thomson's, σ_T, and the optical depth of the source is given by

$$\tau_T = n_e \sigma_T R \qquad (3.19)$$

where n_e is the (average) density of electrons in the source. The quantity

$$l = \frac{1}{n_e \sigma_T} \qquad (3.20)$$

is obviously the *mean free path* of the photon, that is, the average distance covered by the photon between two scattering events.

Let us now consider the case $\tau_T \gg 1$. From elementary courses we know that the average distance the photon has covered along any direction, after N scattering events, is $\propto \sqrt{N}$. Therefore, if $\tau_T \gg 1$, an average of τ_T^2 scattering are needed for the given photon to escape from the source. On the other hand, when $\tau_T \ll 1$, the average number of scattering events is $\approx \tau_T$. In the estimates, it is sufficient to use as the number of scattering events $N \approx \tau + \tau^2$, which obviously has the correct behavior within the two limits $\tau_T \ll 1$ and $\tau_T \gg 1$.

Finally, we note that it is possible to make a rough estimate of the luminosity of an object with a large optical depth to absorption, $\tau_a \gg 1$, as for example a star with a radius R, in the following way. Most photons emitted within the source will not manage to reach the surface, just because $\tau_a \gg 1$; the only ones that succeed in this are (more or less) produced within a mean free path from the surface, namely, within a distance $R/\tau_{a(\mathrm{eff})}$ (see in this regard problem 1). The emissivity of this material is $\alpha_\nu B_\nu(T)$, the surface $4\pi R^2$, whence $L = 4\pi R^2 R/\tau_{a(\mathrm{eff})} \alpha_\nu B_\nu(T)$.

3.1.1   Radiation Transport

In the following, we describe the transport of radiation in a weakly inhomogeneous medium. In this case, the problem allows a simplified treatment, which should already be known to readers from courses on stellar evolution; we will nonetheless discuss this problem for sake of completeness.

Let us go back to the transport equation, to the case in which there are absorption, characterized by a coefficient α_ν, and scattering (generally due to nonrelativistic electrons) with a coefficient σ_ν. We must now make the dependence of j_ν from I_ν explicit. Let us start by noting that, in the presence of absorption only, we already know that

$$\frac{dI_\nu}{ds} = \alpha_\nu B_\nu - \alpha_\nu I_\nu \qquad (3.21)$$

On the other hand, if we consider the case of pure scattering, we must identify the emissivity j_ν. However, it is obvious that when we come across the case of *pure scattering*, the emissivity must exclusively be given by a fraction of the incident intensity, since this latter is immediately reemitted, provided two conditions are satisfied. First, the energy of the scattered photon must equal that of the incident photon; this case is called *coherent scattering*. This is more or less true, for example, for scattering off Newtonian electrons, for which we can neglect the Compton effects.[1] Second, there must be no correlation between the directions of motion of incoming and outgoing photons; in this case we talk of *isotropic scattering*. We have, therefore;

$$j_\nu = \sigma_\nu J_\nu \equiv \frac{\sigma_\nu}{4\pi} \int I_\nu d\Omega \qquad (3.22)$$

where J_ν is the average intensity. This equation holds because isotropic scattering randomizes incident radiation.

Therefore, in the presence of scattering only, we find

$$\frac{dI_\nu}{ds} = \sigma_\nu J_\nu - \sigma_\nu I_\nu \qquad (3.23)$$

---

[1] Please note, however (see further on in this chapter), that repeated scatterings from Newtonian electrons may lead to important effects.

Combining this with equation 3.21, we finally find

$$\frac{dI_\nu}{ds} = -\alpha_\nu(I_\nu - B_\nu) - \sigma_\nu(I_\nu - J_\nu) = -(\alpha_\nu + \sigma_\nu)(I_\nu - S_\nu),$$

$$S_\nu \equiv \frac{\alpha_\nu B_\nu + \sigma_\nu J_\nu}{\alpha_\nu + \sigma_\nu} \qquad (3.24)$$

Here S_ν is, of course, the new source function.

In many problems, such as stellar interiors, scattering and absorption are both present, which of course makes the problem harder. There is, however, also a simplifying circumstance, namely, under many conditions it happens that the medium is only weakly inhomogeneous, and therefore spatial gradients are small and can be treated as perturbations.

Let us consider, for the sake of simplicity, a plane-symmetric system, that is, a system in which all quantities exclusively depend on the coordinate z, which indicates depth from the surface; also, physical quantities are assumed to be independent of the coordinates x and y. The problem of transport also depends on the angle θ between the direction in consideration and the z axis, and we therefore expect that intensity be a function of z and $\mu \equiv \cos\theta$. Moreover, since $dz = \mu ds$, we have

$$\mu\frac{\partial I_\nu(\mu, z)}{\partial z} = -(\alpha_\nu + \sigma_\nu)(I_\nu - S_\nu) \qquad (3.25)$$

which can be rewritten as

$$I(\mu, z) = S_\nu - \frac{\mu}{\alpha_\nu + \sigma_\nu}\frac{\partial I_\nu}{\partial z} \qquad (3.26)$$

We are considering the case of weakly inhomogeneous systems, where the term containing the derivative is small. Let us then take as a zero-order approximation

$$I_\nu^{(0)} = S_\nu \qquad (3.27)$$

Now we notice that S_ν does not depend on μ (see eq. 3.24), so that

$$J_\nu^{(0)} \equiv \frac{1}{4\pi}\int I_\nu^{(0)} d\Omega = S_\nu \qquad (3.28)$$

When this happens, we see from the definition of S_ν, equation 3.24, that $S_\nu = B_\nu$, and thus

$$I_\nu^{(0)} = B_\nu(T) \tag{3.29}$$

Let us now refine our estimate by reintroducing this estimate of $I_\nu^{(0)}$ into equation 3.26 to obtain a higher-order approximation:

$$I_\nu^{(1)} = B_\nu - \frac{\mu}{\alpha_\nu + \sigma_\nu} \frac{\partial B_\nu}{\partial z} \tag{3.30}$$

This is the approximate solution, when the medium is weakly inhomogeneous; in this case, T depends on z, and therefore $B_\nu(T)$ also depends on z. In principle, we could continue to look for more and more accurate approximate solutions by reintroducing this approximation into equation, 3.26 but normally the first approximation is sufficient. In general, all diffusion terms are linear in the gradient (why?).

Now we calculate the flux:

$$F_R(\nu) \equiv \int I_\nu^{(1)} \cos\theta d\Omega \tag{3.31}$$

If we now introduce equation 3.30 into equation 3.31, we can see that the first term on the right-hand side of equation 3.30 gives a vanishing contribution, and we are left with

$$F_R(\nu) = -\frac{2\pi}{\alpha_\nu + \sigma_\nu} \frac{\partial B_\nu}{\partial z} \int_{-1}^{1} \mu^2 d\mu = -\frac{4\pi}{3(\alpha_\nu + \sigma_\nu)} \frac{\partial B_\nu}{\partial T} \frac{\partial T}{\partial z} \tag{3.32}$$

This equation can be further simplified by integrating over all frequencies:

$$F_R \equiv \int F_R(\nu) d\nu = -\frac{4\pi}{3} \frac{\partial T}{\partial z} \frac{\partial}{\partial T} \int \frac{d\nu}{\alpha_\nu + \sigma_\nu} B_\nu \tag{3.33}$$

We now define an average absorption coefficient α_R, called the *Rosseland coefficient*, from the name of the astrophysicist who solved this problem in this approximation:

$$\frac{1}{\alpha_R} \equiv \frac{\int \frac{d\nu}{\alpha_\nu + \sigma_\nu} \frac{\partial B_\nu}{\partial T}}{\int d\nu \frac{\partial B_\nu}{\partial T}} \tag{3.34}$$

which finally gives

$$F_{\mathrm{R}}(z) = -\frac{ac}{3\alpha_{\mathrm{R}}}\frac{\partial T^4}{\partial z} \qquad (3.35)$$

This is called the *radiation transport equation* and tells us how the radiation flux is transported in the presence of a weak temperature gradient.

3.2   Low-Temperature Thermal Emission

We have ordered the radiative processes in this way: first we discuss the thermal, low-temperature ones, followed by bremsstrahlung, then the nonthermal ones, synchrotron and inverse Compton. Let us start with a very brief discussion of the low-temperature phenomena; in a high-energy astrophysics course, they are obviously the least important. Interested readers can find this topic discussed at length by Spitzer (1978).

Low-temperature thermal processes are due to the radiative deexcitation of lines of ions, excited by collisions with electrons, and with neutral atoms when ionization is low. The relative abundance of ions at any given temperature is determined by the equilibrium between photoionization and radiative recombination.

Energy loss takes place through collisional excitation of lines deexcited by photon emission. This term must obviously be proportional to R_{jk}, the rate at which the level k is excited from the level j. This rate is given by the product $n_e n_i$ of the electron and ion densities, times the coefficients $\gamma_{jk}^{(i)}$, which depend on the chemical species, on the initial j and the final k energy levels, and on the temperature, and can be measured in a laboratory or be calculated through quantum mechanics. Once we know the frequency $n_e n_i \gamma_{jk}^{(i)}$ with which higher energy levels are excited, we find that energy loss is

$$\rho\Lambda = n_e \sum_i n_i \sum_j \sum_k P_{kj}\gamma_{jk}^{(i)}\hbar\omega_{kj} \qquad (3.36)$$

where the last term on the right is the energy of the emitted photon, and P_{kj} is the probability of radiative (rather than collisional) deexcitation, and we assumed that the rate of downward transitions balances the rate of upward ones. In order to calculate P_{kj}, the number of radiative transitions, per unit of time, by n_i atoms is $n_i A_{kj}$, whereas the one by collisional deexcitations is $n_e n_i \gamma_{kj}$, so the fraction of radiative transitions is

$$P_{kj} = \frac{A_{kj}}{A_{kj} + n_e \gamma_{kj}} \tag{3.37}$$

Two comments are warranted here. In a laboratory, density is large enough to make the frequency of collisional deexcitation usually much larger than the radiative one; therefore, most collisions do not cause thermal energy losses, because the kinetic energy that has been used to excite an electron to a higher energy level is restored in the following collision: $P_{kj} \ll 1$. In astrophysical conditions, where densities are very low, radiative deexcitations are much more common. However, even here density may occasionally be so large as to make some radiative transitions too slow.

The most important radiative transitions, in astrophysics, are the forbidden or semiforbidden lines of the various stages of ionization of C, N, and O. These lines are called forbidden or semiforbidden because they are very weak in the laboratory; the term dates back to the classification of spectral lines made by spectroscopists at the end of the nineteenth century. The reason why these lines are forbidden or semiforbidden is nowadays well known: they violate one or more quantum-mechanic selection rules for the emission of photons, so that they are not simple dipole transitions, but higher-order transitions.[2] Since radiative deexcitation is slow, there is time for a collisional de-excitation without emission of photons, which explains the weakness of the lines, and their name. Strange

---

[2] Therefore, their coefficients A are not proportional to the first power of the fine-structure constant $e^2/\hbar c \approx 1/137$, but to one of its higher powers, and are therefore suppressed.

as it may appear, these lines are actually responsible for most cooling of totally or partially ionized gas at densities around $n \approx 1$ particle cm^{-3}, and temperatures $T \approx 10^4$ K. Indeed, the photons emitted in this process will never be reabsorbed, since their reabsorption, just like their emission, is not likely to take place; therefore these photons, once emitted, are lost forever.

For this reason, the cooling capacity of a gas is a very sensitive function of the abundance of these elements (in astronomers' parlance, of *metallicity*, a term with which astronomers indicate the sum of the abundances of all elements apart from H and He, from Li onward, even though, in chemistry, they are far from being metals). This function is also rather sensitive to density, because these transitions are forbidden or semiforbidden, so that even a small density increase may lead to the suppression of radiative deexcitation in favor of collisional deexcitation.

At slightly higher temperatures, $T \approx 10^5$ K, the collisional excitation of hydrogen dominates cooling. The reason why this occurs only at moderately large temperatures is that the first excited level of hydrogen is separated from the fundamental level by ≈ 10 eV, and the coefficients γ_{jk} all contain a factor $e^{-E_{jk}/kT}$, where E_{jk} is the difference of energy between the two levels. Since 10 eV $\approx 10^5$ K, the collisional excitation of excited levels of hydrogen becomes possible only at these relatively high temperatures.

The total cooling function $\rho\Lambda$, (i.e., energy lost per unit time and volume) is given by

$$\rho\Lambda = n_{\mathrm{e}} n_{\mathrm{p}} \Lambda_{\mathrm{c}} \qquad (3.38)$$

The coefficient Λ_{c} is called the cooling function and is tabulated for different values of n_{p} (to take into account collisional deexcitations), and of chemical abundances (to take into account the dependence on metallicity). In figure 3.1, we give the cooling function for the range of density and (solar) chemical abundances that are more relevant for this book. Spitzer

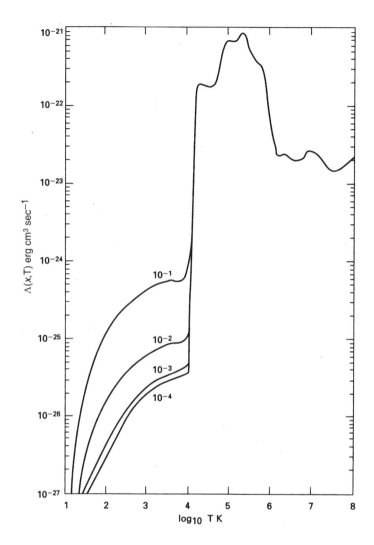

Figure 3.1. Cooling function as a function of temperature, for solar abundances of metals. The curves at low temperatures represent different assumptions about the residual ionization, which is not in equilibrium and thus needs to be parametrized. Dalgarno and McCray 1972.

(1978) also gives references for finding the cooling function under other conditions.

At low temperatures, the cooling function depends on the residual ionization, which is due to the conditions of irradiation (and therefore depends on local conditions) and is sometimes even out of thermal equilibrium because of the very long recombination times. The peak at $T \approx 10^5$ K is given by the excitation of hydrogen (dominant effect); the forbidden and semiforbidden lines of C, N, and O give the prevailing contribution for $T \approx 10^4$ K. The decrease beyond 10^5 K is due to the disappearance of neutral hydrogen, and thus of the possibility of exciting its higher states by collision. The flattening at $T \approx 10^6$ K is due to the contribution of bremsstrahlung, which becomes important just because hydrogen is no longer neutral, and protons can freely collide with electrons. Bremsstrahlung will be treated in the next section.

Here we should define the cooling time t_c:

$$t_c \equiv \frac{3}{2} \frac{NkT}{n_e n_p \Lambda} \tag{3.39}$$

The term in the numerator is the kinetic energy of gas, where N is the total number of free particles per unit volume, $N = n_e + n_p + n_H + n_{He} + \ldots$ On the other hand, the term in the denominator is the total energy loss per unit volume and time. Therefore t_c is simply the time scale on which the gas changes its temperature, in the absence of heating processes. If these processes are instead present, and if we indicate with $\rho\Gamma$ the energy gain per unit time and volume, we have

$$t_c = \frac{3}{2} \frac{NkT}{n_e n_p \Lambda - \rho\Gamma} \tag{3.40}$$

In equilibrium, this time is obviously infinite.

3.3   Bremsstrahlung

Bremsstrahlung, or *braking radiation*, is the radiation emitted during the collision between two particles. Please note the following fact: if the colliding particles are identical, their dipole moment, $q\vec{r_1} + q\vec{r_2}$, is proportional to the position of their center of mass, $m\vec{r_1} + m\vec{r_2}$, which is strictly constant in a collision between two bodies. It follows that there cannot be bremsstrahlung in the collision between two identical particles.

In a gas composed of ions and electrons, the main source of bremsstrahlung is given by the collision between electrons and ions, since the ion-ion collisions, because of the heavier masses involved, generate much smaller accelerations and, consequently, negligible quantities of emitted radiation. In thermal conditions, we can integrate the emission by collision for arbitrary relative velocities on the distribution of relative velocities due to two Maxwell-Boltzmann thermal distributions of velocities. The rate of energy emission ϵ_{br} per unit volume, time, and frequency is obviously proportional to the product of the electrons' density n_e, the ions' density n_i, and the square of the ions' charge Z_i, because acceleration is proportional to charge, but the emitted energy is proportional to the square of acceleration. We have

$$\epsilon_{br}(\nu) = \frac{32\pi e^6}{3m_e c^3} \left(\frac{2\pi}{3km_e T} \right)^{1/2} Z_i^2 n_e n_i e^{-h\nu/kT} g_{br} \qquad (3.41)$$

It is often useful to have a numerical version of this equation:

$$\epsilon_{br}(\nu) = 6.8 \times 10^{-38} Z_i^2 T^{-1/2} n_e n_i e^{-h\nu/kT} g_{br} \text{ erg s}^{-1} \text{ cm}^{-3} \text{ Hz}^{-1}$$

$$(3.42)$$

The factor $g_{br}(\nu, T)$ is called the Gaunt factor and is a dimensionless numerical factor that contains the quantum corrections to classical formulae; it varies very slowly with ν and T and, for most astrophysical situations, ranges between 0.2 and 5.5. It can be found tabulated, for example, by Karzas and Latter (1961).

We should notice the meaning of two factors. The first one is the term $e^{-h\nu/kT}$, which gives most of the dependence of ϵ_{br} on frequency. This term is present because, in a gas with temperature T, there are few particles with energy larger than kT, the only ones that can emit photons with an energy $h\nu \gtrsim kT$. This factor directly mirrors the lack of high-energy particles in Maxwell-Boltzmann statistics.

The other interesting term is $T^{-1/2}$. This is due to the fact that in single collisions, the slower a particle, the more it radiates. Indeed, as electrons pass near ions, they are subject to ever smaller deflections as their speed increases, and accelerations also last correspondingly shorter times and emit less. Please note, however, that when $T \to 0$, $\epsilon_{br} \to 0$, because the dominant term is $e^{-h\nu/kT}$: the cooler the gas, the less it can emit.

Integrating over frequencies in equation 3.41, we obtain the total emissivity:

$$
\begin{aligned}
\epsilon_{br} &= \frac{32\pi e^6}{3hm_ec^3}\left(\frac{2\pi kT}{3m_e}\right)^{1/2} Z_i^2 n_e n_i \bar{g}_{br} \\
&= 1.4 \times 10^{-27}\, T^{1/2} Z_i^2 n_e n_i \bar{g}_{br} \text{ erg s}^{-1}\text{ cm}^{-3} \quad (3.43)
\end{aligned}
$$

The coefficient $\bar{g}_{br}(T)$ is the Gaunt coefficient averaged on the frequency, which varies between 1.1 and 1.5 and is therefore almost completely independent from T.

There is also an absorption coefficient connected to the emissivity, which represents nothing but the opposite process: a collision between three bodies (i.e., electron, ion, and photon) in which the electron absorbs the photon in the electrostatic field of the ion. The connection between the two is given by equation. 3.15, which expresses the fact that the bremsstrahlung is balanced, in thermal equilibrium, by the absorption of the photon in a three-body collision, if photons are distributed like a blackbody. It should not be surprising that a two-body process (i.e., the bremsstrahlung emission) is balanced, in thermal equilibrium, by a three-body process. An extremely general principle, called the *principle of detailed balance*, says that in thermal equilibrium, each process is balanced by its opposite. If this were not so, it would be easy to invent violations of the second principle of thermo-dynamics.

We shall give once again the relevant formulae for the co-efficient of absorption: before integrating on the frequencies, we have

$$\alpha_\nu^{(\mathrm{br})} = \frac{4e^6}{3m_e hc} \left(\frac{2\pi}{3km_e T}\right)^{1/2} Z_i^2 n_e n_i \frac{1 - e^{-h\nu/kT}}{\nu^3} g_{\mathrm{br}}$$

$$= 3.7 \times 10^8\, T^{-1/2} Z_i n_e n_i \frac{1 - e^{-h\nu/kT}}{\nu^3} g_{\mathrm{br}}\ \mathrm{cm}^{-1} \quad (3.44)$$

Here the term $1 - e^{-h\nu/kT}$ represents the correction due to stimulated emission. At high frequencies, this term ≈ 1, and therefore $\alpha_\nu \propto \nu^{-3}$, whereas, at low frequencies $h\nu \ll kT$,

$$\alpha_\nu^{(\mathrm{br})} \approx 0.02\, T^{-3/2} Z_i^2 n_e n_i \nu^{-2} g_{\mathrm{br}}\ \mathrm{cm}^{-1} \quad (3.45)$$

All of the above refers to nonrelativistic particles. The case in which ions and electrons are both relativistic and thermal finds no application in astrophysics, at least so far. However, electrons and ions may be thermal, but electrons, because of their small mass, may be relativistic, even though the motion of ions is still Newtonian. In this case, the emis-sivity of equation 3.43 must be corrected; the new emissivity $\epsilon_{\mathrm{br,rel}}$ is given by

$$\epsilon_{\mathrm{br,rel}} = \epsilon_{\mathrm{br}} \left(1 + \frac{T}{2.3 \times 10^9\ \mathrm{K}}\right) \quad (3.46)$$

which shows that these corrections acquire importance only at very high temperatures. When electrons become relativis-tic, however, other radiative processes, such as synchrotron and inverse compton, become more important.

3.4 Synchrotron

3.4.1 Power Radiated by a Single Particle

Let us consider the equation of motion of a relativistic particle with a charge q in a magnetic field constant in modulus and direction, $\vec{B} = B_0 \hat{z}$:

$$\frac{dp^\mu}{ds} = \frac{q}{c} F^{\mu\nu} u_\nu \qquad (3.47)$$

where $F^{\mu\nu}$ is the tensor that represents the electromagnetic field in special relativity. This equation can be decomposed into its space and time components:

$$\frac{d}{dt}(\gamma m \vec{v}) = \frac{q}{c}\vec{v} \wedge \vec{B}, \qquad \frac{d}{dt}\gamma m c^2 = q\vec{v} \cdot \vec{E} = 0 \qquad (3.48)$$

where the last equality holds because we have assumed zero electric field. From the second one, we have $d\gamma/dt = 0$, and therefore from the first one we find

$$\gamma m \frac{d\vec{v}}{dt} = \frac{q}{c}\vec{v} \wedge \vec{B} \qquad (3.49)$$

Let us multiply this equation times $\vec{B}$: we see that the right-hand side vanishes, and therefore

$$\frac{dv_\parallel}{dt} = 0 \qquad (3.50)$$

The velocity component along the magnetic field is not changed by the presence of the magnetic field. On the other hand, for the perpendicular component $\vec{v}_\perp$ we find

$$\frac{d\vec{v}_\perp}{dt} = \frac{q\vec{v}_\perp}{\gamma mc} \wedge \vec{B} \qquad (3.51)$$

from which we see immediately (just multiply by $\vec{v}_\perp$) that the modulus of $\vec{v}_\perp$ is constant. This equation simply describes a pure rotation with frequency

$$\omega_c = \frac{qB_0}{\gamma mc} \qquad (3.52)$$

called the *Larmor*, or *cyclotron*, or *gyration* frequency.

The total radiated power can be easily inferred, thanks to the Larmor formula:

$$P = \frac{2q^2}{3c^3} a^2 \qquad (3.53)$$

where $\vec{a}$ is the particle acceleration. This formula was inferred, in elementary courses, in the Newtonian limit, but its generalization to relativistic particles can be obtained as follows. *First*, notice that P is the ratio between the time components of two four-vectors, namely, energy and time. Therefore, it is a relativistic invariant. We must now write it in an explicitly invariant form, with the condition that when the speed is low with respect to c, it must reduce to the preceding expression. Obviously, we have to write

$$P = -\frac{2q^2}{3c^3}a^\mu a_\mu \tag{3.54}$$

where a^μ is the four-acceleration. In order to see that for low speeds this formula reduces to the preceding one, we remember that the four-velocity u^μ obeys the equation $u^\mu u_\mu = 1$, and therefore, by derivation with respect to s, $a^\mu u_\mu = 0$. The quantity $a^\mu a_\mu$ is a relativistic invariant, and we can thus compute it in any reference frame. Let us take the one in which the particle is momentarily at rest, where $u^\mu = (1,0,0,0)$. The equation $a^\mu u_\mu = 0$ therefore reduces to $a^0 = 0$. So, by calculating $a^\mu a_\mu$ in the reference frame where the particle is momentarily at rest (which is also the reference frame in which we expect eq. 3.53 to hold), we find $a^\mu a_\mu = a^0 a_0 - \vec{a} \cdot \vec{a} = -\vec{a} \cdot \vec{a}$, which shows that expression 3.54 reduces to equation 3.53, within the limit $v \ll c$.

As we know, the acceleration $\vec{a}_q$, in the reference frame in which the particle is momentarily at rest, is given, in terms of acceleration in the reference frame in which the particle has a Lorentz factor γ, by

$$a_{q\parallel} = \gamma^3 a_\parallel \qquad a_{q\perp} = \gamma^2 a_\perp \tag{3.55}$$

where $\parallel$ and $\perp$ give the components of a parallel to and perpendicular to the speed μ, respectively. This is useful because, in the motion within the magnetic field, the acceleration is always perpendicular to velocity (cf. eq. 3.51), $a_\perp = \omega_c v_\perp$, so that we find

$$P = \frac{2q^2}{3c^3}\gamma^4 \omega_c^2 v_\perp^2 = \frac{2q^4 \gamma^2 B_0^2}{3m^2 c^5}v^2 \sin\theta^2 \tag{3.56}$$

where θ is the angle between the particle velocity and the magnetic field (note that θ is constant with time: why?). If we have a series of particles with isotropically distributed velocities, it is necessary to replace $\sin\theta^2$ with its value averaged on an isotropic distribution:

$$\frac{1}{4\pi} \int \sin\theta^2 d\cos\theta d\phi = \frac{2}{3} \qquad (3.57)$$

Thus we find our final formula:

$$P = \frac{4}{3}\sigma_T c \left(\frac{v}{c}\right)^2 \gamma^2 \epsilon_B \qquad (3.58)$$

where $\epsilon_B = B^2/8\pi$ is the energy density in the magnetic field, and $\sigma_T = 8\pi(q^2/mc^2)^2/3$ is the Thomson cross section for the particle. From this we see that in the Newtonian limit ($v \ll c, \gamma - 1 \ll 1$) the energy lost per unit time is proportional to v^2, and therefore to the particle energy ($=1/2mv^2$), while in the relativistic limit ($c - v \ll c, \gamma \gg 1$), the energy lost per unit time is proportional to γ^2, and therefore to the square of the particle energy ($=\gamma mc^2$). Moreover, the energy lost is proportional to the magnetic field energy density and to the Thomson cross section and is therefore large for the electrons and almost negligible for ions.

For the single electron, we can define a cooling time as

$$t_c \equiv \frac{E}{P} = \frac{\gamma m_e c^2}{P} \propto \frac{1}{\gamma} \qquad (3.59)$$

3.4.2   The Spectrum of a Single Particle

The motion of a single particle is strictly periodic, with a frequency given by equation 3.52, and therefore, the total spectrum will be given by a series of spectral lines at the fundamental frequency (eq. 3.52) and its harmonics. However, the separation between the lines, being the same as the fundamental frequency, becomes smaller and smaller as the particle energy increases: $\omega_c \to 0$ when $\gamma \to +\infty$. Therefore, every time we have an observing instrument with a limited spectral resolution, the spectrum will no longer appear like the linear combination of infinitely sharp lines, but like a continuum. This is especially true in the limit of large Lorentz factors. Since this the most useful limit in astrophysics, we shall simply discuss the spectrum in this continuum limit. We note that what is traditionally called *cyclotron* emission is the one in which lines are well separated and distinct (and thus, the one due to Newtonian particles), while *synchrotron* emission is the opposite limit, where each particle emits a quasi continuum, with a negligible separation between the lines and is due to relativistic particles.

Landau and Lifshitz (1987a) give the spectrum as

$$dP = d\nu \frac{\sqrt{3}e^3 B_0 \sin\theta}{mc^2} F\left(\frac{\nu}{\nu_{\rm s}}\right) \qquad (3.60)$$

where

$$\nu_{\rm s} \equiv \frac{3eB_0}{4\pi mc}\gamma^2 \sin\theta \qquad (3.61)$$

and the function F is defined as

$$F(x) \equiv x \int_x^\infty K_{5/3}(y)dy \qquad (3.62)$$

where $K_{5/3}$ is a modified Bessel function. The limits of F are much more useful than the exact expression. The function $F(x)$ has a maximum for $x \approx 0.29$, corresponding to $F(0.29) \approx 0.92$, and the asymptotic expressions

$$F(x) \approx \begin{cases} \frac{4\pi}{\sqrt{3}\Gamma(1/3)}\left(\frac{x}{2}\right)^{1/3} & x \ll 1 \\ \left(\frac{\pi}{2}\right)^{1/2} e^{-x} x^{1/2} & x \gg 1 \end{cases} \qquad (3.63)$$

whence we can see that the spectrum of a single particle grows slowly (like $\nu^{1/3}$), until $\nu \lesssim \nu_{\rm s}$, reaches a maximum, and is then exponentially cut off.

The surprising part of this spectrum is that the maximum is reached at the frequency in equation 3.61, rather than at

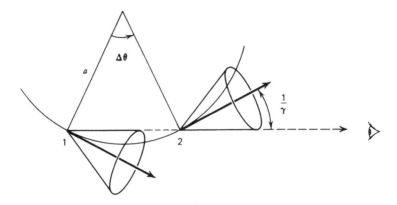

Figure 3.2. Geometry for the explanation of the shape of the synchrotron spectrum. Rybicki and Lightman 1979.

the frequency given in equation 3.52; the two differ by a factor of γ^3! The reason for this phenomenon is the so-called *relativistic beaming*, which takes place every time a particle has a Lorentz factor $\gamma \gg 1$. Even though the particle has, in its reference frame, a totally isotropic emission, the observer sees the emission confined within a small cone of half-opening $\approx 1/\gamma$. For the proof, see problem 4.

Let us now consider an electron rotating around a magnetic field line (fig. 3.2). Most of the time, we cannot see it, because its small emission cone, with a half-opening $1/\gamma$, is not directed toward us. Therefore, it is visible for only a small fraction of its orbit. It is easy to see that this fraction is $2/\gamma$, so that taking the rotation period from equation 3.52, we realize that we see the electron for only a short time $T = 2\pi/\gamma\omega_c = 2\pi mc/eB$. However, the electron emission as we perceive it lasts even less time than this, because the source is relativistic and therefore, so to speak, runs after the photons it has just emitted. Let us call t_0 the moment in which the emission cone is oriented to reach us for the first time, and $t_0 + T$ the moment in which it is oriented to reach us for the last time. The electron emits the first photon at time t_0, and the last one reaching us is emitted at time $t_0 + T$.

However, the last photon is not emitted at the same location of the first one, but at a distance vT closer to the observer. In the interval T the first photon has covered a distance cT, so that when the last photon is emitted, it is located a distance $(c - v)T$ ahead of the last photon. This distance is covered by the second photon at a speed c, so that the two photons are separated in time by $\triangle t = (1 - v/c)T$. Since we assume $\gamma \gg 1$, $v/c \approx 1 - 1/2\gamma^2$, the time interval separating the two photons is

$$\triangle t = \frac{T}{2\gamma^2} = \frac{\pi m c}{e B \gamma^2} \tag{3.64}$$

Therefore, at each rotation of the electron, on the long time scale $\gamma m c / e B$, it is visible for only the (short!) time $\triangle t$. However, we know from Fourier's analysis that when a signal changes by a significant amount on a time scale $\triangle t$, its Fourier transform contains frequencies up to the frequency $1/\triangle t$, which is of the same order of magnitude as ν_s. That is the reason why there are such high frequencies in the synchrotron spectrum of relativistic particles.

The spectrum given by equation 3.60 is the instantaneous spectrum; if we observe all the photons emitted in a time interval dt in which the particle energy does not vary appreciably, they follow equation 3.60. However, in many astrophysical situations, our observing instruments are not very efficient, and the exposures of astrophysical objects require hours and sometimes even weeks. In these cases, it often happens that the particle energy varies significantly during an exposure, and therefore the instrument will pick up *all* the photons emitted by the particle, while its energy changes. We speak then of an integrated spectrum, which differs significantly from the instantaneous spectrum. If P is the power radiated instantaneously (eq. 3.58) by a particle of energy E, we call cooling time the quantity

$$t_c \equiv \frac{E}{P} \tag{3.65}$$

Because of the above argument, the instantaneous spectrum, equation 3.60, applies when observations last $\ll t_c$, whereas, when they last $\gg t_c$, we must change our argument as follows. Let us idealize the instantaneous spectrum emitted as a Dirac's delta at a frequency given by equation 3.61:

$$dP = P\delta(\nu - \nu_s)d\nu \qquad (3.66)$$

where the P (eq. 3.58) appears to give the correct normalization. The amount of energy emitted in photons with a frequency ν, in a time dt is obviously

$$d\mathcal{E}_\nu = P\delta(\nu - \nu_s)d\nu dt \qquad (3.67)$$

But we also remember that $Pdt = dE = mc^2 d\gamma$, so that the total amount of energy emitted in photons with frequencies between ν and $\nu + d\nu$ during the particle cooling from an initial energy E_i to a final $E_f \ll E_i$ is given by

$$\mathcal{E}_\nu = \int d\mathcal{E}_\nu = \int_{E_f}^{E_i} \delta(\nu - \nu_s)dE = \int_{\gamma_f}^{\gamma_i} mc^2 \delta(\nu - \nu_s)d\gamma \quad (3.68)$$

The integral can be calculated with elementary methods:

$$\mathcal{E}_\nu = \left(\frac{\pi m^3 c^5}{3eB_0\nu}\right)^{1/2}, \text{ for } \nu_{min} \leq \nu \leq \nu_{max}; \nu_{min} \equiv \frac{3eB_0\gamma_f^2}{2mc};$$

$$\nu_{max} \equiv \frac{3eB_0\gamma_i^2}{2mc} \quad (3.69)$$

which shows the very important fact that in this case, the spectrum is proportional to $\nu^{-1/2}$, within the interval of frequencies covered by the particles during their cooling.

3.4.3   The Spectrum of a Group of Nonthermal Particles

In many astrophysical situations, a population of relativistic particles is generated, with the following energy distribution:

$$dN = AE^{-\alpha}dE, \quad 2 \lesssim \alpha \lesssim 3, \quad E_{min} \leq E \leq E_{max} \quad (3.70)$$

with $E_{\min} \ll E_{\max}$. The origin of these nonthermal particles will be discussed in the next chapter. Here we shall simply study their synchrotron emission and, in particular, the instantaneously emitted spectrum.

The spectrum of energy instantaneously emitted by synchrotron photons is given by

$$j_\nu = \frac{\sqrt{3}e^3 B_0 \cos\theta}{mc^2} A \int_{E_{\min}}^{E_{\max}} E^{-\alpha} dE F\left(\frac{\nu}{\nu_{\rm s}}\right) \qquad (3.71)$$

where we must remember that $\nu_{\rm s}$ and E are connected by $E = mc^2\gamma$, $\nu_{\rm s} \propto \gamma^2$. This equation immediately gives the exact result, but we wish to find an approximate expression in order to understand the spectral shape.

Let us define

$$\nu_1 \equiv \frac{3eB_0}{4\pi mc}\gamma_{\min}^2, \qquad \nu_2 \equiv \frac{3eB_0}{4\pi mc}\gamma_{\max}^2 \qquad (3.72)$$

where $\gamma_{\min}$ and $\gamma_{\max}$ are the Lorentz factors of the particles of energy $E_{\min}$ and $E_{\max}$, respectively, and ν_1 and ν_2 are the synchrotron frequencies at which peaks the emission of particles also of energy $E_{\min}$ and $E_{\max}$, respectively.

We can then consider three spectral regions. The first one has $\nu < \nu_1$; in this case, we see that all the particles are in the regime $F(x) \propto \nu^{1/3}$ (see eq. 3.63), and therefore the spectrum must have the shape $j_\nu \propto \nu^{1/3}$. The third region has $\nu > \nu_2$, in which case we see that we are considering the emission at such a frequency that $F(x) \approx e^{-x}$, and thus the emission is negligible: $j_\nu \approx 0$. Finally, the intermediate region is the most interesting one: $\nu_1 \leq \nu \leq \nu_2$. Here we shall approximate once again the emission of each particle as

$$dP = P\delta(\nu - \nu_{\rm s})d\nu \qquad (3.73)$$

and we use this expression in equation 3.71, thus finding

$$j_\nu = A \int_{E_{\min}}^{E_{\max}} P E^{-\alpha} dE \delta(\nu - \nu_{\rm s}) \qquad (3.74)$$

which can be calculated with elementary methods, thus obtaining

$$j_\nu \propto \nu^{-s} \qquad (3.75)$$

where

$$s \equiv \frac{\alpha - 1}{2} \qquad (3.76)$$

which is the fundamental relation connecting the spectral index of the synchrotron spectrum with the spectral slope of the energy distribution of nonthermal particles.

Summing up, we find for the spectrum the following, approximate shape:

$$j_\nu \approx \begin{cases} b\left(\frac{\nu}{\nu_1}\right)^{1/3} & \nu < \nu_1 \\ b\left(\frac{\nu}{\nu_1}\right)^{-s} & \nu_1 \leq \nu \leq \nu_2 \\ 0 & \nu > \nu_2 \end{cases} \qquad (3.77)$$

The dimensional quantity $b\nu_1^s$ can be calculated with precision (Jones, O'Dell, and Stein 1974):

$$b\nu_1^s = 2.71 \times 10^{-22}(2.80 \times 10^6)^s q n_0 B^{1+s} \text{ erg s}^{-1} \text{ cm}^{-3} \text{ Hz}^{s-1}$$

$$(3.78)$$

Here n_0 is the density of emitting particles, which is linked to the coefficient A in equation 3.70 by

$$n_0 = A(mc^2)^{1-\gamma} \qquad (3.79)$$

whereas the numerical factor $q \approx \mathcal{O}(1)$ is also tabulated by Jones, O'Dell, and Stein.

3.4.4 Quantum Corrections

In calculating the spectrum of the instantaneous emission of a single particle, we have assumed that its motion does not change as a consequence of the emission. However, we know from quantum mechanics that photons possess a momentum, and therefore the emitting particle must recoil because of the emission. This effect will be negligible, until the momentum of typical emitted photons, with a frequency ν_s given by equation 3.61, is small compared to the momentum of the particle. However, when $h\nu_s \approx \gamma mc^2$, we must take into account the changes imposed by quantum mechanics on particle motion. Since $\nu_s \propto \gamma^2$, we see that for sufficiently large Lorentz factors, computations must be done quantistically, not classically. We define a parameter χ:

$$\chi \equiv \frac{B_0}{B_q}\frac{p}{mc} \approx \frac{h\nu_s}{mc^2\gamma} \qquad (3.80)$$

where p is the particle momentum, and $B_q \equiv m^2c^3/e\hbar \approx 4.4 \times 10^{13}$ G[3] is sometimes improperly called the *quantum magnetic field*. The parameter χ measures the deviations from Newtonian motion: for $\chi \ll 1$, the motion is Newtonian, and the above computations are valid, whereas for $\chi \gg 1$ we expect changes to the spectrum, as well as to the total radiated power.

It is worth noting that these changes can take place in astrophysical situations only when the particle has a Lorentz factor $\gamma \gg 1$, except near pulsars, which have magnetic fields so intense (as high as $\approx 10^{15}$ G) that we need to take into account quantum corrections even at relatively low values of momentum.

We give here, without demonstration, the power radiated in the limit $\chi \gg 1$, referring the interested reader to Landau and Lifshitz (1981b) for the details. The total radiated power is

$$P_q \approx \frac{32 \times 3^{2/3}\Gamma(2/3)e^2m^2c^3}{243\hbar^2}\left(\frac{B_0\gamma}{B_q}\right)^{2/3} \quad \text{for } \chi \gg 1 \quad (3.81)$$

whence we see that, when $\chi \gtrsim 1$, the radiated power stops growing as γ^2 and grows more moderately as $\gamma^{2/3}$.

---

[3]The numerical value was obtained by using the electron mass.

As far as the spectrum is concerned, it has a wide, flat peak around the frequency ν, which satisfies

$$\frac{h\nu}{E - h\nu} \approx \chi \qquad (3.82)$$

It is easy to see that, in the case $\chi \ll 1$, the peak is at $\nu \approx \nu_s$ (eq. 3.61), whereas for $\chi \gg 1$, the above equation gives, for the particle energy *after* emission,

$$E' \approx mc^2 \frac{B_q}{B_0} \qquad (3.83)$$

In general, $E' \ll E$, so we see that the particle energy does not change smoothly, but through emission of photons, which carry away a large fraction of its energy at each emission. Therefore, we cannot imagine the energy loss as a continuous, smooth process; this process is intrinsically discrete, and the energy loss cannot be followed analytically, but only numerically. In fact, it takes place in big jumps. In particular, in all cases in which $B_q/B_0 \lesssim 1$, we see that particles become Newtonian even after the emission of a single photon.

3.4.5   Self-absorption

The opposite process of the emission of a photon through cyclotron emission is the absorption of a photon by a particle within a magnetic field. This process corresponds to an absorption coefficient α_ν, which, however, we cannot calculate from the relation

$$j_\nu = \alpha_\nu B_\nu(T) \qquad (3.84)$$

This relation is valid in thermal equilibrium, whereas we want absorption by the relativistic, nonthermal particles we discussed above. These are the only ones for which we calculated j_ν, but they are certainly nonthermal. When thermal equilibrium does not necessarily hold, we describe particles through their *distribution function f*: the number dN of particles contained within a volume dV with momenta included in a small volume d^3p is

$$dN = f(p)dV d^3p \qquad (3.85)$$

where f is the particles' phase space density. When thermal equilibrium holds, for classical and therefore *distinct* particles, we know from statistical mechanics that $f = Ae^{-E/kT}$, where A is a normalization constant. Away from thermal equilibrium, we can assume an isotropic situation, in which case the energy distribution of particles is given by

$$n(E)dE = 4\pi p^2 f(p)dp \qquad (3.86)$$

In the presence of a magnetic field, the assumption of isotropy implies that we are observing a situation in which the magnetic field assumes all possible orientations, so that we are averaging over a region large enough that the average (vectorial) value of the magnetic field vanishes.

Taking into account stimulated emission, and considering only the case in which the energy of the photon $h\nu \ll E$, the energy of the emitting particle, the absorption coefficient (Rybicki and Lightman 1979) is

$$\alpha_\nu = -\frac{c^2}{8\pi\nu^2} \int dE E^2 \frac{d}{dE} \left(\frac{n(E)}{E^2} \right) P(\nu, E) \qquad (3.87)$$

Here, $P(\nu, E)$ is the energy radiated by a particle of energy E per unit time in photons of frequency ν (eq 3.60).

We can now calculate α_ν for the distribution of nonthermal particles, equation 3.70, by straightforward but long computations. We obtain

$$\alpha_\nu = \frac{\sqrt{3}e^3}{8\pi m} \left(\frac{3e}{2\pi m^3 c^5} \right)^{\alpha/2}$$

$$\times A(B_0 \cos\theta)^{(\alpha+2)/2} \Gamma \left(\frac{3\alpha + 2}{12} \right) \Gamma \left(\frac{3p + 22}{12} \right) \nu^{-\beta}$$

$$\beta \equiv \frac{\alpha + 4}{2} \qquad (3.88)$$

The numerical coefficient can also be expressed as

$$\alpha_\nu = 1.98 \times 10^7 (2.80 \times 10^6)^{\alpha/2} B_0^{(\alpha+2)/2} n_0 t \nu^{-\beta} \text{ cm}^{-1} \quad (3.89)$$

where, once again, the quantity $t \approx \mathcal{O}(1)$ was tabulated by Jones, O'Dell, and Stein (1974), after averaging over the angle θ.

From this we see that the sources will be optically thick $(\tau_\nu \gg 1)$ at low frequencies and, after a certain frequency characteristic of each source, they become optically thin.

As we shall soon see, this result is very important. We know that when the source we are considering is optically thick $(\tau_\nu \gg 1)$, we see $I_\nu \approx j_\nu/\alpha_\nu$ (cf. eq. 3.17). Since $j_\nu \propto \nu^{-(\alpha-1)/2}$ and $\alpha_\nu \propto \nu^{-(\alpha+4)/2}$, we find that $j_\nu/\alpha_\nu \propto \nu^{5/2}$. Therefore, the emission from an optically thick region is proportional to $\nu^{5/2}$. The spectrum of a synchrotron source is thus extremely characteristic: at low frequencies, where the source is optically thick, its emission grows with the frequency as $\nu^{5/2}$, until a frequency is reached where $\tau_\nu \lesssim 1$. From there on, the source is optically thin, $I_\nu \approx j_\nu$, and therefore the spectrum becomes $\propto \nu^{-(p-1)/2}$. This behavior is typical of every absorbed source, but the part of the spectrum going as $\nu^{5/2}$ is typical of synchrotron self-absorption and constitutes a unique signature of the presence and relevance of this process. Indeed, it differs from its purely thermal analogue, for which $\alpha_\nu \propto \nu^2$.

3.4.6　Cyclotron Lines

In an intense magnetic field, it is possible to see Newtonian electrons emitting at the cyclotron frequency and at its first harmonics. This effect is also known in white dwarfs, but we shall concentrate on pulsars, when the magnetic field is at least $B \gtrsim 10^{12}$ G. In this case, the energy of the emitted photon is very high:

$$\hbar\omega_c = \hbar \frac{eB}{m_e c} = 12 \text{ keV} \frac{B}{10^{12} \text{ G}} \quad (3.90)$$

which is a remarkable fraction of the electron's energy at rest, $m_e c^2 \approx 511$ keV. It follows that the problem must be treated in quantum mechanics. From elementary courses, we know that the energy of the electron motion perpendicular to the field is quantized, and the energy levels, called *Landau levels*, are separated by the energy $\hbar\omega_c$. As a consequence, when the electrons pass from a given energy level to a lower one, a photon should be emitted with energy given by an integer multiple of $\hbar\omega_c$.

The best-known case in which this phenomenon is observed is the X-ray pulsar Her X-1, in which $\hbar\omega \approx 60$ keV. The interpretation of this spectral feature is not unequivocal; it can be seen as an emission line at 60 keV, and as an absorption line at 40 keV. However, this last interpretation is made less plausible by the fact that no element, among those abundant on the pulsar surface, has lines in this region of the spectrum. Moreover, in the spectrum of Her X-1, a further line has been discovered at ≈ 110 keV, which corresponds neatly to the first harmonic of the fundamental. Furthermore, the surface magnetic field inferred from this interpretation, $\approx 5 \times 10^{12}$ G, is exactly what we can expect from a pulsar, but note also that the magnetic field—thusly measured—is the surface field, rather than just the dipole component, which is inferred from the slow-down rate of radio pulsars.

An observation that could confirm this interpretation of the line is the presence of polarization. This is exactly what happens in magnetic white dwarfs of the type AM Her, which, exactly because they display up to a 30% polarization, are called *polar*. However, in the relatively weak magnetic fields of white dwarfs, $\approx 10^7 - 10^8$ G, cyclotron emission takes place in the optical region of the spectrum, where polarization is easily measured. On the other hand, in the case of pulsars, cyclotron emission takes place in the X region of the spectrum, where polarization measures are much more difficult.

Since polarization in the X band is not detectable as of now, theorists have mainly tried to predict the shape of the spectrum of the various lines, their relative normalization, and their dependence on the rotational characteristics of the star (phase and period), but the results are not yet conclusive.

3.4.7   Processes in an Intense Magnetic Field

There are two further important processes when the magnetic field is particularly strong: curvature radiation and the process of conversion

$$\gamma + B \rightarrow \gamma' + B + e^- + e^+ \qquad (3.91)$$

Let us consider curvature radiation. Since synchrotron emission is so effective, a particle within an intense magnetic field dissipates the component of its energy perpendicular to the line of the magnetic field and follows with a uniform motion the line itself. In many astrophysical situations, such as around pulsars and black holes, the magnetic lines are not straight, but curved, with curvature radius ρ: in this case, the force that the particle feels as it follows a curved line also makes it radiate. It is possible to derive all quantities related to this process exploiting the analogy with synchrotron radiation. Let us therefore assume that the velocity $v_\parallel$ parallel to the curvature radius of the magnetic line vanishes, and consider a particle moving along the magnetic line (and thus, perpendicular to the curvature radius) with a velocity $v_\perp$. We have, obviously,

$$a_\perp = \frac{v_\perp^2}{\rho} = \omega^2 \rho \qquad (3.92)$$

which can be introduced into equation 3.56 to obtain

$$P = \frac{2e^2}{3c^3} \gamma^4 \frac{v_\perp^4}{\rho^2} \qquad (3.93)$$

Therefore we see that the curvature radiation is perfectly identical to the cyclothron/synchrotron one, except for the replacement $\omega_c \rightarrow v_\perp/\rho$.

We are particularly interested in the hyperrelativistic case, with an eye to future developments. If we replace

$\omega_{\rm c} \to v_\perp/\rho \approx c/\rho$, we obtain from equation 3.60

$$\frac{dP}{d\nu} = \sqrt{3}e^2\gamma\frac{1}{\rho}F(x) \tag{3.94}$$

where

$$x \equiv \frac{4\pi\nu\rho}{3\gamma^3 c} \tag{3.95}$$

and the function F is once again given, in exact form by equation 3.62, and in approximate form by equation 3.63. In particular, we see that the spectrum of a single particle peaks for $x \approx 0.29$, namely, for

$$\omega \approx 0.43\gamma^3\frac{c}{\rho} \tag{3.96}$$

The process

$$\gamma + B \to \gamma' + B + e^- + e^+ \tag{3.97}$$

describes the creation of pairs by a photon in a magnetic field. This was studied by Erber (1966), to whom we refer for the derivation of the following formulae, which are included as reference. If the photon propagates parallel to the magnetic field, there is obviously no effect; however, if the photon propagates for a length d perpendicular to a magnetic field B, it is subject to an attenuation. Thus defined, the average number of pairs $n_{\rm p}$ produced by a single photon is given by

$$n_{\rm p} = 1 - \exp(-\alpha(\chi)d) \tag{3.98}$$

where α is the *coefficient of attenuation per unit length*. The parameter χ is defined by

$$\chi \equiv \frac{h\nu}{2m_e c^2}\frac{Be\hbar}{m_e^2 c^3} \tag{3.99}$$

Erber (1966) shows that the coefficient of attenuation is given by

$$\alpha(\chi) = \frac{e^3 B}{2m_e c^3 \hbar}T(\chi) \tag{3.100}$$

where an approximate form for $T(\chi)$ is

$$T(\chi) \approx \begin{cases} 0.46 \exp(-4/(3\chi)) & \chi \ll 1 \\ 0.60\chi^{-1/3} & \chi \gg 1 \end{cases} \qquad (3.101)$$

and an excellent approximation (even though less transparent) is

$$T(\chi) \approx \frac{0.16}{\chi} K_{1/3}^2(2/(\chi 3)) \qquad (3.102)$$

3.4.8   The Razin-Tsytovich Effect

In appendix A we show that plasma effects lead to an absorption of all frequencies

$$\omega \lesssim \omega_{\mathrm{p}} \equiv \left(\frac{4\pi n_e e^2}{m_e}\right)^{1/2} \qquad (3.103)$$

However, plasma is also responsible for a different absorption process called the *Razin-Tsytovich effect*, for strongly beamed radiation, like the synchrotron one, which leads to the absorption of all frequencies

$$\omega \lesssim \gamma\omega_{\mathrm{p}} \gg \omega_{\mathrm{p}} \qquad (3.104)$$

Therefore, the Razin-Tsytovich effect greatly increases the interval of absorbed frequencies.

In order to understand the origin of this effect, we remember that synchrotron radiation is strongly beamed, being concentrated in a cone of half-opening angle

$$\theta \approx \frac{1}{\gamma} = \sqrt{1 - v^2/c^2} \qquad (3.105)$$

When this radiation propagates in a plasma with a dielectric constant $\epsilon = 1 - \omega_{\mathrm{p}}^2/\omega^2$, the equations to be solved differ from Maxwell's in vacuum. They are

$$\nabla \cdot \vec{E} = \frac{4\pi\rho}{\epsilon} \ , \quad \nabla \cdot \vec{B} = 0$$

$$\nabla \wedge \vec{E} = -\frac{1}{c}\frac{\partial \vec{B}}{\partial t} \ , \quad \nabla \wedge \vec{B} = \frac{4\pi\vec{j}}{c} + \frac{\epsilon}{c}\frac{\partial \vec{E}}{\partial t} \qquad (3.106)$$

If we assume ϵ to be a constant, we can see that this system can be transformed into the vacuum Maxwell equations by means of the following substitutions:

$$
\begin{aligned}
\vec{E}' &= \sqrt{\epsilon}\vec{E}\,, & c' &= c/\sqrt{\epsilon} \\
\vec{B}' &= \vec{B}\,, & \phi' &= \sqrt{\epsilon}\phi \\
\vec{A}' &= \vec{A}\,, & e' &= e/\sqrt{\epsilon}
\end{aligned}
\tag{3.107}
$$

Since the new system of Maxwell equations, in primed quantities, is formally identical to that in vacuum, we know that the synchrotron emission will produce a beaming effect, with a half-angle of opening given by

$$
\theta_{RT} \approx \sqrt{1 - v^2/c'^2} = \sqrt{1 - n^2 v^2/c^2}
\tag{3.108}
$$

where I used the well-known fact (see appendix A) that $\epsilon = n^2$, with n the index of refraction. From this we can see that the angle $\theta_{RT} \approx \theta$, except when n^2 differs from 1, which may occur in a plasma because

$$
n^2 = \epsilon = 1 - \frac{\omega_{\mathrm{p}}^2}{\omega^2}
\tag{3.109}
$$

For a relativistic particle, $v^2/c^2 \approx 1 - 1/(2\gamma^2)$, so that

$$
\theta_{RT} \approx \sqrt{\omega_{\mathrm{p}}^2/\omega^2 + 1/(2\gamma^2)}
\tag{3.110}
$$

whence we can see that the beaming angle is

$$
\theta_{RT} \approx \frac{\omega_{\mathrm{p}}}{\omega}
\tag{3.111}
$$

when

$$
\omega \lesssim \gamma\omega_{\mathrm{p}}
\tag{3.112}
$$

while it reduces to its usual value, $\theta \approx 1/\gamma$ for $\omega \gtrsim \gamma\omega_{\mathrm{p}}$. Of course, the suppression of beaming suppresses the propagation, not the emission of waves with frequencies $\omega \lesssim \gamma\omega_{\mathrm{p}}$; synchrotron propagation is damped at these frequencies.

3.5   Compton Processes

We call Compton processes those processes in which we take into account the recoil that photons impart to the electrons that scatter them. In classical physics the only scattering process is Thomson's, in which the energy of the photons is not affected by scattering event; in this case, we talk of *coherent* or *elastic* scattering. As a consequence, there can be no exchange of energy between the two. However, since photons have energy and momentum, we are forced to consider the recoil of electrons; there must then be an exchange of energy, which can be transferred from photons to electrons or vice versa. We talk of *direct* Compton effect when the photon transfers energy to the electron, and *inverse* Compton effect (IC) in the opposite case.

For the sake of completeness, we shall give here the formulae relevant to Thomson scattering. In this case, the energy of the photon after the scattering process is identical to the one before the process:

$$\hbar\omega_{\text{f}} = \hbar\omega_{\text{i}} \qquad (3.113)$$

where, obviously, ω_{i} and ω_{f} are the initial and final frequency of the photon, respectively. The differential and integrated cross sections, for nonpolarized incident radiation, are

$$\frac{d\sigma_{\text{T}}}{d\Omega} = \frac{r_0^2}{2}(1 + \cos^2\theta) \qquad (3.114)$$

$$\sigma_{\text{T}} = \tfrac{8\pi}{3}r_0^2 \qquad (3.115)$$

Here, once again, $r_0 \equiv e^2/m_e c^2$ is the classical radius of the electron, and θ is the photon deflection angle.

Let us now consider Compton processes. We shall consider a photon scattered by an electron in the reference frame in which the electron is initially at rest. Its initial four-momentum is then $p_{\text{i}\mu}^{\text{e}} = (m_e c, 0)$, whereas the photon's, before and after the scattering, is $p_{\text{i}\mu}^{\text{f}} = \hbar\omega_{\text{i}}/c(1, \vec{n}_{\text{i}})$, $p_{\text{f}\mu}^{\text{f}} = \hbar\omega_{\text{f}}/c(1, \vec{n}_{\text{f}})$. Conservation of the total four-momentum yields

$$p^e_{f\mu} = p^f_{i\mu} + p^e_{i\mu} - p^f_{f\mu} \qquad (3.116)$$

Taking the modulus of the two sides of this equation, we can eliminate the final four-momentum of the electron, thus finding the photon's final energy as

$$\hbar\omega_f = \frac{\hbar\omega_i}{1 + \frac{\hbar\omega_i}{m_e c^2}(1 - \cos\theta)} \qquad (3.117)$$

where θ is the photon's angle of deflection from its initial direction of motion. The cross section for this process, called the Klein-Nishina process, is given, as usual, by Landau and Lifshitz (1981b):

$$\frac{d\sigma}{d\Omega} = \frac{r_0^2 \, \omega_f^2}{2 \, \omega_i^2}\left(\frac{\omega_i}{\omega_f} + \frac{\omega_f}{\omega_i} - \sin^2\theta\right) \qquad (3.118)$$

reducing to the classical Thomson form for $\omega_i \approx \omega_f$. The whole cross section is conveniently expressed in terms of $x \equiv \hbar\omega_i/m_e c^2$:

$$\begin{aligned}
\sigma = \sigma_T \frac{3}{4} &\left(\frac{1+x}{x^3}\left(\frac{2x(1+x)}{1+2x} - \ln(1+2x)\right)\right. \\
&\left. + \frac{1}{2x}\ln(1+2x) - \frac{1+3x}{(1+2x)^2}\right)
\end{aligned} \qquad (3.119)$$

As usual, this has simple limits when the photon's energy is small compared with the rest mass of the electron ($x \ll 1$), and when it is large ($x \gg 1$). We have

$$\sigma \approx \sigma_T \begin{cases} (1 - 2x) & x \ll 1 \\ \frac{3}{8x}(\ln x + 1/2 + \ln 2) & x \gg 1 \end{cases} \qquad (3.120)$$

The regime $x \gg 1$ is called the *Klein-Nishina regime*, from the name of the two physicists who first derived the quantum corrections to the classical Thomson cross section.

3.5.1 Physical Mechanism of the Inverse Compton

So far, we have talked exclusively of energy transfer from photons to the electron, but there must also obviously be the opposite process. In order to have an immediate impression of the energy transfer from the electron to the photon, it is convenient to consider, as an exemplary case, a hyper-relativistic electron, $\gamma \gg 1$, and a photon with initial energy $\hbar\omega_i$, in the reference frame of the laboratory. In the electron rest frame, the photon has energy before the collision,

$$\hbar\omega_i' = \hbar\omega_i\gamma(1 - \beta\cos\theta) \qquad (3.121)$$

where θ is the angle between the directions of motion of photon and electron, in the laboratory system, and $\beta = v/c$ is the speed of the electron in dimensionless units. Let us now assume that this energy $\hbar\omega_i' \ll m_e c^2$. We can then apply Thomson scattering, according to which, in the electron frame, the photon's energy remains almost unchanged, so that

$$\hbar\omega_f' \approx \hbar\omega_i'\left(1 - \frac{\hbar\omega_i'}{m_e c^2}(1 - \cos\Theta)\right) \qquad (3.122)$$

This is just equation 3.117, expanded to the first order in the small quantity $\hbar\omega_i/m_e c^2$. Θ is the angle of deflection, which, in terms of the angles of the momentum direction *before* (θ', ϕ') and *after* (θ_1, ϕ_1) the scattering, still in the electron frame, is given by

$$\cos\Theta = \cos\theta_1\cos\theta' + \sin\theta'\sin\theta_1\cos(\phi' - \phi_1) \qquad (3.123)$$

However, in the laboratory frame, the photon now has an energy

$$\hbar\omega_f = \hbar\omega_i'\gamma(1 + \beta\cos\theta_1) \qquad (3.124)$$

When comparing the two equations above, we can see that the photon now has an energy γ^2 times the initial one, except for a very small range of very unlucky angles, namely, those for which $\gamma(1-\beta\cos\theta) \lesssim 1$. As a consequence, if both photons and electrons have isotropically distributed velocities, when we average over all initial angles, the *unlucky* angles will not weigh much, and the average energy gain is $\propto \gamma^2$. We note in passing that the photon energy in the electron frame is γ times the initial one; therefore, the initial photon energy, in the lab frame and in the electron frame, and the final photon energy in the lab frame, respectively, are in the ratio

$$1 : \gamma : \gamma^2 \qquad (3.125)$$

For the validity of this argument, the photon energy in the electron frame must satisfy $\approx \gamma \hbar \omega_i \ll m_e c^2$.

The mechanism underlying this energy transfer is obvious, since it is the relativistic *beaming* we discussed earlier in this chapter. This argument makes the energy transfer from the electron to the photon perfectly clear when the electron is hyperrelativistic, but it applies in toto also in the case of Newtonian electrons. However, obviously, in the case of hyperrelativistic electrons $\gamma \gg 1$; the photons can acquire huge amounts of energy in just one jump, thus making the IC process the most efficient for the generation of very high energy photons.

In order to understand what happens in detail, we should ask which is the variation of the photon's energy in the Newtonian case. By using the above formulae, we can write the photon's energy in the laboratory frame as

$$\hbar \omega_f = \hbar \omega_i \gamma^2 (1 + \beta \cos \theta_1)(1 - \beta \cos \theta)$$
$$\times \left(1 - \frac{\hbar \omega_i \gamma}{m_e c^2}(1 - \beta \cos \theta)(1 - \cos \Theta) \right) \quad (3.126)$$

We now keep only the most significant terms in β and in $\hbar \omega / m_e c^2$:

$$\hbar \omega_f \approx \hbar \omega_i \gamma^2 \left(1 + \beta \cos \theta_1 - \beta \cos \theta - \beta^2 \cos \theta \cos \theta_1 - \frac{\hbar \omega_i \gamma}{m_e c^2}(1 - \cos \Theta) \right).$$
$$(3.127)$$

Let us now average this expression over all scattering events. Calling fd^3p the number of photons with momentum between $\vec{p}$ and $\vec{p} + d\vec{p}$, the flux seen by the electron is $|v_{\text{rel}}|fd^3p$, where

$$v_{\text{rel}} = c\frac{\cos\theta - \beta}{1 - \beta\cos\theta} \tag{3.128}$$

is the relative velocity between electron and photon. For this flux, the probability of scattering within a solid angle $d\Omega_\Delta$ is given by

$$dP = A|v_{\text{rel}}|fp^2dpd\phi d\mu\frac{d\sigma_{\text{T}}}{d\Omega_\Delta}d\Omega_\Delta \tag{3.129}$$

Here A is a normalization constant. Taking into account the fact that we are considering photons with a given initial energy, and that f does not depend on the angle because of the assumed isotropy, we have

$$dP = A|v_{\text{rel}}|d\phi d\mu\frac{d\sigma_{\text{T}}}{d\Omega_\Delta} \tag{3.130}$$

In the Newtonian limit, $\beta \ll 1$, we can show that $A =$ constant $+ \mathcal{O}(\beta^2)$, and this correction can be easily neglected (see eq. 3.127). Now the average of equation 3.127 can be easily made term by term. Indeed, the angle θ' along which the photon is moving, in the electron's reference frame of rest *before* the scattering event, is related to the same angle in the laboratory system θ by

$$\cos\theta' = \frac{\cos\theta - \beta}{1 - \beta\cos\theta} \tag{3.131}$$

whereas the angle θ_1 of the photon's motion after the scattering event, still in the electron's reference frame at rest, is connected to the initial direction of motion θ', as well as to the scattering angle Δ, by

$$\theta_1 = \theta' + \Delta \tag{3.132}$$

and the distribution of the scattering angle Δ is given by equation 3.114, which is valid within the Thomson limit.

Thus we find

$$\overline{\hbar\omega_{\rm f} - \hbar\omega_{\rm i}} \approx -\frac{\overline{\hbar^2\omega_{\rm i}^2}}{m_{\rm e}c^2} + 4\beta^2\overline{\hbar\omega_{\rm i}} \qquad (3.133)$$

where $\overline{X}$ indicates the mean value of the quantity X averaged over the angular part of the photons' distribution. This useful equation tells us that there are two effects that concur, with opposite signs, to determine the change of energy in the photon. The term with the negative sign is due to the energy loss of the photon, which transfers energy to the electron recoil. The term with the positive sign is the increase of the photon energy through the relativistic beaming effect, which was discussed above. Thence we can see that photons can only *acquire* energy through relativistic beaming.

Averaging over the distribution of the electrons' velocity, we find

$$\overline{\hbar\omega_{\rm f} - \hbar\omega_{\rm i}} \approx -\frac{\overline{\hbar^2\omega_{\rm i}^2}}{m_{\rm e}c^2} + 4\overline{\beta}^2\overline{\hbar\omega_{\rm i}} \qquad (3.134)$$

where $\overline{\beta}^2$ indicates the mean quadratic value of velocity (in units of c) of the electrons. Note that in some texts, you will find $\beta^2/3$, instead of β^2, given above; the reason is that here, β^2 is the dispersion in one-dimensional velocity, whereas other authors indicate with the same symbol the dispersion of three-dimensional velocity, which is in fact three times larger.

If electrons have a thermal distribution, the mean of β^2 must be computed using the Maxwell-Boltzmann distribution, thus obtaining

$$\overline{\hbar\omega_{\rm f} - \hbar\omega_{\rm i}} \approx -\frac{\overline{\hbar^2\omega_{\rm i}^2}}{m_{\rm e}c^2} + 4\frac{kT}{m_{\rm e}c^2}\overline{\hbar\omega_{\rm i}} \qquad (3.135)$$

If this is correct, then we should find that if photons have a a thermal distribution at the same temperature of electrons, there must be—on average—no net energy gain for photons. From the Planck distribution, we find that

$$\overline{\hbar\omega} = 3kT_\gamma, \qquad \overline{\hbar^2\omega^2} = 12(kT_\gamma)^2 \qquad (3.136)$$

where T_γ is the gas temperature of photons. Introducing these results in the equation above, indeed, we find that

$$\hbar\omega_{\rm f} - \hbar\omega_{\rm i} = 0 \Leftrightarrow T_\gamma = T \qquad (3.137)$$

We have checked that in thermal equilibrium at the same temperature, photons and electrons do not exchange energy.

In a similar way, we can calculate the total energy lost by the electron per unit time, by averaging over angles (Blumenthal and Gould 1970). We find that the average energy lost per unit time by the electron, taking into account its recoil, is given by

$$P = \frac{4}{3}\sigma_{\rm T} c \gamma^2 \beta^2 \epsilon_\gamma \left(1 - \frac{63}{10} \frac{\gamma \overline{\hbar^2 \omega^2}}{m_e c^2 \overline{\hbar\omega}} \right) \qquad (3.138)$$

where ϵ_γ is the energy density in photons, when the scattering process takes place in the Thomson regime. These are the photons for which $\hbar\omega_{\rm i} \lesssim m_e c^2/\gamma$, so that

$$\epsilon_\gamma \approx 4\pi c \int_0^{m_e c^2/\gamma} f(x) x^3 dx \qquad (3.139)$$

Notice that the second term in parentheses gives the correction due to the recoil. This correction is always small, because, in order to derive this formula, it has been assumed that the energy of the photon within the electron's reference frame at rest, $\approx \gamma \hbar\omega$, is $\ll m_e c^2$. However, apart from this constraint, this equation is valid for any velocity of the electron, both $v \ll c$ and $\gamma \gg 1$. This shows that the electron's energy losses are *always positive*: the beaming effect always leads to an energy loss.

We can define, once again, the cooling time for the IC of the single electron $t_{\rm c}$ as

$$t_{\rm c} \equiv \frac{E}{P} = \frac{\gamma m_e c^2}{P} \propto \frac{1}{\gamma} \qquad (3.140)$$

exactly like synchrotron (but the coefficient is different here).

Equation 3.138 also represents the solution of problem 4 in chapter 2. In fact, if equation 3.138 represents the power emitted by an electron, the force to which the electron is subject, in the Newtonian regime, is obviously given by

$$\vec{F}_{\rm C} = -\frac{4}{3}\sigma_{\rm T}\epsilon_\gamma \frac{\vec{v}}{c} \tag{3.141}$$

where we have neglected the small correction due to recoil. Now we can easily understand the signs. This force is called *Compton drag*, and the version we have given applies to each particle moving with a velocity $\vec{v}$ in the reference frame in which the photon gas appears isotropic, even if the photon gas spectrum is not thermal. Once again, the only true constraint is that $\gamma\hbar\omega \ll m_e c^2$.

Comparing equation 3.138 with equation 3.58, we remark that the ratio between the power emitted through synchrotron and IC is simply given (neglecting the correction due to recoil of the electron) by

$$\frac{P_{\rm s}}{P_{\rm IC}} = \frac{\epsilon_B}{\epsilon_\gamma} \tag{3.142}$$

the ratio between the energy densities in the magnetic field and in photons.

When many photons have energy $>m_e c^2$, in the electron's reference frame at rest, Blumenthal and Gould (1970) give an expression equivalent to equation 3.138, but approximated, in the so-called Klein-Nishina regime:

$$P_{\rm KN} = \frac{3\pi}{2}\sigma_{\rm T} m_e c^3 \int_{m_e c^2/\gamma}^{\infty} x^2 f(x)dx \left(\ln\frac{4\gamma x}{m_e c^2} - \frac{11}{6}\right) \tag{3.143}$$

3.5.2 The Spectrum of Inverse Compton Processes

The processes of Compton scattering, direct and inverse, can transfer energy, as we have seen, from photons to electrons, and vice versa. At this point, obviously, the spectrum of the emerging radiation depends on how many times photons are scattered. Indeed, the answer will be very different depending on whether photons are scattered once only, or many times. Therefore, we distinguish several distinct cases.

The Newtonian Regime for Small Optical Depth

In the Newtonian regime, we have already shown that the average fractional increase of photon energy in each scattering event is given by equation 3.135. The energy transfer vanishes when the electron temperature reaches a value called the *Compton temperature*, T_C:

$$kT_C \equiv \frac{\overline{(\hbar\omega)^2}}{4\overline{\hbar\omega}} \qquad (3.144)$$

When $T \gg T_C$, as in the case of the well-known Sunyaev-Zel'dovich effect, there is a net transfer of energy from electrons to photons, which are pushed toward energies larger than the initial ones by a factor $1 + 4kT/m_ec^2$.

The Relativistic Regime for Small Optical Depth

When electrons are hyperrelativistic, we can determine the average fractional energy gain by photons in the following way. In its motion, at a velocity $\approx c$, the electron scatters per unit time a quantity of energy in photons given by $c\sigma_T\epsilon_\gamma$; comparing this with equation 3.138 (where we neglect the small correction due to recoil), we can see that, on average, scattered photons increase their energy by a factor $4/3\gamma^2\beta^2$. Therefore,

$$\frac{\overline{\hbar\omega_f - \hbar\omega_i}}{\hbar\omega_i} = \frac{4}{3}\beta^2\gamma^2 \qquad (3.145)$$

which is obviously the equivalent of equation 3.134. The spectrum that emerges after *one* scattering is given by Blumenthal and Gould (1970). It is convenient to visualize it as follows. Let us consider an isotropic, monochromatic distribution of photons with an energy $\hbar\omega_i$ before the scattering process, with F_o indicating the number of photons crossing one unit of area, per unit time and solid angle. The specific emissivity j_{IC}, due to the scattering process of IC, is given by

$$j_{IC}(\hbar\omega_f) \approx \frac{3n_e\sigma_T F_o}{4\gamma^2\hbar\omega_i} f(x) \qquad (3.146)$$

where

$$x \equiv \frac{\hbar\omega_f}{4\gamma^2\hbar\omega_i}; \qquad f(x) \equiv 2x\ln x + x + 1 - 2x^2, \qquad x \le 1$$

$$(3.147)$$

Physically, the result is very simple. There is a maximum gain of energy for the photon, occurring when the initial photon moves directly toward the electron and is scattered exactly in the electron's direction of motion. This can easily be seen from equations 3.121 and 3.124. The maximum energy gain is $4\gamma^2$, so the scattering of monochromatic photons can never produce photons with an energy larger than $4\gamma^2\hbar\omega_i$, which is where the spectrum ends: $f(1) = 0$. The various angles possible between the electron's and the photon's directions of motion, before and after scattering, generate a range of amplifying factors, from low values of x, $x \ll 1$, up to $x = 1$. An excellent approximation for $f(x)$ is given by

$$f(x) \approx \frac{2}{3}(1 - x) \qquad (3.148)$$

We can now ask about the spectrum of photons produced by the IC, when the population of electrons, responsible for the scattering is distributed in energy as

$$dN = AE^{-\alpha}dE, \qquad E_{min} \le E \le E_{max} \qquad (3.149)$$

It is given by the convolution of the preceding equation with equation 3.146:

$$J_{IC} = A \int E^{-\alpha}dE j_{IC}(\hbar\omega_f) \qquad (3.150)$$

where the term to be integrated depends on E through the Lorentz factor γ of the electron. It is convenient to transform the integral by changing the integration variable to x. Thus we obtain

$$J_{\mathrm{IC}} = \frac{3}{8}\sigma_{\mathrm{T}}cA\big(m_ec^2\big)^{1-\alpha}Q(\alpha)\left(\frac{\omega_{\mathrm{i}}}{\omega}\right)^{(\alpha-1)/2} \qquad (3.151)$$

The dimensionless quantity $Q(\alpha)$ is given by

$$Q(\alpha) \equiv 2^{\alpha+3}\,\frac{\alpha^2 + 4\alpha + 11}{(\alpha+3)^2(\alpha+5)(\alpha+1)} \qquad (3.152)$$

Strictly speaking, this equation is valid only for a limited range of photon energies. Since the electrons have Lorentz factors from γ_{min} to γ_{max}, corresponding to energies E_{min} and E_{max}, respectively, the spectrum certainly does not extend beyond $4\gamma_{\mathrm{max}}^2\hbar\omega_{\mathrm{i}}$, but, in fact, it does not even go below $4\gamma_{\mathrm{min}}^2\hbar\omega_{\mathrm{i}}$.

Apart from this constraint, we can see that the spectrum due to one scattering is identical to that due to synchrotron processes, because the spectral index α_{IC} and the particles's index α obey the same relation as the synchrotron index:

$$\alpha_{\mathrm{IC}} = \frac{\alpha - 1}{2} \qquad (3.153)$$

Multiple Scattering Regime for Small Optical Depth

Occasionally, multiple scattering events can become important even in the limit of small optical depth, $\tau_{\mathrm{T}} \ll 1$. This may happen when, in the band we are considering, there are no other emission mechanisms (including Compton scattering with a smaller number of events). In this case, the emerging spectrum is once again a power law, but without a universal spectral index.

The following argument, from Zel'dovich, determines the spectrum. Let us consider an initial spectrum of photons with average energy $\hbar\omega_{\mathrm{i}} \ll kT_{\mathrm{e}}$ or $\hbar\omega_i \ll \gamma m_ec^2$, according to whether the electrons are thermally distributed or not. We have already seen that, after a scattering event, the photon's energy increases by a factor $A = \overline{\omega_f}/\omega_i$, which we have calculated separately for the Newtonian and hyperrelativistic limits. After n scattering events, the new photon energy will be

$$\hbar\omega_n \approx A^n \hbar\omega_i \qquad (3.154)$$

whereas the spectrum amplitude, which is obviously proportional to the probability of n scattering events $(P_n \propto \tau_T^n)$ decreases by a factor τ_T^n. It follows that the radiation intensity, at the frequency ω_n is approximately given by

$$I(\hbar\omega_n) \approx I(\hbar\omega_i)\tau_T^n \approx I(\hbar\omega_i)\left(\frac{\omega_n}{\omega_i}\right)^{-\alpha_{IC}} \qquad (3.155)$$

where the spectral index, in this case, is

$$\alpha_{IC} \equiv -\frac{\ln \tau_T}{\ln A} \qquad (3.156)$$

We have seen that $A > 1$, whereas, by assumption, $\tau_T < 1$, and thus $\alpha_{IC} > 0$. Moreover, in the relativistic case, $A \gg 1$, and if $\tau_T \approx \mathcal{O}(1)$, $\alpha_{IC} \ll 1$; in this regime, extremely flat spectra are produced. On the other hand, in the Newtonian limit, $A = 1 + \mathcal{O}(\beta^2)$, and the spectra will be correspondingly steep. This obviously happens because electrons are assumed to be Newtonian, or, better, cold, and it is difficult to steal from them their their small amounts of energy.

It should be noted that, if we want to treat this problem in this quasi-continuum limit, the photons coming from two, three, and more scattering events must overlap smoothly, to give the impression of a continuum. Of course, this is much easier in the Newtonian case, when $A - 1 \ll 1$, rather than in the relativistic case, where a better approximation consists in identifying a series of distinct bumps.

Large Optical Depths for Newtonian Electrons

This is the most common, and best studied case. In this paragraph, we shall assume that electrons are distributed thermally, with $kT \ll m_e c^2$. Let us assume that at first, photons have very low energies in comparison with the thermal energy of the electrons. In this case, the net energy gain by collision is given by equation 3.135, where we can neglect the recoil term

$$\frac{\overline{\triangle\hbar\omega}}{\hbar\omega_{\mathrm{i}}} \approx 4\frac{kT}{m_{\mathrm{e}}c^2} \ll 1 \qquad (3.157)$$

At the same time, however, the number of collisions is assumed to be large. If $\tau_{\mathrm{T}} \gg 1$, the number of effective collisions in order to get out of the source is τ_{T}^2, which is even larger. In order to know whether the spectrum is significantly changed, it proves convenient to define in this regime a parameter y, called the *Compton parameter*, thus determined:

$$y \equiv 4\frac{kT}{m_{\mathrm{e}}c^2}\tau_{\mathrm{T}}^2 \qquad (3.158)$$

Here we should note that, even when $\tau_{\mathrm{T}} \ll 1$, it is possible to define a Compton parameter as

$$y \equiv 4\frac{kT}{m_{\mathrm{e}}c^2}\tau_{\mathrm{T}} \qquad (3.159)$$

which has, obviously, the same physical meaning. Therefore, the correct definition of the Compton parameter, in the non-relativistic regime, is

$$y \equiv 4\frac{kT}{m_{\mathrm{e}}c^2}\max\left(\tau_{\mathrm{T}}, \tau_{\mathrm{T}}^2\right) \qquad (3.160)$$

This parameter describes the cumulative effect of many scattering events on the photon's energy. Indeed, the photon's energy changes with the number of N collisions as

$$\frac{d\hbar\omega}{dN} = \hbar\omega\frac{4kT}{m_{\mathrm{e}}c^2} \qquad (3.161)$$

which has the easy solution

$$\hbar\omega_{\mathrm{f}} = \hbar\omega_{\mathrm{i}}e^{4kTN/m_{\mathrm{e}}c^2} \qquad (3.162)$$

However, since the total number of collisions is τ_{T}^2, we find

$$\hbar\omega_{\mathrm{f}} = \hbar\omega_{\mathrm{i}} e^y \qquad (3.163)$$

which illustrates the importance of the Compton parameter: if $y \ll 1$, the source is not thick enough to allow sufficiently many collisions, and the spectrum does not change appreciably; photons do not manage to steal much energy from the electrons. If, on the other hand, $y \gg 1$, the spectrum is significantly changed, and the photons can swallow up a lot of energy from the electrons. However, we must recall that the energy transfer process from the electrons saturates when the photons' average energy is comparable to that of the electrons (see equation 3.135 and discussion thereof), which occurs at the Compton temperature. In this case, the photon's maximum energy is not that of the above equation, but

$$\hbar\omega_{\mathrm{f}} \approx kT \qquad (3.164)$$

The complete evolution of the spectrum is described by the so-called Kompaneets equation, which will not be derived here (see Rybicki and Lightman 1979). In order to describe a photon gas, we normally use its phase density, or the so-called occupation number for the energy levels $\mathcal{N}$. The term *occupation number of the energy levels* reminds us that there must be terms for stimulated emission, since the photon obeys Bose-Einstein statistics.

We define a photon's dimensionless energy as

$$x \equiv \frac{\hbar\omega}{kT} \qquad (3.165)$$

The Kompaneets equation states:

$$\frac{\partial\mathcal{N}}{\partial t_{\mathrm{c}}} = \frac{kT}{m_{\mathrm{e}}c^2 x^2} \frac{\partial}{\partial x}\left(x^4\left(\frac{\partial\mathcal{N}}{\partial x} + \mathcal{N} + \mathcal{N}^2\right)\right) + Q \qquad (3.166)$$

The term Q takes into account the injection of low-energy photons, which are later energized by collisions with the electrons, and their possible escape from the source, which occurs necessarily because the sources we consider always have finite dimensions. The dimensionless time t_c is the so-called *Compton-unit* time:

$$t_c \equiv n_e \sigma_T ct \qquad (3.167)$$

It is easy to see that, for a photon crossing a spherical-symmetric source from the center to the surface, $t_c = y$. The Compton parameter is therefore identical to the Compton time necessary to get out of the source. The term $\mathcal{N}^2$ is responsible for stimulated emission, for particles subject to Bose-Einstein statistics. It is rarely comparable to the term $\mathcal{N}$ and is often neglected.

In order to get acquainted with Kompaneets's equation, we now find its solution in the regime of thermal equilibrium. In equilibrium nothing should vary with time, so both sides of equation 3.166 must vanish. Obviously there can be neither escape nor injection of new photons: $Q = 0$. Therefore, the solution is

$$\mathcal{N} = \frac{1}{e^{\alpha + x} - 1} \qquad (3.168)$$

where α is an arbitrary constant. This solution is naturally very similar to the occupation number for the Planck distribution, which, however, has $\alpha = 0$. Whence the difference? You will remember from courses in statistical mechanics that in thermal equilibrium, the number of occupation for bosons is

$$\mathcal{N} = \frac{1}{e^{x - \mu / kT} - 1} \qquad (3.169)$$

where μ is the chemical potential. You will also remember that the chemical potential is indispensable in order to conserve the total number of particles of a given species. Therefore, the solution of the Kompaneets equation behaves as if it had a chemical potential and should conserve the number of photons. But this is absolutely true: since here we are considering only scattering processes, the total number of photons can neither increase nor decrease, which explains the presence of α in the solution. Why, instead, does the Planck equation have $\alpha = 0$? Because, in thermal equilibrium, photons are scattered but also emitted and absorbed, and their number can vary freely (remember that there is no law for the conservation of photon number) until thermal equilibrium is reached. This is the physical meaning of equation 3.168.

In order to understand what happens in the case of a source out of thermal equilibrium, with finite dimensions, we shall not introduce a scattering term, which takes into account the source's lack of homogeneity (which would be the correct approach), but use an approximation introduced by Shapiro, Lightman, and Eardley (1976). We shall assume that Q contains a loss term, $Q = -\mathcal{N}/\tau_T^2$, where the coefficient τ_T^{-2} is suggested by the fact that the escape process from the source is, in our hypothesis, of a scattering type, and this requires τ_T^2 scattering events. In this case, we can easily find an approximate solution:

$$\mathcal{N} \approx \frac{x^w}{e^{\alpha+x} - 1} \tag{3.170}$$

The exponential term takes into account the fact that photons rarely get more energy than the electrons with which they are in thermal contact. The parameter w is determined through the introduction of the approximate solution in the Kompaneets equation (eq. 3.166), neglecting the term of stimulated emission $\mathcal{N}^2$, and using the above-discussed form for Q. We find

$$w = \frac{3}{2}\left(-1 \pm \sqrt{1 + \frac{16}{9y}}\right) \tag{3.171}$$

When $y \gg 1$, the solution with the positive sign has $0 \lesssim w \lesssim 1/2$, and therefore the energy distribution $x^3\mathcal{N}$ is almost thermal:

$$x^3\mathcal{N} \propto \frac{x^{3+w}}{e^{\alpha+x} - 1} \tag{3.172}$$

To obtain the Planck distribution, we should have $w = 0$, which is recovered in the $y \to \infty$ limit. This is obviously the limit in which the photon gas is well thermalized and is led to a distribution and a temperature close to that of the electrons, always taking into account that the number of scattering events is large, but not infinite (namely, w is small, but not actually zero). In order to understand why we have neglected the solution with the negative sign, solve problem 11 now.

In the opposite limit, $y \ll 1$, we find $w \approx -3/2 - 2/\sqrt{y}$, so the energy distribution is

$$x^3 \mathcal{N} \propto \frac{x^{3/2 - 2/\sqrt{y}}}{e^{\alpha + x} - 1} \qquad (3.173)$$

which shows that in this case, the distribution is a steep power law, up to the electrons' thermal energy, beyond which the usual exponential cut intervenes.

In the intermediate regime, Zdziarski (1985) has proposed the following approximate form:

$$x^3 \mathcal{N} \propto \left(x^{3+w} + qx^3 \right) \frac{1}{e^{\alpha + x} - 1} \qquad (3.174)$$

The ratio q between the two components is given by

$$q = \frac{\Gamma(-3 - w)}{\Gamma(-3 - 2w)} (1 - p(\tau_{\mathrm{T}})) \qquad (3.175)$$

where $\Gamma(x)$ is the usual gamma function, w is the solution corresponding to the minus sign in equation 3.171 and $p(\tau_{\mathrm{T}})$ is the probability of escape from a spherical cloud, given in problem 2.

Analytical solutions for the Kompaneets equations have been found for inhomogeneous sources with the introduction of a scattering term (Sunyaev and Titarchuk 1980), and numerical solutions have also been found when account is taken of the Klein-Nishina cross section, rather than the Thomson one (Ross, Weaver, and McKray 1978).

Large Optical Depths for Relativistic Electrons

When electrons are relativistic, the simple approach sketched in the paragraph above cannot be used, because, as we have seen, the average energy amplification for each photon is

$4/3\gamma^2 \gg 1$, and thus we cannot use a formalism of the Fokker-Planck kind, like the Kompaneets equation above, which is derived strictly under the hypothesis that the energy gain is small when compared with the photon energy. The only possible approach then is through numerical simulations. However, so far this problem has received little attention in the literature; therefore we refer here to the original literature. All the numerical codes suitable for this problem come essentially from the one created by Pozd'nyakov, Sobol, and Sunyaev (1977); see, for example, work by Zdziarski (1985). Other enlightening works, which, however, use a Monte Carlo technique, are those by Stern et al. (1995) and by Poutanen and Svensson (1996).

3.5.3 About the Compton Parameter

The diligent reader will have certainly noticed that we have not defined a Compton parameter y in the relativistic regime. It is obvious, from what we have just said, that its usefulness, is, at most, limited, since the spectral evolution cannot be described by a Fokker-Planck-type equation like Kompaneets', because the energy gain in each scattering event can be $\gg 1$. However, there are fundamental reasons that make y useless in the relativistic or even transrelativistic limits.

In the Newtonian limit, when $\tau_T \gg 1$, the Kompaneets equation describes the spectrum evolution in terms of y and nothing else, as described above (see also Illarionov and Sunyaev 1975; Sunyaev and Titarchuk 1985). However, in the relativistic limit, a Compton parameter with the same properties cannot be defined.

We can obtain the average energy gain, also in the relativistic regime, which is $\mathcal{O}(\gamma^2)$ for a single electron with a Lorentz factor γ. This gain can be averaged on a *thermal* distribution of electrons (Svensson 1984):

$$A_1 = 4\frac{kT}{m_e c^2}\frac{K_3\left(kT/m_e c^2\right)}{K_2\left(kT/m_e c^2\right)} + 16\left(\frac{kT}{m_e c^2}\right)^2 \tag{3.176}$$

where K_2 and K_3 are modified Bessel functions; this expression can be approximated as

$$A - 1 \approx 4\frac{kT}{m_e c^2} + 16 \left(\frac{kT}{m_e c^2}\right)^2 \qquad (3.177)$$

However, it is not obvious which is the generalization of y. For example, Loeb, McKee, and Lahav (1991) suppose that the distribution of collisions follows a Poisson distribution; then they calculate the relationship between incoming and outgoing photon energy, and, imposing that their ratio is once again e^y as in the Newtonian case, they find

$$y = (A - 1)\langle N \rangle \qquad (3.178)$$

where $\langle N \rangle$ is the average number of collisions, thus obtaining

$$y = (A - 1)\tau_T(1 + \tau_T/3) \qquad (3.179)$$

On the other hand, Zdziarski, Coppi, and Lamb (1990) treat all photons as if they suffered only the average number of collisions and find, in the same manner of Loeb, McKee, and Lahav, that

$$y = (\ln A)\tau_T(1 + \tau_T/3) \qquad (3.180)$$

We must remember that, in the Newtonian limit, these two definitions of y coincide; the fact that they do not coincide, in the relativistic limit, indicates the uselessness of the definition of y. Finally, we may add that in pair plasmas illuminated by an outside source, Pietrini and Krolik (1995) find another result still, thus showing essentially that the Compton parameter in the relativistic regime is useless.

3.5.4 Self-synchro-Compton and Compton Limit

It happens sometimes that sources are so compact that there is a high probability that photons produced by the synchrotron processes are used in the inverse compton process by the same electrons that produced them. We speak then of *self-synchro-Compton*, or SSC. In this case, we can obtain the spectrum of the radiation emitted assuming that each electron with a Lorentz γ factor scatters incident photons, with an energy $\hbar\omega$, to a new energy $\hbar\omega \times 4\gamma^2/3$. In other words, we neglect the dependence of the photon's energy gain from the angle of incidence. Thus we find

$$j_\nu \propto \int d\nu'\nu'^{-\alpha} \int d\gamma\gamma^{-p}\delta(\nu - 4\gamma^2\nu'/3) \propto \int d\nu'\nu'^{-\alpha-1} \propto \nu^{-\alpha}$$

$$(3.181)$$

In other words, the SSC spectral index is identical to that of synchrotron, $\alpha = (p-1)/2$. We can easily see that this spectral form holds between $\omega_{\min} = \gamma_{\min}^4\omega_L$ and $\omega_{\max} = \gamma_{\max}^4\omega_L$, where $\omega_L = eB/(m_e c)$ is the Larmor angular frequency, and I neglected the factor $4/3$ as inessential.

This equation applies only to photons that have undergone one IC process; if photons have also undergone $2, 3, \ldots N, \ldots$ scattering events, the spectrum will extend to energies of the order $\gamma_{\max}^{2+2N}\omega_L$. As a general rule, these further Compton peaks are not important, except in one case. Comparing equations 3.58 and 3.138, we see that the total power emitted by a particle scales as

$$\eta \equiv \frac{P_{IC}}{P_s} = \frac{\epsilon_\gamma}{\epsilon_B} \qquad (3.182)$$

where I neglected the recoil terms in P_{IC}, and ϵ_γ and ϵ_B are the energy densities in photons and in the magnetic field, respectively. Thus we see that, when $\eta > 1$, particles lose more energy through the IC effect than through synchrotron; besides, the photons that underwent two collisions will absorb much more energy than those that underwent only one. On the other hand, the photons that underwent N collisions will contain less energy than those that underwent $N + 1$ collisions, and so on. As a consequence, $\eta = 1$ represents the limit beyond which a thermal catastrophe takes place, which is called the *Compton catastrophe*.

Let us now verify when $\eta = 1$ occurs. In order to do this, we must calculate ϵ_γ; let us assume that the photons' spectrum is self-absorbed up to a frequency ν_m. In this case, we know that $B_\nu = j_\nu/\alpha_\nu$ and, therefore, that it reaches a strong peak around ν_m, as we saw above (see the discussion following equation 3.89). Therefore,

$$\epsilon_\gamma = \int B_\nu d\nu \approx B_{\nu_m} \nu_m \tag{3.183}$$

The quantity B_ν is directly observable; radioastronomers express it in terms of *brightness temperature* T_b, thus defined:

$$kT_b \equiv \frac{B_\nu c^2}{2\nu^2} \tag{3.184}$$

This is the temperature that a blackbody with the same specific intensity B_ν would have at the given frequency. It is useful because it is a directly observed quantity. Note that kT_b is also the average energy of the process emitting at a frequency ν, a remark that will soon come handy. Putting together the two preceding equations, we thus find

$$\epsilon_\gamma \propto T_b \nu_m^3 \tag{3.185}$$

Since kT_b is the average energy of the system emitting at a frequency ν_m, we know that, this being a synchrotron emission,

$$\nu_m = \nu_L \left(\frac{kT_b}{m_e c^2} \right)^2 \tag{3.186}$$

where $\nu_L = eB/(2\pi m_e c)$ is the Larmor frequency. It is now easy to check the value of the energy density in the magnetic field:

$$\frac{B^2}{8\pi} = m_e c^2 \frac{\pi \nu_L^2}{2 r_e c^2} \tag{3.187}$$

where $r_e \equiv e^2/m_e c^2$ is the usual classic radius of the electron. Thence we find that

$$B^2 \propto \frac{\nu_m^2}{T_b^4} \tag{3.188}$$

which, introduced into equation 3.182, gives

$$\eta \propto T_{\mathrm{b}}^5 \nu_{\mathrm{m}} \tag{3.189}$$

The exact coefficient of proportionality, whose derivation is now left to the reader, is

$$\eta = \left(\frac{324 e^2}{\pi m_{\mathrm{e}}^6 c^{13}}\right) \nu_{\mathrm{m}} T_{\mathrm{b}}^5 \tag{3.190}$$

Thus we find that the critical brightness temperature ($\eta = 1$) is

$$T_{\mathrm{c}} = 10^{12} \left(\frac{1\ \mathrm{GHz}}{\nu_{\mathrm{m}}}\right)^{1/5} \mathrm{K} \tag{3.191}$$

The physical meaning of this quantity is very simple: there should be no source beyond this brightness temperature, since for $T_{\mathrm{b}} > T_{\mathrm{c}}$ the losses due to the IC process are catastrophically rapid.

3.5.5 Compton Broadening

There is another very significant effect due to the Compton processes, which has interesting observational consequences. Let us consider equation 3.127, in the Newtonian limit (i.e., neglecting all terms $\mathcal{O}(\beta^2)$). Thus we find

$$\frac{\hbar \omega_{\mathrm{f}}}{\hbar \omega_{\mathrm{i}}} \approx 1 + \vec{\beta} \cdot (\hat{n}_{\mathrm{f}} - \hat{n}_{\mathrm{i}}) \tag{3.192}$$

where $\hat{n}_{\mathrm{i}}$, $\hat{n}_{\mathrm{f}}$ are unit vectors directed along the initial and final momenta of the photon. We see that the photon energy after the collision differs from the initial one, not because of the electron's recoil, which we neglected, but because of the electron's velocity. When we average over an isotropic distribution of electrons, the average effect obviously disappears, but the quadratic one remains. An initially monochromatic distribution of photons acquires, after a collision, a spectral width given by $\Delta \nu / \nu \approx (\overline{\beta^2})^{1/2}$.

A simple computation (Pozd'nyakov, Sobol, and Sunyaev 1983) gives, for the width $\triangle E$ of a spectral line,

$$\frac{\triangle E}{E} = \left(\frac{2kT}{m_e c^2}\right)^{1/2} = 0.063 \left(\frac{kT}{1 \text{ keV}}\right)^{1/2} \tag{3.193}$$

The meaning of this equation is pretty obvious: at sufficiently high temperatures ($kT \approx 1$ keV $\rightarrow 10^7$ K), the average quadratic change in the photon energy is large enough to push the photon outside the natural width of the line. Furthermore, for photons with an energy ≈ 1 keV, this remains true even for scattering by very cold electrons, because in that case, the electron's recoil modifies the photon's energy by about $(\hbar\omega)^2/m_e c^2 \approx 100$ eV, enough to remove it from the line.

Moreover, we note that if we saw spectral lines with a width smaller than this, they should be wholly generated within the cortex of the source, that is, that part of the source in which $\tau_T \ll 1$. In all the other parts of the source (namely, the internal ones) the line is inevitably broadened, sometimes so much so that it merges with the underlying continuum.

This effect, first discussed by Pozd'nyakov, Sobol, and Sunyaev (1983), is of critical relevance for the detection of the iron recombination lines in the soft X band.

3.6　Relativistic Effects

There are at least two excellent reasons to consider what happens when there are relativistic *bulk* motions, namely, when the Lorentz factor $\gg 1$ pushes not single particles, but macroscopic (and detectable) amounts of matter. First of all, such motions are directly observed in at least two classes of sources: blazars and gamma ray bursts. Second, some radio sources show brightness temperatures $\gg T_c \approx 10^{12}$ K, up to values $T_b \approx 10^{18}$ K, which are prohibited by the argument just made about the Compton catastrophe.

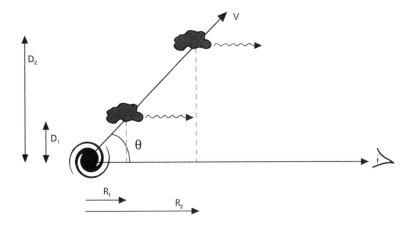

Figure 3.3. Geometry for the Rees model (1966) for superluminal motions.

3.6.1   Superluminal Motions

It is possible to observe plasma clouds detaching themselves from the main component of blazars, in the radio region, and moving away from them. If we know the distance of the object hosting the blazar, thanks to optical observations, we can then calculate the apparent velocity of the cloud, perpendicular to the distance (see fig. 3.3).

These velocities often appear larger than the speed of light, some even by a factor of 10. The explanation of this phenomenon (Rees 1966) shows that this paradox can be explained within relativistic kinematics.

We assume that V is the velocity of the plasma cloud, moving along a direction that makes an angle θ with the line of sight. Let us suppose that the cloud starts at an instant t_0; the photons emitted at this time reach Earth at an instant $t'_0 = t_0 + D/c$, where, obviously, D is the distance of the source. At a following instant t_1, the cloud is at a distance $V(t_1 - t_0)$ from the source from which it moved off; its distance from the source projected on the plane of the sky is $D_\perp = V(t_1 - t_0) \sin \theta$, whereas its distance from Earth is $D - V(t_1 -$

t_0) $\cos\theta$. Therefore, the photons that left at time t_1 reach Earth at the time $t_1' = t_1 + (D - Vt_1\cos\theta)/c$. Thus, to an observer on Earth, the cloud appears to have covered the distance $D_\perp$ in the time $t_1' - t_0'$, and the resulting apparent velocity v_{app} is

$$v_{app} = \frac{V\sin\theta}{1 - V\cos\theta/c} \qquad (3.194)$$

This apparent velocity vanishes for $\theta = 0, \pi$ and therefore has its peak in between these extremes. A simple computation shows that this is realized for $\cos\theta = V/c$, and thus we have a peak:

$$v_{app}^{(max)} = \gamma V \qquad (3.195)$$

From this we see that if θ has a favorable inclination, we can easily have $v_{app} \gg c$, with no contradiction with the theory of special relativity as we know it. In fact, the so-called *superluminal velocities* are now normally interpreted as a demonstration that the motion of the sources in question is hyper-relativistic, $\gamma \gg 1$.

3.6.2   Emission Properties of Relativistic Sources

When a source with macroscopic dimensions is in relativistic motion ($\gamma \gg 1$) with respect to the observer, the quantities we have used to characterize radiation, such as specific intensity I, flux J, and so on, must be transformed into the observer's rest frame. Here we derive how these macroscopic quantities are transformed through the Lorentz transformations.

Let us consider, once again, a source moving with speed V along a direction that makes an angle θ with the line of sight to the observer. We can easily see that if the source emits photons with a frequency ν_0, the frequency of the received photons ν is

$$\nu = \frac{\nu_0}{\gamma(1 - V\cos\theta/c)} \equiv q\nu_0 \qquad (3.196)$$

If the factor is $q > 1$, we say that the photon has been *blueshifted*; if $q < 1$, it has been *redshifted*.

The transformation of frequencies immediately allows us to derive the transformation of times; indeed, since $T = 1/\nu$ and $T_0 = 1/\nu_0$, times transform as

$$\frac{T}{T_0} = \frac{\nu_0}{\nu} = \frac{1}{q} \tag{3.197}$$

It is also easy to derive the law of transformation of solid angles:

$$\sin\theta d\theta d\phi = \frac{1}{q^2} \sin\theta_0 d\theta_0 d\phi_0 \tag{3.198}$$

or

$$d\Omega = \frac{1}{q^2} d\Omega_0 \tag{3.199}$$

We are now ready to transform the specific intensity I between two reference frames. As a reminder to the reader, I is the energy that arrives per unit time, frequency, and solid angle. Therefore, as seen in the source's reference frame, the energy that arrives is

$$h\nu_0 N(\nu_0) d\nu_0 dt_0 d\Omega_0 \tag{3.200}$$

Here $N(\nu)$ is the number of photons, obviously a relativistic invariant, which must be the same in both the source frame and in the observer's. On the other hand, in the reference frame of the observer, the quantity of energy received is

$$h\nu N(\nu) d\nu dt d\Omega = q h\nu_0 N(\nu_0) q d\nu_0 \frac{dt_0}{q} q^2 d\Omega_0$$
$$= q^3 h\nu_0 N(\nu_0) d\nu_0 dt_0 d\Omega_0 \tag{3.201}$$

which immediately implies

$$I(\nu) = q^3 I(\nu_0) \tag{3.202}$$

We can immediately apply this result to an object whose spectrum is that of a blackbody, $I(\nu_0) = 2h\nu_0^3/c^2$ $(\exp(h\nu_0/kT_0) - 1)^{-1}$. We find

$$I(\nu) = \frac{2q^3 h\nu_0^3}{c^2} \frac{1}{e^{h\nu_0/kT_0} - 1} = \frac{2h\nu^3}{c^2} \frac{1}{e^{h\nu/kT} - 1} , \quad T = qT_0$$

$$(3.203)$$

Therefore, an object moving in a relativistic way toward us seems hotter than it actually is in its reference frame, by a factor q.

If, on the other hand, the source has a nonthermal spectrum, $I \propto \nu^{-\alpha}$, it is easy to see that the observed flux,

$$F(\nu) = \frac{L(\nu_0)}{4\pi D^2} q^3 \qquad (3.204)$$

gives us

$$F(\nu) = \frac{L(\nu)}{4\pi D^2} q^{3+\alpha} \qquad (3.205)$$

It has been observed that many sources accreting material from the disk emit material along the rotation axis of the disk itself, in two jets going in opposite directions. This means that q is different for the two jets, and therefore the luminosity ratio between the two is given by

$$\frac{F_+}{F_-} = \left(\frac{1 + V\cos\theta/c}{1 - V\cos\theta/c}\right)^{3+\alpha} \qquad (3.206)$$

This number can be large: for an ideal orientation ($\cos\theta = V/c$) and large Lorentz factors $\gamma \gg 1$, we have

$$\frac{F_+}{F_-} \approx (2\gamma^2)^{3+\alpha} \qquad (3.207)$$

For typical synchrotron spectra, $\alpha \approx 1/2$; for $\gamma = 10$, the ratio between the values of luminosity is $\approx 10^8$, which explains why some sources showing superluminal expansions have only one jet: the other one is so weak that it is below the detection threshold.

However, we should notice that the two jets are not observed simultaneously. In fact, since the jet moving away is further away from the observer, the photons we receive from the far-away jet must have left much earlier than those of the approaching jet, which we receive simultaneously. This difference obviously tends to *reduce* the luminosity contrast F_+/F_-, unless the source is truly stationary in its energy output.

Finally, we can say that the relativistic expansion helps us to understand that the observations of brightness temperatures larger than the Compton limit T_c, equation 3.190, do not represent true violations. In fact, we have seen that both temperature and frequency measured by the observer are higher than those measured in the reference frame of the emitting plasma by a factor q. As a consequence, if we use the observed quantities ν_m and T_b in equation 3.190, we overestimate the parameter η by a factor q^6. Therefore, observations of $T_b = 10^{18}$ K are consistent with the Compton limit, equation 3.191, provided they refer to material moving with respect to us, with a factor $q = 1/(\gamma(1 - V \cos\theta/c)) \gtrsim 10$.

3.7 Pair Creation and Annihilation

Only three of all processes of creation and annihilation of particles have so far played a role in high energy astrophysics. The first one concerns the creation, absorption, and scattering of neutrinos and is indispensable for an understanding of supernova formation, as well as of the collapse toward a neutron star or a black hole within the nucleus of a massive star. We shall not discuss these processes here. The second one is the process of pion photoproduction in collisions between a nucleon and a photon, which will be treated in the next section. Here we shall discuss the third relevant process, namely, the creation and annihilation of electron-positron pairs (from now on, we shall not specify again that they are e^+/e^- pairs).

The fundamental process of pair creation is

$$\gamma + \gamma \leftrightarrows e^+ + e^- \qquad (3.208)$$

The reason is that the other processes that may involve pairs, such as

$$\gamma + e^\pm \leftrightarrows e^\pm + e^+ + e^-, \gamma + Z \leftrightarrows Z + e^+ + e^-,$$

$$e^\pm + e^\pm \leftrightarrows e^\pm + e^\pm + e^+ + e^-, e^\pm + Z \leftrightarrows e^\pm e^\pm + Z + e^+ + e^-$$

$$(3.209)$$

have smaller cross sections in comparison with the process in equation 3.208. Indeed, the cross section for equation 3.208 is, in order of magnitude, $\approx \sigma_T = 8\pi(e^2/m_e c^2)^2/3 = 0.66 \times 10^{-24}$ cm^2, whereas the other processes, which have one or two more vertices, must have cross sections smaller by one factor α, or α^2, where $\alpha = e^2/\hbar c \approx 1/137$ is the fine-structure constant. That is why we neglect processes in equation 3.209, for which you can find references in work by Landau and Lifshitz (1981b) or Svensson (1982).

The cross section for the process

$$e^+ + e^- \rightarrow \gamma + \gamma \qquad (3.210)$$

can be easily expressed in terms of the relativistic invariant

$$\tau \equiv \frac{(p_-^\mu + p_+^\mu)(p_{-\mu} + p_{+\mu})}{m_e^2 c^4} \qquad (3.211)$$

and has the value (Landau and Lifshitz 1981b)

$$\sigma_{\rm ann} = \frac{2\pi r_0^2}{\tau^2(\tau-4)} \left((\tau^2 + 4\tau - 8) \ln \frac{\sqrt{\tau} + \sqrt{\tau-4}}{\sqrt{\tau} - \sqrt{\tau-4}} - (\tau+4)\sqrt{\tau(\tau-4)} \right)$$

$$(3.212)$$

where $r_0 \equiv e^2/m_e c^2$ is the classical electron radius. It has two simple limits, the Newtonian one ($\tau \rightarrow 4$) and the hyper-relativistic one ($\tau \gg 1$):

$$\sigma_{\rm ann} \approx \begin{cases} \frac{\pi r_0^2}{\sqrt{\tau-4}} & \tau - 4 \ll 1 \\ \frac{2\pi r_0^2}{\tau}(\ln \tau - 1) & \tau \gg 1 \end{cases} \qquad (3.213)$$

When one of the two particles is at rest, and the other one is in motion with a Lorentz factor γ, it is easy to find that

$$\tau = 2(1 + \gamma) \tag{3.214}$$

introducing this value into the equation 3.212, we find the cross section in function of energy in the laboratory system. In this case, the Newtonian limit assumes the form

$$\sigma_{\text{ann}} \approx \frac{\pi c r_0^2}{v_{\text{rel}}} = \frac{3 c \sigma_T}{8 v_{\text{rel}}} \tag{3.215}$$

where v_{rel} is the relative velocity between the particles. This dependence of σ_{ann} on v_{rel}^{-1} implies that the annihilation rate, in other words, the probability, per time unit, of a particle annihilating while moving in a background of antiparticles with a numerical density n cm^{-3}, is given by

$$\dot{P} \approx \frac{3}{8} \sigma_T c n \tag{3.216}$$

independent of the temperature, provided the relative motion is subrelativistic. On the other hand, at very high temperatures, that is, for

$$\mathcal{T} \equiv \frac{kT}{m_e c^2} \gg 1 \tag{3.217}$$

we can show that (Svensson 1982)

$$\dot{P} \approx \frac{3}{16} \sigma_T c n \mathcal{T}^{-2} \ln(2\mathcal{T}) \tag{3.218}$$

The coefficient of Newtonian cooling is obviously given by

$$\Lambda_{\text{ann}} = m_e c^2 \dot{P} n_- n_+ \tag{3.219}$$

where n_- and n_+ are the densities of the electrons and positrons, respectively.

The spectrum of the annihilation process is obviously much simpler in the reference frame of the center of mass, because here it is given by two photons each with an energy $\sqrt{p^2 c^2 + m_e^2 c^4}$. When the pairs are Newtonian, as is the case in the one example observed so far, the diffuse source located in our galactic center, the line is essentially monochromatic and centered at $m_e c^2 = 511$ keV. The width of the line will be about v/c, where v is the typical thermal velocity.

The inverse process,

$$\gamma + \gamma \to e^+ + e^- \tag{3.220}$$

obviously has a threshold. In the reference frame of the center of mass, since the energy of each photon must equal that of each lepton, we must have $h\nu \geq \sqrt{p^2 c^2 + m_e^2 c^4}$, and therefore, in any case $h\nu > m_e c^2$. If the two photons have a collinear momentum (namely, $\vec{k}_1 = k_1 \hat{n}_1, \vec{k}_2 = k_2 \hat{n}_2, \hat{n}_1 = -\hat{n}_2$), we can reason as follows. If we consider the four-vectors $k^\mu = (\omega/c, \vec{k})$ for each photon, we know that $k_1^\mu k_{2\mu}$ is a relativistic invariant. If we calculate it in the system of the center of mass, we find $k_1^\mu k_{2\mu} = \omega^2/c^2 - \vec{k}_1 \cdot \vec{k}_2 = \omega^2/c^2 + k^2 = 2\omega^2/c^2$, because in the center of mass frame, $\omega_1 = \omega_2$ and $\vec{k}_1 = -\vec{k}_2$. But if we now calculate the same invariant in the reference frame in which photons have opposite momenta, we find $k_1^\mu k_{2\mu} = \omega_1 \omega_2/c^2 + k_1 k_2 = 2\omega_1 \omega_2/c^2$, and, by equating these two results, we find $\hbar\omega_1 \hbar\omega_2 = p^2 c^2 + m_e^2 c^4$. Finally, the threshold condition is

$$\hbar\omega_1 \hbar\omega_2 \geq m_e^2 c^4 \tag{3.221}$$

The pair creation cross section can be directly inferred from the principle of detailed balance (but one must pay attention to a crucial factor of 2 (Landau and Lifshitz 1981b):

$$\sigma_{cr} = 2 \frac{v^2}{c^2} \sigma_{ann} = \frac{3\sigma_T}{8t} \left(\left(2 + \frac{2}{t} - \frac{1}{t^2} \right) \ln(t^{1/2} + \sqrt{t-1}) \right.$$

$$\left. - \left(1 + \frac{1}{t} \right) \left(1 - \frac{1}{t} \right)^{1/2} \right) \tag{3.222}$$

where $t \equiv (k_1^\mu + k_2^\mu)(k_{1\mu} + k_{2\mu})/m_e^2 c^4$. Here k_1 and k_2 are the photons' four-wave vectors before the creation of the pair, and t automatically turns out equal, in the center of mass frame, to the square of the photon's energy in dimensionless units.

The most interesting limits of this cross section are the one just above the threshold, $t - 1 \ll 1$, and the very high energy one, $t \to \infty$. The two corresponding limits are

$$\sigma_{\text{cr}} \approx \frac{3\sigma_{\text{T}}}{8} \begin{cases} \sqrt{t - 1} & t - 1 \ll 1 \\ \frac{\ln t}{t} & t \gg 1 \end{cases} \tag{3.223}$$

The most important quantity inferred from this cross section is the absorption coefficient for pair creation, for a photon with a given energy ϵ. It can be easily shown that

$$\alpha_{\text{cr}} = \frac{2m_e^3 c^6}{\epsilon^2} \int_1^\infty dt\sigma_{\text{cr}}(t)t \int_{tm_e^2 c^4/\epsilon}^\infty d\epsilon' \frac{n(\epsilon')}{\epsilon'^2} \tag{3.224}$$

3.8 Cosmological Attenuations

3.8.1 Protons

The reaction (Greisen 1967, Zatspein and Kuzmin 1966)

$$p + \gamma \to \begin{cases} n + \pi^+ \\ p + \pi^0 \end{cases} \tag{3.225}$$

has recently assumed an extraordinary importance in cosmic-ray physics, since it limits drastically the propagation of very high energy particles.

This is a threshold reaction. Indeed, in order for it to take place, energy in the frame of the center of mass must exceed the rest mass of the nucleon-pion system. We shall call $p^{\text{P}}, p^\gamma, p^{\text{N}}$, and p^π the four-momenta of proton and initial photon, as well as of nucleon (p or n) and final pion. In the frame of the center of mass, the three-dimensional components (namely, the normal momenta) of $p^{\text{P}} + p^\gamma$ cancel, by definition of the center of mass frame. Therefore, the quantity

$$s \equiv (p_\mu^{\text{P}} + p_\mu^\gamma)(p^{\text{P}\mu} + p^{\gamma\mu}) \tag{3.226}$$

simply takes the value of the square of the total energy E_c^2. On the other hand, s is a Lorentz invariant; therefore, no matter the reference frame in which we calculate it, we always get the same result, that is, the square of the collision energy in the center of mass frame. The proton has a four-momentum $p^p = (E, c\vec{p})$, whereas the photon $p^\gamma = \hbar\omega(1, \vec{n})$. By assuming, for convenience, that the proton is ultrarelativistic (thus $E \approx pc$), we find

$$E_c^2 = m_p^2 c^4 + 2\hbar\omega E(1 - \cos\theta) \qquad (3.227)$$

and we must therefore have

$$E_c^2 \geq (m_N + m_\pi)^2 c^4 \qquad (3.228)$$

which can now be rewritten, using the above equation, as

$$E(1 - \cos\theta) \geq \frac{m_\pi^2 + 2m_\pi m_N}{2\hbar\omega} c^4 \approx \frac{m_\pi m_p}{\hbar\omega} c^4 \qquad (3.229)$$

The important case is that of hyperrelativistic protons, which move in intergalactic space. In this case, the photons that lead to the production of pions are the ones in the cosmic microwave background, which have a temperature $T \approx 2.73\ K$. Thus we easily find that the threshold energy E_{th} is

$$E_{th} \approx 10^{20} \text{eV} \qquad (3.230)$$

The threshold that appears sharp in the above computation is less definite in nature. In fact, the threshold appears at different energies, according to the angle between the directions in which photon and proton are moving. Furthermore, the photons of the cosmic microwave background are certainly not monochromatic, but follow a Planck distribution, so that a proton can be below threshold for some photons and above it for others. That is why, in equation 3.230, there is no exact equality.

Once the proton is above threshold for the production of photo-pions with most cosmic microwave background photons, its energy losses increase dramatically compared to below threshold, for a number of reasons. When there is no photo-pion production, the proton loses energy through the reaction

$$p + \gamma \rightarrow p + e^+ + e^- \tag{3.231}$$

which has a much lower threshold energy. However, this reaction causes, for each event, a much lower energy loss, because the rest mass of the electron-positron pair is much smaller than that of the pion ($m_e = 511$ keV against $m_\pi = 135$ MeV and $m_\pi = 139$ MeV for the neutral and charged pion, respectively). This quantity is called *anelasticity* ξ. In the case of the reaction 3.225, it reaches, on average, about 5% of the proton before the event, $\xi = \triangle E/E \approx 0.05$.

There is, however, a second reason why equation 3.231 produces smaller losses than equation 3.225, namely, the second one has only one electromagnetic vertex, whereas the first one has two. Therefore, equation 3.231 is of the second order in the small parameter $e^2/\hbar c \approx 1/137$, also called the *fine structure constant*, whereas equation 3.225 is only of the first order in $e^2/\hbar c$.

In fact, it is even worse (or better?) than this. The cross section for equation 3.225 is simply measured in a laboratory, bombarding a target made of protons with photons of suitably high energy. In this reference frame, obviously called the *laboratory system*, the photon's threshold energy is slightly larger than $\approx m_\pi c^2 \approx 140$ MeV, but, above all, it reaches a peak just above the threshold, which roughly corresponds to about $\sigma \approx 0.5$ mb $= 0.5 \times 10^{-27}$ cm^2.

We can now calculate the typical length on which the proton loses energy, in order of magnitude. In the cosmic microwave background at $T = 2.73$ K there are about $n_\gamma \approx 400$ photons cm^{-3}. The mean free path between collisions is thus given by the usual relation:

$$l = \frac{1}{n_\gamma \sigma} \tag{3.232}$$

where, in each collision, a fraction $\xi \approx 0.05$ of the total energy is lost. As a consequence,

$$\frac{dE}{dx} = -\xi \frac{E}{l} \tag{3.233}$$

Therefore, the attenuation path $L \equiv l/\xi$, defined as the typical one on which energy is lost, is

$$L = \frac{l}{\xi} \approx 10^{25} \text{ cm} \approx 3 \text{ Mpc} \tag{3.234}$$

Thence we can see that $L \ll c/H_\circ$, the Hubble radius. This very fast energy loss continues until the proton energy is above threshold, equation 3.230. When the proton energy has decreased enough that it is below threshold for most photons, equation 3.231 is solely responsible for energy losses, but it has an inelasticity that since the electron's mass is smaller than the pion's one by about a factor of 270, is smaller by a factor of 135 (an electron and a positron are created); therefore, $\xi \lesssim 4 \times 10^{-4}$. This fact alone makes the attenuation path longer than the Hubble radius, $c/H_\circ$; but we must also take into account the reduced cross section, which further increases the attenuation path.

The total result of this exercise is to demonstrate that the attenuation path of very high energy protons varies from $\gg c/H_\circ$, immediately below threshold, to $\ll c/H_\circ$ immediately above the threshold. The exact position of the threshold, as well as its width, can be accurately calculated with numerical simulations. Figure 3.4 shows the attenuation path as a function of the energy; here we can see how L changes by almost four orders of magnitude while the proton's energy changes only by a factor of 2, around 6×10^{19} eV. Correspondingly, the spectrum of very high energy particles measured at Earth must undergo a sharp drop exactly at the threshold energy, since, by varying the particle energy by just a factor of 2, the visible part of the universe is dramatically reduced to $L \ll c/H_\circ$. This is called the GZK effect, from the name of the physicists (Greisen 1967; Zat'sepin, and Kuzmin

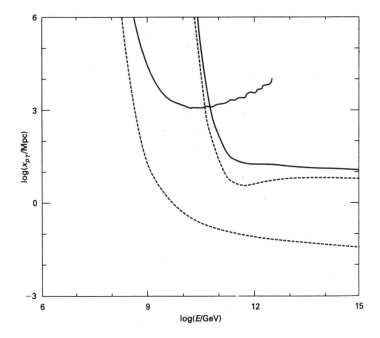

Figure 3.4. Interaction length (broken lines) and energy loss length ($E/(dE/dx)$, solid lines) for the protons in the cosmic microwave background, for pair production e^-/e^+ (lines on the left) and pions (lines on the right). From Protheroe and Johnson 1996.

1967) who first discussed it, immediately after the discovery of the cosmic microwave background.

As a further corollary, note that the reaction of pion photoproduction, equation 3.225, also generates neutrinos from the decay of charged pions (there are also negative pions; see problem 12). Since the other particles produced in the pion's decay are hyperrelativistic, too, and thus almost massless, neutrinos will take away the same fraction of the pion's energy. Therefore, typically, if the inelasticity of the reaction is ξ, each neutrino will have a fraction $s \approx 0.05\xi/3$ of the primary proton's energy. It follows that we expect huge fluxes of neutrinos on Earth, with energies up to $\approx 10^{18}$ eV.

3.8.2   Photons

The reaction of pair production

$$\gamma + \gamma \rightarrow e^+ + e^- \tag{3.235}$$

limits the maximum energy of the photons that can reach us from cosmological distances. The only uncertainty in this computation is which photon background to use as the target for high-energy photons. We find that the dominant background, in this case, is not the cosmic microwave background, which has photons with too low energy and comes therefore into play at energies that are too high, but the infrared background.

An approximate, quantitative estimate can be made as follows. First, the luminosity density in the universe is well known:

$$n_{\rm L} = 1.6 \times 10^8 L_\odot \; {\rm Mpc}^{-3} \tag{3.236}$$

This light is due to the galaxies, which emit a large fraction (≈ 0.1) of their whole luminosity in infrared photons, with $\hbar\omega \approx 0.1$ eV. As a consequence, the density of infrared photons in the universe, assuming that galaxies have maintained a constant luminosity since their formation $\approx 10^{10}$ years ago, is $n_{\rm IR} = 0.01$ cm^{-3}. Taking the cross section of the order of the Thomson one, we find the mean free path:

$$l = \frac{1}{n_{\rm IR}\sigma_{\rm T}} = 30 \; {\rm Mpc} \tag{3.237}$$

As we saw above, the condition for pair creation is

$$\hbar\omega \, \hbar\omega_{\rm IR} \gtrsim \left(m_e c^2\right)^2 \tag{3.238}$$

this tells us that if the target photon is in the infrared ($\hbar\omega_{\rm IR} \approx 0.1$ eV), the high-energy photon must have an energy of about

$$\hbar\omega \gtrsim 10 \; {\rm TeV} \tag{3.239}$$

On the other hand, if we had used as a target the cosmic microwave background photons ($\hbar\omega \approx kT \lesssim 10^{-3}$ eV), we

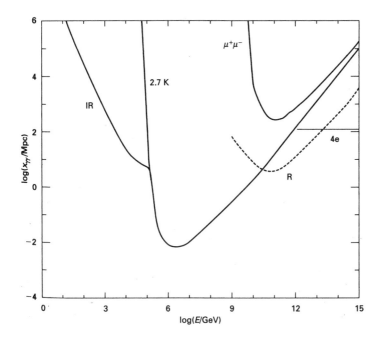

Figure 3.5. Interaction length for photons for pair production against photons of the cosmic microwave background (2.7 K), the infrared background (IR), and the radio background (R); also, for the creation of $\mu^+\mu^-$ pairs and for double pair production e^-e^+ in the cosmic background. Protheroe and Johnson 1996.

would have discovered that they are much more numerous, but the production of pairs becomes possible only when the sample photon has an energy $\gtrsim 10^{15}$ eV. Thus, we expect that the spectrum of photons arriving at Earth from any source farther than ≈ 30 Mpc must be cut off beyond ≈ 10 TeV.

Once again, since we are discussing the absorption due to a threshold process, the mean free path must be very short for photons with energies immediately above the threshold, but practically infinite for those with an energy immediately below the threshold. A precise computation for the mean free path is shown in fig. 3.5.

3.9   Problems

1. Show that, when a source has optical depth for Thomson scattering $\tau_T \gg 1$, the true optical depth to absorption is given by $\tau_a^{(\text{eff})} = \sqrt{\tau_a \tau_T}$, where τ_a is the optical depth naively calculated from the definition.

2. Consider a spherical cloud, with constant density, and optical depth by absorption (calculated from the center to the surface) τ. Neglecting scattering, show that the probability for a photon to escape from the cloud is given by

$$ p(\tau) = \frac{3}{4\tau} \left(1 - \frac{1}{2\tau^2} + \left(\frac{1}{\tau} + \frac{1}{2\tau^2} \right) e^{-2\tau} \right) $$

It is necessary to average over photons emitted in different parts of the cloud, and over all possible directions. For the answer, see Osterbrock (Astrophysics of gaseous nebulae).

3. Consider a gas made of electrons only. What is the main radiative mechanism of this gas? How does this argument change if the gas is composed of electrons and positrons? Could you distinguish the two systems if they were in thermal equilibrium at the same temperature? When can you distinguish them?

4. Consider a gas in which photons have isotropically distributed momenta; now consider the same distribution seen from a reference frame S' in motion with velocity V relative to the gas. If we call θ the azimuthal angle, with the z axis directed along the axis of the relative motion of the two systems, measured in the reference frame S', show that the photons' distribution, in the system S', is given by $dN \propto (1 - V/c)/(1 - V \cos\theta/c)^2 \sin\theta \, d\theta$. Deduce the effect of relativistic beaming, and show that the aperture semiangle is $d\theta = 1/\gamma$, in the hyperrelativistic limit $\gamma \gg 1$. Now show that this result also applies to nonisotropic distributions, provided anisotropy is not too extreme (what does "too extreme" actually mean?).

5. A hyperrelativistic electron is immersed in a uniform magnetic field. How does its energy change with time? And if the electron is Newtonian?

6. The spectrum of a source, which you suspect is due to synchrotron, varies as $\nu^{5/2}$ at the low frequencies, then as ν^2, and finally, after reaching a peak, decreases as $\nu^{-0.7}$. How can you interpret this spectrum? Is it consistent with the hypothesis that this is synchrotron emission from relativistic electrons?

7. In the text we did not discuss the emission of cyclotron by a thermal Newtonian gas; in fact, we never observe this, in astrophysics. Why? Hint: Consider self-absorption.

8. Derive equation 3.224. In order to check your computations, see Svensson (1982).

9. A jet of matter of mass M moves with Lorentz factor $\gamma \gg 1$ within a gas of photons with thermal spectrum and density, at a temperature T. Explain why, and describe how, the jet slows down.

10. In equation 3.151, we deduced the spectrum due to a power-law distribution of relativistic electrons, when the seed photons are strictly monochromatic. Should you change this result when the photons' distribution is thermal, with $kT \ll m_e c^2$? When should you change this result?

11. In discussing the solutions of the Kompaneets equation (eq. 3.166), in the limit of finite source, but $y \gg 1$, we neglected the solution for the index w with the negative sign (eq. 3.171). Why? (Hint: If the total number of low-energy photons injected per unit time is fixed, is the contribution of this solution at photon energies much higher than the injection's one large or small?)

12. Equation 3.225 does not produce π^-. Is there a way to produce it? (Hint: Consider what happens to a neutron produced in eg. 3.225.)

13. The Kompaneets equation (eq. 3.166), when the injection-loss term $Q = 0$, has the form of a conservation equation:

$$\frac{\partial}{\partial t} x^2 \mathcal{N} = -\frac{\partial}{\partial x} x^2 j(x) \qquad (3.240)$$

Can you explain why? (Hint: Is there a total quantity that is conserved? Why, physically, is it conserved?)

Chapter 4

Nonthermal Particles

One of the peculiarities of high-energy astrophysics is the existence of a class of particles, known as *nonthermal*, whose energy distribution does not follow the usual Maxwell-Boltzmann law; they follow instead a power law:

$$dN \propto E^{-p} dE \qquad (4.1)$$

where E is the particle kinetic energy. From a certain point of view, this is exactly what allows us to carry out observations: if the whole universe were in thermal equilibrium, and at the same temperature, it would be impossible to distinguish the sources from the background.

The existence of these particles is revealed by two distinct phenomena. On one hand, we have the ubiquity of nonthermal spectra, with a power-law behavior, $I_\nu \propto \nu^{-\alpha}$, easily distinguishable from the thermal spectra. We saw in chapter 3 that these spectra are easily generated by processes such as synchrotron and inverse Compton, provided there is a family of nonthermal particles, distributed according to the law given in the equation above. On the other hand, Earth is hit by *cosmic rays*, which are particles of extraterrestrial origin, with individual energies from GeV to 3×10^{20} eV, with a spectrum given, with a good approximation, by the equation above, with $p \approx 3$. Obviously, the particles we see emitting nonthermal spectra in distant sources (even extragalactic ones) are the same we see on the Earth, whose existence and distribution according to equation 4.1 cannot be doubted.

The fundamental result of the Newtonian shock theory, equation 4.22 (Bell 1978; Blandford and Ostriker 1978), is that the spectrum of accelerated particles is universal, in other words, it can be determined despite our ignorance of a large number of (apparently relevant) details. This theory shows that neglecting losses, the spectrum index is very close to $p = 2$ for strong shocks.

In this chapter, we shall discuss, first of all, the genesis of these particles, in particular how they are separated from thermal particles, and how they are accelerated, presenting the classical theory of Fermi acceleration around a Newtonian shock. Second, we shall discuss which processes limit the largest attainable energy. Then we shall examine in detail the interaction between thermal and nonthermal particles; here we shall also refer to a few important problems that still have no solution. We shall later introduce, in very general form, the theory for the acceleration around a shock with arbitrary (not only Newtonian) Lorentz factors. After that, we shall briefly discuss the acceleration from the unipolar inductor, the only occurrence in which an electric field plays an important role in Astrophysics, and which constitutes the only valid alternative to shock acceleration, in order to explain the properties of cosmic rays (but not necessarily of the nonthermal emission, observed in situ).

4.1 The Classic Theory of Acceleration

The question of acceleration of nonthermal particles divides naturally into two separate questions: the first one is how it is possible to separate a group of particles (destined to become nonthermal) from the group of thermal particles (this problem is called *injection*), whereas the second one is how it is possible to accelerate these same particles to the observed high energies. Strange as it may appear, we know the answer to the second question much better, and that is why we shall start from the problem of *acceleration.*

4.1.1 Acceleration

Let us consider, for convenience, a particle already separated from its thermal brethren and accelerated up to marginally relativistic energies, namely, with $v \approx c$. From chapter 2 (eq. 2.97), we know that the time scale t_E on which it exchanges its kinetic energy with thermal particles tends to $+\infty$ when $E = mv^2/2 \gg kT$. In order to be definite, using $n_f = 1 \text{ cm}^{-3}$, we easily discover that $t_E \gg t_H$, the age of the universe, for protons and for electrons; therefore, if a particle has become marginally relativistic, it can no longer effectively exchange its kinetic energy with the thermal particles. Moreover, if there is a process that further accelerates it, and drives it further away from the thermal energy kT, it becomes ever harder to exchange energy.

In the same way, we can see that the deflection time t_d, equation 2.94, for the same energies, leads to very long times: for protons, $t_d \gg t_H$. On the other hand, for electrons this is not true, since $t_d \approx 10^5$ yr, but the deflection of electrons or protons does not take place through collisions with other particles. In fact, all charged particles are deflected by the irregular magnetic field that is always present in the astrophysical plasma. The deflection time is comparable to the time employed by the particle to cross a Larmor radius, so that we find

$$t_L = \frac{r_L}{v} = \frac{m_e v}{eB} \approx 0.1 \; s \frac{1 \mu G}{B} \tag{4.2}$$

for marginally relativistic electrons and a typical interstellar magnetic field. In short, we have seen that marginally relativistic particles do not suffer anelastic collisions (namely, collisions in which they lose their energy) with thermal particles, and thus are deflected by the irregular magnetic field of the interstellar medium, conserving all the while their energy.

The deflection process of electrically charged particles by interstellar magnetic fields is far from trivial; see, for example, the fine book by Schlickheiser (2001). First of all, the particles may move in stationary but spatially disordered magnetic fields; also, magnetic fields may be turbulent, in others words, they may vary with time, with an average value and deviation independent from time. Besides, nonthermal particles also emit (Alfvén) waves, which can significantly contribute to the deflection of other nonthermal particles. However, it is important that, at least in the Newtonian limit, we can determine the spectrum of nonthermal particles without any reference to all these details. This is why we shall not discuss them, but we shall rather concentrate on what we can deduce while leaving all details aside.

The Fermi Mechanism

There is a particle *acceleration* mechanism, originally proposed by Fermi (1954). How can we *gain* energy, if the magnetic field conserves energy? The idea is very simple: let us consider a relativistic particle, with energy E and momentum p, in the laboratory reference frame, which moves toward a galactic cloud, moving at a speed $-V$ along the x axis, and θ is the angle that the particles' speed forms with the x axis. The galactic cloud has a mass $\gg m$, the mass of the particle, so that the collision is entirely elastic; indeed, we can imagine that the cloud contains a magnetic field deflecting the particle, so that the particle emerges from the cloud moving in a new direction. In the cloud's reference frame, the particle has an energy

$$E' = \gamma(E + pV\cos\theta) , \qquad \gamma \equiv \left(1 - \frac{V^2}{c^2}\right)^{-1/2} \qquad (4.3)$$

whereas the momentum p'_x along the cloud's velocity, the only element changing with the Lorentz transformation, is given by

$$p'_x = \gamma\left(p\cos\theta + \frac{VE}{c^2}\right) \qquad (4.4)$$

Since the particle collides against a much more massive object, we can assume that the collision is elastic, so that the

particle energy after the collision is identical to the one before the collision, $E'_{after} = E'$, whereas its momentum just
changes sign, $p'_{x,after} = -p'_x$. These relationships are valid in
the cloud's reference frame, whereas, in the laboratory reference frame, its energy is

$$E'' = \gamma(E'_{after} - Vp'_{x,after}) = \gamma(E' + Vp'_x) \qquad (4.5)$$

where we have taken into account the elasticity of the collision. Replacing the above equation with equations 4.3 and
4.4, we find

$$E'' = \gamma^2 E \left(1 + \frac{2Vv\cos\theta}{c^2} + \frac{V^2}{c^2}\right) \qquad (4.6)$$

where we used $v\cos\theta/c^2 = p_x/E$, the particle speed in the
laboratory system. Since, in general, $V \ll c$, we may expand the above equation in the small parameter V/c, thus
obtaining

$$\frac{\Delta E}{E} \approx \frac{2Vv\cos\theta}{c^2} + 2\frac{V^2}{c^2} \qquad (4.7)$$

If we had only head-on collisions in our system, $\cos\theta = 1$;
the energy gain would be of the first order in the small parameter V/c. Unfortunately, clouds' velocities are distributed
randomly, and we expect that if we calculate the average of
all possible values of $\cos\theta$, the first–order term will vanish,
leaving only second-order terms $\propto (V/c)^2$. This expectation is
actually realized, as we now show.

The number of collisions per unit time is proportional to
the relative velocity, $(V\cos\theta + v)/(1 + vV\cos\theta/c^2)$, and to the
solid angle, $d\Omega/4\pi$; taking advantage of the fact that $v \approx c$
and keeping only terms linear in V/c, we find

$$P(\cos\theta)d\cos\theta = \frac{1}{2}\left(1 + \frac{V\cos\theta}{c}\right)d\cos\theta \qquad (4.8)$$

where we also introduced the correct normalization. As we
can see, head-on collisions ($\cos\theta = 1$) are slightly more common than tail collisions ($\cos\theta = -1$) for the same reason that

when we run in the rain, we get wetter in the front than in the back.

We can now use this probability in order to obtain the average energy gain per collision. Using the distribution we obtain from equation 4.7, we find

$$\left\langle \frac{\triangle E}{E} \right\rangle = \frac{8}{3} \frac{V^2}{c^2} \tag{4.9}$$

which is the well-known Fermi result: on average, particles gain energy, even if only at the second order in V/c, which is why this process is called a *second-order Fermi process*.

The second-order nature of the process has at least two disadvantages. First of all, obviously, it is very slow: even around supernovae, where $V \approx 10^4$ km s^{-1}, $(V/c)^2 \ll 1$. Second, it leads to spectra (eq. 4.1) with indices p that can assume *any* value. On the other hand, the observations of synchrotron spectra in external sources and the cosmic ray spectrum suggest that $p \approx 2 - 3$, a property that is sometimes called *universality*. These results may be corrected by the theory, which will be exposed in the next paragraph. The fact remains that Fermi's theory constitutes the basis of this new theory too; the derivation of the nonthermal particles' spectrum by means of second-order Fermi theory, instead, has been almost completely forgotten; you can find it in volume 2 of the book by Longair (1999).

It is worth noting a paradoxical aspect of this acceleration process: in the clouds' reference frame, there is only a magnetic field, which seems therefore capable of accelerating particles, even though it obviously conserves energy. The paradox is solved if we notice that the particles entering the cloud also feel the presence of an electric field, because of the Lorentz transformations. In the cloud's reference frame there is only a magnetic field, whereas in the observer's and particles' reference frames, there is also an electric field, which is truly responsible for the acceleration.

Acceleration at Shocks

For many years, the *second-order* mechanism was considered the only possible one, until physicists (Bell 1978; Blandford and Ostriker 1978) determined that there is a situation in which we can realize a *first-order* mechanism: this occurs around shock waves.

The idea can be visualized very easily. Let us consider a shock, and let us place ourselves in its reference frame; namely, we imagine the shock at $x = 0$. Matter enters the shock with velocity u_u from left to right, emerging with velocity u_d toward the right. Let us consider a nonthermal relativistic particle with an energy E in the reference frame of the preshock fluid, just as it is about to enter the postshock zone. The preshock fluid moves toward the shocked material with velocity $u_u - u_d$; the situation is totally analogous to the galactic cloud running toward the nonthermal particle (fig. 4.3). Because of the Lorentz transformation, in the postshock fluid the particle has a different energy, E'; if the particle reenters the preshock fluid, it will have conserved exactly the same energy in the postshock reference frame, E'. The energy of the particle E'' in the preshock reference frame is once again linked to E by equation 4.6, but with the constraint $\cos\theta \geq 0$; in fact, the velocity of the particle must allow it to enter the postshock fluid! As a consequence, when we average over all possible directions, the first-order term *cannot* vanish, and $\langle \triangle E/E \rangle \propto V/c$: this is now a first-order process!

Let us now compute the average energy gain per shock crossing. As we noticed before, equation 4.6 still holds, but we must find a new probability distribution, $P(\cos\theta)$. In fact, in the paragraph above, we had many clouds with different velocities along the x axis, whereas here the postshock fluid always has a given velocity, and the distribution of probability is created by the particles that cross the shock with different angles θ. The flux of particles crossing the shock, per unit time and surface, is given by $nv_x = nc\cos\theta$, provided

$\cos\theta \geq 0$. The total number of particles crossing the surface is

$$J = \int_{\cos\theta \geq 0} \frac{d\Omega}{4\pi} nc\cos\theta = \frac{nc}{4} \qquad (4.10)$$

thus the probability of finding a particle crossing along an angle θ is

$$P(\cos\theta)\frac{d\cos\theta}{2} = \frac{nv_x}{2J}d\cos\theta = 2\cos\theta\,d\cos\theta, \quad \cos\theta \geq 0$$

$$(4.11)$$

where the constraint $\cos\theta \geq 0$ restricts us to particles crossing the shock from left to right, not vice versa. If we use this distribution in order to take the average of equation 4.6, we find

$$\left\langle \frac{\triangle E}{E} \right\rangle = \frac{2}{3}\frac{V}{c} = \frac{2}{3}\frac{u_u - u_d}{c} \qquad (4.12)$$

where we neglected the terms $\mathcal{O}(u/c)^2$, and the last equality comes from the fact that V is the velocity of the postshock fluid with respect to the preshock one. It is now easy to realize that the problem is totally symmetric in the sense that if we consider the particles crossing the shock from the postshock to the preshock fluid, they also gain the very same factor in energy! We can therefore say that in a whole cycle around the shock, for example, from the postshock to the preshock fluid, and then once again to the postshock one, the average energy gain is given by

$$\left\langle \frac{\triangle E}{E} \right\rangle_{\text{cycle}} = \frac{4}{3}\frac{u_u - u_d}{c} \qquad (4.13)$$

In the derivation of this equation, we have completely neglected any effect of the shock on nonthermal particles, but it is easy to demonstrate that this effect is wholly negligible. Indeed, in our discussion about the thickness of noncollisional shocks, in chapter 1, we mentioned that the thickness of the shock is roughly equal to the Larmor radius of a thermal proton. Since the nonthermal particles have energies several orders of magnitude larger than those of thermal particles, and, consequently, their Larmor radii are much larger, the effect of the shock on the nonthermal particles is in any case infinitesimal, since it corresponds to a deflection at most by an angle kT/E, where E is the energy of the thermal particle. Therefore the shock is essentially invisible to these particles.

We have *assumed* up to now that particles may be deflected back to the shock. This would never be possible, of course, if these particles were confined by collisions to a pure Brownian motion, since this is too slow to allow them to return to the shock. We showed above, however, that these particles are not subject to collisions with other particles, thermal or otherwise. This, however, does not necessarily imply that the nonthermal particles will automatically manage to return to the shock, and we should now ask ourselves how likely it is that nonthermal particles go back to the shock a given number of times. Naively, we might suspect that this is a very difficult question, depending on the detailed properties of the deflection mechanism, the properties of the magnetic field and its turbulence, and so on, but luckily for us this is not so. The answer we are going to prove shortly is that in each cycle, each particle has a finite probability of returning to the shock, a probability given by $P = 1 - 4u_{\rm d}/c$. This result will also help us to determine the particles' spectrum.

However, we would like to understand first of all why particles will not necessarily go back to the shock. If the magnetic field were totally independent of time and were spatially homogeneous, each particle would have an extremely simple motion, sliding along the direction of the field while spinning around it. Therefore, since, as we studied in chapter 2, field lines are frozen in the plasma, they are also subject to advection, that is, they are dragged by the fluid in its motion; therefore, a nonthermal particle is brought along by the fluid in its motion. The presence of magnetic fields, variable in time or inhomogeneous in space, changes the situation

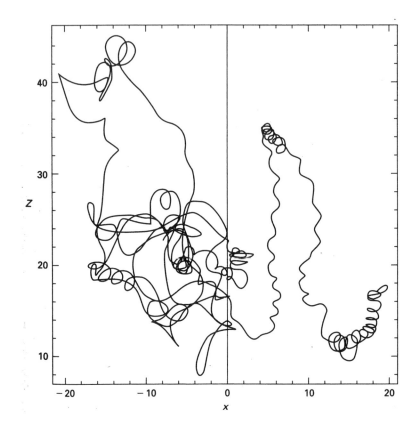

Figure 4.1. Motion of a nonthermal particle around the shock.
Ballard and Heavens 1992.

because this effect causes a drift from a magnetic field line to
another one (fig. 4.1). This process is obviously very similar
to *diffusion,* and we know from the elementary treatment of
diffusion that after a time t, an average particle will find itself
a distance $D \propto t^{1/2}$ from its point of origin; if the particle
now leaves from the shock in the upstream direction, it moves
as $D \propto t^{1/2}$, whereas simultaneously the motion of the fluid
takes it back to the shock by an amount $d = u_{\mathrm{u}}t$. In the long
run, the advective part $(d = u_{\mathrm{u}}t)$ prevails over the scattering part of the motion, so that a particle leaving the shock

against the flow (namely, toward upstream) *always* returns to the shock. However, if the particle leaves the shock downstream, that is, following the flow in the postshock fluid, the fluid tends to carry it away from the shock and prevents it from returning to the shock. Thus, either the particle manages to return to the shock in the first instants of its motion, when $D > d$, or the probability of returning to the shock decreases as t grows, because for $t \to \infty$ $d \gg D$. Therefore, particles always complete the first half of the acceleration cycle in which they are in the upstream section, but they may or may not complete the second half of the cycle in the downstream section. How can we calculate the probability P to return to the shock, thus completing an acceleration cycle?

The answer to this question is possible thanks to an elegant argument by Bell (1978). The total flux of particles entering and exiting the postshock area must vanish because we are considering a stationary situation, where nothing varies with time, and in particular the number of particles in the postshock area cannot vary. A flux of particles J_+ enters the downstream section through the surface of the shock per unit time and area, a flux J_- leaves the downstream region through the same surface, and a flux J_∞ leaves the postshock area at downstream infinity. J_∞ represents the particles that are dragged away by advection. Obviously, in a stationary situation, we must have

$$J_+ = J_- + J_\infty \tag{4.14}$$

which only expresses the fact that, as many particles enter the postshock area, so many exit it. The probability that they return to the shock is obviously given by

$$P = \frac{J_-}{J_+} = \frac{J_-}{J_- + J_\infty} \tag{4.15}$$

which requires that we calculate J_- and J_∞.

Let n_0 be the density of nonthermal particles in the postshock area; n_0 is obviously a constant, because on this side of the shock there is of course no spatial variation. The number of particles crossing the shock toward the preshock area per unit time and surface is given by equation 4.10:

$$J_- = \frac{n_0 c}{4} \tag{4.16}$$

where, for the sake of simplicity, we considered only relativistic particles. On the other end, the number of particles removed per unit time and surface is given by

$$J_\infty = n_0 u_{\mathrm{d}} \tag{4.17}$$

It is easy to be convinced of this result if we consider that the particles being removed are tied to the magnetic field lines, frozen in the fluid, and scattering is too slow for them to be taken back. Thus we find

$$P = \frac{c}{c + 4u_{\mathrm{d}}} \approx 1 - \frac{4u_{\mathrm{d}}}{c} \tag{4.18}$$

where the last equality follows from the fact that in the Newtonian limit, $u_{\mathrm{d}} \ll c$. This is the simple result we announced above.

At this point, we are ready to compute the spectrum of nonthermal particles. Let E_0 and N_0 be the energy and the initial number of particles respectively; after k cycles, there will only be $N_k = N_0 P^k$ particles left, with an energy $E_k = E_0 A^k$, where

$$A = 1 + \frac{\triangle E}{E} = 1 + \frac{4}{3} \frac{u_{\mathrm{u}} - u_{\mathrm{d}}}{c} \tag{4.19}$$

is the average energy amplification of each particle for each cycle. Eliminating the useless factor k between N_k and E_k, we find

$$N = \text{constant} \times E^s, \qquad s \equiv \frac{\ln P}{\ln A} \tag{4.20}$$

However, this is an integral distribution, that is, the number of all the particles that completed k *or more* cycles, whereas we are interested in the differential distribution, which is simply obtained:

$$dN = \text{constant} \times E^{-p}dE, \qquad p \equiv 1 - \frac{\ln P}{\ln A} \qquad (4.21)$$

We can simplify the computation of the exponent p as follows:

$$p = 1 - \frac{\ln P}{\ln A} = 1 - \frac{\ln(1 - 4u_{\mathrm{d}}/c)}{\ln(1 + 4(u_{\mathrm{u}} - u_{\mathrm{d}}/3c)} \approx 1 + \frac{3u_{\mathrm{d}}}{u_{\mathrm{u}} - u_{\mathrm{d}}}$$
$$(4.22)$$

which is the expression we were looking for. In the important case of the strong shock for $\gamma = 5/3$, we know that $u_{\mathrm{d}} = u_{\mathrm{u}}/4$, so that we find

$$p = 2 \qquad (4.23)$$

As if by magic, all references to the scattering properties of the medium, before and after the shock, have disappeared from this equation. In fact, we did not even specify them. The only quantity left is the speed jump, $u_{\mathrm{d}}/u_{\mathrm{u}}$; the spectral slope of particles accelerated at the shock depends on nothing else. This is the result we anticipated in the introduction of this chapter.

The spectrum with $p = 2$ has the following feature: the number of particles is dominated by the low-energy ones, but the total energy is dominated by the high-energy ones, even if only logarithmically. It is, however, worth noticing that any small loss (be it a radiative loss, or a leak of particles from the acceleration area) will make the spectrum steeper than $p = 2$, so that both the total number of particles and the energetics will be dominated by low-energy particles.

At this point, it becomes obvious that the true spectrum is a superimposition of the spectra of particles that have completed $0, 1, 2, \ldots N, \ldots$ cycles around the shock. In figure 4.2 we show the result of numerical simulations for *relativistic* shocks, for which the energy gain A is of the order of ≈ 2, so that the peaks are well separated and easier to distinguish. This figure shows how to obtain a power law superposing a number of spectra that are not necessarily power laws.

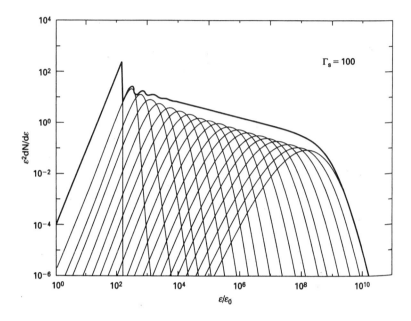

Figure 4.2. Spectrum of nonthermal particles accelerated around a relativistic shock (with $\Gamma = 100$); thin lines indicate the spectrum of particles that completed $1, 2, 3, \ldots$ cycles around the shock. Lemoine and Pelletier 2003.

Constraints

It is important to note that this result has been obtained under the assumption that nonthermal particles are distributed in an *isotropic* way, namely, independently of the direction of motion. In fact, this assumption has been used to obtain equations 4.11 and 4.16, and therefore the value of A and of P derive directly from it. This is a correct assumption, but only in the limit $u_{\mathrm{u}} \ll c$. In order to see it, let us argue as follows. If there were no shock wave, the problem would obviously be isotropic. However, the presence of a shock leads to the existence of three different reference frames (upstream, downstream, and comoving with the shock), and there is no distribution that can be simultaneously isotropic in all three of them. The existence of multiple reference frames naturally creates an anisotropy, which is simply due to the fact that the components of the momentum perpendicular to the relative speed among the various reference systems do not change, whereas the Lorentz transformations change the component of the momentum parallel to the relative speed. Therefore, even though the distribution is isotropic in one reference frame, it will not be in any other system.

However, as long as the velocities in question are small with respect to the speed of light, obviously the angular distribution of nonthermal particles differs but slightly from isotropy, namely,

$$f(\cos\theta) = f_0 + \frac{u_{\mathrm{u}}}{c}f_1(\cos\theta) + \cdots \approx f_0 \qquad (4.24)$$

with f_0 independent of direction. If we included the next term f_1 (which depends on the angle), we would find corrections to the spectral index of particles, p, of the order of $\mathcal{O}(u_{\mathrm{u}}/c)$, which are negligible in the Newtonian limit.

We can also measure the deviation from isotropy by considering the flux of particles from one side of the shock to the other. In the presence of *exact* isotropy, we should have $J_+ = J_-$, which, combined with equation 4.14, implies $J_\infty = 0$, which is possible only for $u_{\mathrm{d}} = 0$, and this, in its turn, is possible only for $u_{\mathrm{u}} = 0$, and thus no shock (there are no subsonic shocks, and for $u_{\mathrm{u}} = 0$ the shock is *surely* subsonic!). This justifies our expectation, namely, in the absence of a shock, the angular distribution function of particles at the shock is isotropic. However, if $u_{\mathrm{d}} > 0$, $J_+ \neq J_-$, and the difference (or, in other words, anisotropy) can once again be obtained by equation 4.14,

$$\frac{J_+ - J_-}{J_+} = \frac{J_\infty}{J_+} = \frac{4u_{\mathrm{d}}}{c} \ll 1 \qquad (4.25)$$

This shows that anisotropy at the shock is of the order of u_{u}/c, as argued above. Please note furthermore, that all these arguments are related to anisotropy *at the shock*: for $x \to +\infty$, anisotropy, in the reference frame moving with the fluid, must always tend to zero, even for hyperrelativistic situations (why?).

On the other hand, in the relativistic case, these corrections become large. In this case, the angular distribution function f is not known a priori, and therefore we cannot calculate P and A, as we did above; rather, we should consider $f(\cos\theta)$ as one of the unknown quantities of the problem, which must be found through a different analysis. The solution of this problem will be discussed further on in this chapter.

4.1.2   Injection

We have seen above that the particles with a sufficiently high energy do not lose their energy through collisions against thermal particles and are not even deflected by them, but by the magnetic field. This happens for sufficiently high energies; in order to play safe, in the preceding paragraph we considered marginally relativistic particles. Therefore, these particles are decoupled from thermal ones and can thus start their cycle of multiple crossings of the shock, which will lead them to possess hyperrelativistic energies. The problem of *injection* consists in finding these particles, which have an energy, not yet relativistic but large enough to start the acceleration cycle.

The usual solution to this problem is the following: Any collision tends to produce a Maxwell-Boltzmann distribution, which extends to very high energies. The particles within the extreme tail of such a distribution might constitute injection particles. There are at least two types of processes populating the tail of a Maxwell-Boltzmann distribution. First, there are normal collisions between thermal particles; second, there are accelerations from electromagnetic transient fields, which take place inside noncollisional shocks, as described in chapter 1. As we have already explained, the physics inside noncollisional shocks is, at the moment, little known, so that we can specify neither the rate nor the energy of the particles coming out of them; according to common wisdom, the injection of particles to be later accelerated through the first-order Fermi process takes place exactly within the shocks, which thus act as injectors of non-thermal particles.

Why do these particles separate from the rest of the fluid? After all, normal particles are subject to Brownian motion, and they leave their initial positions with correspondingly small speeds. In other words, in order for these particles to separate from the others, their mean free path between collisions must be much longer than that of ordinary particles. This is exactly what happens; indeed, the cross section for scattering of unpolarized particles, with charge e, momentum p and energy E by nuclei with charge Ze initially at rest (often called *Mott scattering* [Sakurai 1967]) is given by

$$\frac{d\sigma_{\mathrm{M}}}{d\Omega} = \left(\frac{Za\hbar}{2c}\right)^2 \frac{E^2}{p^4} \frac{1 - \beta^2 \sin^2(\theta/2)}{\sin^4(\theta/2)} \qquad (4.26)$$

which reduces, in the Newtonian limit, to the Rutherford formula:

$$\frac{d\sigma_{\mathrm{M}}}{d\Omega} = \frac{(Za\hbar c)^2}{4m^2 v^4 \sin^4(\theta/2)} \qquad (4.27)$$

Here $\alpha \equiv e^2/(\hbar c) \approx 1/137$ is the fine-structure constant. From these equations, we can see that the scattering cross section, at every angle, is a decreasing function of the particle energy, in the Newtonian limit and in the relativistic one. As a consequence, particles that we consider for injection tend to have longer mean free paths in comparison with particles with an energy $\approx kT$, and thus they tend to separate from the fluid they originally belonged to.

Initially, these particles may be accelerated by secondary processes unlike the Fermi one. For instance, in a hydromagnetic shock in which the magnetic field is not exactly parallel to the direction of the shock's motion, there is an electric field parallel to the shock's surface, which, in the reference frame of the shock itself, has intensity

$$\vec{E} = -\frac{\vec{V}}{c} \wedge \vec{B} \qquad (4.28)$$

In this equation $\vec{B}$ is the magnetic field in the reference frame of the preshock fluid, whereas $\vec{V} = \vec{u}_{\mathrm{u}}$ is the velocity at which the shock propagates in the unperturbed medium. The existence of this electric field is only a consequence of the assumption of ideal magnetohydrodynamics. It was actually suggested (Jokipii 1987) that this electric field is so efficient in accelerating particles that it competes, in certain conditions, with the Fermi mechanism; it can increase the particle energy by a quantity $ZeVBD/c$, where D is the distance covered by the particle before being scattered far away from the shock. The mechanism is obviously limited by the fact that D is a typical mean free path of the particles and, therefore, a rather small quantity, but this does not mean that at least in the first stages of acceleration, this mechanism, called *drift*, may not help particles in the extreme Maxwell-Boltzmann tail to reach energies large enough for the Fermi process to start.

4.2   Constraints on the Maximum Energy

The spectrum deduced above,

$$dN = A \times E^{-p}dE, \qquad p = 1 + \frac{3u_{\mathrm{d}}}{u_{\mathrm{u}} - u_{\mathrm{d}}} \qquad (4.29)$$

formally extends to $E \to \infty$, because the problem has been strongly idealized; we have considered neither that the region around the shock has finite dimensions, nor that the shock propagation may last for only a finite time. Moreover, we have completely neglected any kind of radiative losses on the part of nonthermal particles. When we abandon these idealizations, we find that there is a typical maximum energy above which particles cannot be accelerated.

Let us start by showing that the acceleration process of particles of given energies lasts a finite time, and calculate t_a, the acceleration time. In order to do this, let us calculate the time each particle spends in the region before (and after) the shock, by recurring to the following nice argument. As discussed above, nonthermal particles, which are injected at the shock, cannot reach upstream infinity, because they have to swim against the flow: the advection flux that brings them back to the shock is obviously given by $J_{adv} = nu_u$. At the same time, the distribution of particles in the preshock region is not uniform, and therefore there will be a diffusion flux, identical to the one that carries heat in the presence of a temperature gradient: $J_{diff} = -Ddn/dx$, where x is the distance from the shock $(x = 0)$, and D a scattering coefficient of the particles, which we shall soon examine. In a stationary situation, the number of particles on either side of the surface cannot change, so the total flux of particles, per unit time and surface, through this surface (which is all inside the preshock region) must vanish:

$$J_{adv} + J_{diff} = nu_u - D_u\frac{dn}{dx} = 0 \qquad (4.30)$$

where we have to specify that the scattering coefficient is related to the preshock zone. This is a simple differential equation with solution

$$n(x) = n_0 e^{\frac{u_u x}{D_u}} \qquad (4.31)$$

where n_0 is the density of particles at the shock, $x = 0$. The total number N of particles per unit of shock surface is thus

$$N = \int_{-\infty}^{0} dx n(x) = \frac{n_0 D_u}{u_u} \qquad (4.32)$$

But in the present section we have just computed (eq. 4.16) the number of particles entering the preshock region, per unit time and surface:

$$J_- = \frac{n_0 c}{4} \qquad (4.33)$$

which tells us that the *average* particle remains in the preshock region for a time given by

$$t_{\mathrm{u}} = \frac{N}{J_-} = \frac{4D_{\mathrm{u}}}{u_{\mathrm{u}}c} \qquad (4.34)$$

A similar argument shows also that the time spent in the postshock region is

$$t_{\mathrm{d}} = \frac{4D_{\mathrm{d}}}{u_{\mathrm{d}}c} \qquad (4.35)$$

so that the duration of a cycle is

$$t_{\mathrm{cycle}} = t_{\mathrm{u}} + t_{\mathrm{d}} = \frac{4D_{\mathrm{u}}}{u_{\mathrm{u}}c} + \frac{4D_{\mathrm{d}}}{u_{\mathrm{d}}c} \qquad (4.36)$$

In this interval of time, the particle energy increases by an amount $A - 1 = 4(u_{\mathrm{u}} - u_{\mathrm{d}})/3c$ (eq. 4.19), so that we can write

$$\frac{dE}{dt} = \frac{4(u_{\mathrm{u}} - u_{\mathrm{d}})}{3c}\frac{E}{t_{\mathrm{cycle}}} = \frac{E}{t_{\mathrm{a}}} \qquad (4.37)$$

where the acceleration time is

$$t_{\mathrm{a}} \equiv \frac{3ct_{\mathrm{cycle}}}{4(u_{\mathrm{u}} - u_{\mathrm{d}})} = \frac{3}{u_{\mathrm{u}} - u_{\mathrm{d}}}\left(\frac{D_{\mathrm{u}}}{u_{\mathrm{u}}} + \frac{D_{\mathrm{d}}}{u_{\mathrm{d}}}\right) \qquad (4.38)$$

Now we have to define the scattering coefficient D. In the presence of a concentration gradient, we can show that

$$D = \frac{1}{3}\lambda v \qquad (4.39)$$

where λ is the mean free path of the particle subject to scattering, and v its typical velocity. It is called the *Bohm scattering coefficient*, and it is not wholly trivial to derive (see Landau and Lifshitz 1981a). It can be applied to nonthermal particles as follows. Of course, $v \approx c$ for these relativistic particles; as for λ, we may assume it approximately equal to the Larmor radius of the particle in question, so that

$$D = \frac{Ec}{3eB} \qquad (4.40)$$

This coefficient has often been criticized in the literature because it describes a physical situation different from the one where it is used. In fact, it is derived for conditions with a slight deviation from thermodynamic equilibrium, when the scale length on which all quantities vary (for example, $n/(dn/dx)$) is much larger than all mean free paths. On the other hand, in *our* application we use it to describe a very turbulent interstellar medium, in which scattering is certainly not due to the variation with distance of any quantity, but to the space-time variation of turbulent phenomena, which *do not* depend *on average* on either time or space. However, even more detailed analyses have confirmed that at least *in order of magnitude*, Bohm scattering gives the correct answer.

Now we obtain

$$t_a = \alpha E , \qquad \alpha \equiv \frac{c}{e(u_u - u_d)} \left(\frac{1}{u_u B_u} + \frac{1}{B_d u_d} \right) \qquad (4.41)$$

The most important thing to notice is that bringing a particle to an energy E requires a time that grows linearly with E. This time scale must be compared first of all with the time scale for radiative energy loss. Whether we consider synchrotron, or inverse compton, the rate of energy loss is given by

$$\frac{dE}{dt} = -\beta E^2 \qquad (4.42)$$

thus the cooling time t_c is

$$t_c \equiv \frac{E}{\frac{dE}{dt}} = \frac{1}{\beta E} \qquad (4.43)$$

which *decreases* with the energy; when the two times t_c and t_a become equal, the time necessary to accelerate the particle exceeds the one necessary to radiate away most of its own kinetic energy. Therefore, the maximum energy is determined by the condition

$$t_a = t_c \Rightarrow E \leq \left(\frac{1}{\alpha\beta} \right)^{1/2} \qquad (4.44)$$

which may be evaluated in the different concrete situations that may occur.

The time scale for acceleration that we just derived can also be used to show that an important constraint to the maximum energy derives from the finiteness of the source's lifetime. Here we shall concentrate on the important case of supernovae. We are going to evaluate the maximum energy for nuclei with charge Ze, accelerated in the shocks of supernovae. We know that supernovae do not live forever, but they have a strong time evolution. If a supernovae remnant (SNR) has age T, the maximum energy that can be reached is obviously given by the equation

$$t_{\mathrm{a}} - T \qquad (4.45)$$

which expresses only the fact that I have had a finite time (T) available in order to accelerate the particle. Since supernovae generate strong shocks and typically have $\gamma = 5/3$, I know that $u_{\mathrm{d}}/u_{\mathrm{u}} = 1/4$, whence,

$$E_{\max} = \frac{3}{20}\frac{eB_{\mathrm{u}}}{c}V^2 T \qquad (4.46)$$

for parallel shocks.[1] Here V is the instantaneous speed of the shock, which, as we know, varies as $V \propto T^{-3/5}$ for the Sedov solution. This shows that $E_{\max}$ decreases during the Sedov phase, so that the maximum is obtained at the beginning of the phase itself, namely, at the end of the phase of free expansion, when $T = T_{\mathrm{S}}$. In order to determine T_{S}, we note that the phase of free expansion ends when the mass of the gas swept up by the shock equals that ejected in the explosion, that is, when

$$\frac{4\pi}{3}nm_{\mathrm{p}}(VT_S)^3 = M_{\mathrm{ej}} \qquad (4.47)$$

For typical values, $n = 1$ cm^{-3}, $V = 5000$ km s^{-1}, and $M_{\mathrm{ej}} = 10M_{\odot}$, we find $T_{\mathrm{S}} = 10^3$ yr, and therefore

$$E_{\max} = Z \times 3 \times 10^{13}\,\mathrm{eV} \qquad (4.48)$$

---

[1] Namely, for those shocks in which the magnetic field is parallel to the normal to the shock, for which $B_{\mathrm{u}} = B_{\mathrm{d}}$; see chapter 2.

Finally, a very important constraint is represented by the finite spatial extension of the source. Indeed, any particle moving away from the shock must be deflected in order to return to the shock. If the particle leaves the source, and therefore the region where the magnetic field is (presumably!) strong enough to make it return, it will be lost forever. If we call λ (the confusion with the preceding λ is deliberate!) the typical scale length on which the particle is brought back, and R the linear extension of the source, we must of course have $\lambda < R$. What can we take for λ? Obviously, it is reasonable to choose a multiple (g times) of the Larmor radius: $\lambda \approx g r_{\mathrm{L}}$, therefore the condition $\lambda \leq R$ becomes, solving for the energy,

$$E \leq E_{\mathrm{max}} \equiv \frac{ZeBR}{g} \tag{4.49}$$

4.3 More Details in the Newtonian Limit

We have so far treated nonthermal particles as test particles, thus neglecting the reaction they exert on the fluid. However, since particles shuffle back and forth between the two sides of the shock, we can easily see that they exert forces on the fluid; in particular, they decelerate the incoming flux (they decrease u_{u}) and accelerate the outgoing flux (i.e., they increase u_{d}). If the number of accelerated particles is large, we can expect that the pressure exerted by the particles is significant and must be included in the equation for the evolution of the fluid. In order to assess the importance of the particles, we should compare the flux of energy density in nonthermal particles leaving downstream infinity with the flux of kinetic energy of matter entering the shock, which is obviously the ultimate source of particle acceleration. Thus we define the efficiency of acceleration as

$$\eta \equiv \frac{u_{\mathrm{d}} \int E dN}{\frac{1}{2} \rho u_{\mathrm{u}}^3} \tag{4.50}$$

Of course we expect that, for $\eta \ll 1$, particle reaction on the fluid may be neglected, whereas for $\eta \gtrsim 1$ we must take it into account. From independent considerations on supernovae, we know that, for Newtonian shocks,

$$0.01 \gtrsim \eta \gtrsim 0.1 \qquad (4.51)$$

As we said before, we cannot predict the number of particles injected into the nonthermal component, and therefore we cannot predict whether the acceleration mechanism saturates when $\eta \ll 1$, or vice versa. A possible way to study the problem consists in studying whether there are solutions for the whole problem, in which a counterreaction also acts on the fluid, due to the particles themselves, and, in case they exist, whether they are stable or not. Unfortunately, the study of this problem, which is obviously important, is still in its early stages. In this section, we simply describe the relevant equations.

The dynamic properties of a group of particles are conveniently described by their phase space density, that is, a *distribution function* $f(t, \vec{x}, \vec{p})$, which gives the number of particles contained in a volume of space dV, with momenta in an interval d^3p. Since we know the equations of motion for each particle, if we can evaluate f at a given instant, we can derive f at any other time. The distribution function contains all information about nonthermal particles; for example, the energy distribution can be easily extracted from f through the relation

$$\frac{dN}{dE} = 4\pi p^2 f(p)\frac{dp}{dE} , \quad E^2 = m^2 c^4 + p^2 c^2 \qquad (4.52)$$

which is obviously valid in any regime of motion, Newtonian or relativistic.

It is convenient to have an equation for the evolution of f. Since particles interact only with the magnetic field $\vec{B}$, we apply the well-known *Vlasov equation* (see, e.g., Binney and Tremaine 1987):

$$\frac{\partial f}{\partial t} + \vec{v} \cdot \frac{\partial f}{\partial \vec{x}} + \dot{\vec{p}} \cdot \frac{\partial f}{\partial \vec{p}} = 0 \qquad (4.53)$$

where $\dot{\vec{p}}$ is the Lorentz force acting on the particle:

$$\dot{\vec{p}} = q\frac{\vec{v}}{c} \wedge \vec{B} \tag{4.54}$$

This is a collisionless equation; if collisions dominated deflections, as in a normal gas, we would not need an equation for the distribution function, since we know from elementary statistical mechanics that in that case, the distribution function is the one by Maxwell-Boltzmann (for classical particles). However, since nonthermal particles do not experience collisions, we cannot assume this result, which would make our life much easier. That is why we are looking for an equation suitable for particles.[2]

4.3.1 From the Vlasov Equation to the Convection-Scattering Equation

Let us now examine the change of the function f when the particles move from the element $\vec{x}$ to the element $\vec{x} + \Delta\vec{x} = \vec{x} + \vec{v}dt$. We can write in an implicit way:

$$f(\vec{p}, \vec{x} + \vec{v}dt, t + dt) = \int_{\Delta\vec{p}} \Delta\vec{p}\, P(\vec{p} - \Delta\vec{p}, \Delta\vec{p}) f(\vec{p} - \Delta\vec{p}, \vec{x}, t)$$
$$\tag{4.55}$$

where we have introduced the function $P(\vec{p}, \Delta\vec{p})$, which represents the probability that particles with momentum $\vec{p}$ change their momentum by $\Delta\vec{p}$. Evidently, the correct normalization for P is given by the condition

$$\int_{\Delta\vec{p}} d\Delta\vec{p}\, P(\vec{p}, \Delta\vec{p}) = 1 \tag{4.56}$$

The fact that the changes of the vector $\vec{p}$ are a function of the values of $\vec{p}$ and $\vec{x}$ at the same time *only*, tells us that the process under consideration is a Markov process. We can

---

[2] The remaining paragraphs of this section were written by Pasquale Blasi.

expand in Taylor series the functions f and P in equation 4.55:

$$f(\vec{p}, \vec{x} + \vec{v}dt, t + dt) = f(\vec{p}, \vec{x}, t) + \vec{v} \cdot \nabla_{\mathrm{x}} f dt + \frac{\partial f}{\partial t} dt \quad (4.57)$$

$$f(\vec{p} - \Delta\vec{p}, \vec{x}, t) = f(\vec{p}, \vec{x}, t) + (\nabla_{\mathrm{p}} f)\Delta\vec{p} + \frac{1}{2}(\nabla_{\mathrm{p}}^2 f)\Delta\vec{p}\Delta\vec{p} \quad (4.58)$$

$$P(\vec{p} - \Delta\vec{p}, \Delta\vec{p}) = P(\vec{p}, \Delta\vec{p}) - (\nabla_{\mathrm{p}} P)\Delta\vec{p} + \frac{1}{2}(\nabla_{\mathrm{p}}^2 P)\Delta\vec{p}\Delta\vec{p}. \quad (4.59)$$

By replacement in equation 4.55, we therefore have

$$\frac{\partial f}{\partial t} + \vec{v} \cdot \nabla_{\mathrm{x}} f = -\nabla_{\mathrm{p}}[A_{\mathrm{p}} f] + \frac{1}{2}\nabla_{\mathrm{p}}[\nabla_{\mathrm{p}}(D_{\mathrm{pp}} f)] \quad (4.60)$$

where we have introduced the quantities

$$A_{\mathrm{p}} = \frac{1}{\Delta t} \int d\Delta\vec{p} \, \Delta\vec{p} P(\vec{p}, \Delta\vec{p}) \quad (4.61)$$

$$D_{\mathrm{pp}} = \frac{1}{2\Delta t} \int d\Delta\vec{p} \, \Delta\vec{p}\Delta\vec{p} P(\vec{p}, \Delta\vec{p}) \quad (4.62)$$

At this point, we use the fact that the effects of the increase in momentum modulus are of second order in u/c, where u is the velocity in the magnetic field fluctuations, which implies that at this order, we can assume that the momentum modulus remains constant, whereas the particles' direction of motion changes (*pitch angle scattering*). The principle of detailed balance tells us that

$$P(\vec{p}, -\Delta\vec{p}) = P(\vec{p} - \Delta\vec{p}, \Delta\vec{p}) \quad (4.63)$$

Expanding the second term in this relation in a Taylor series, we obtain

$$P(\vec{p}, -\Delta\vec{p}) = P(\vec{p}, \Delta\vec{p}) - \Delta\vec{p}\nabla_{\mathrm{p}} P + \frac{1}{2}\Delta\vec{p}\Delta\vec{p}\nabla_{\mathrm{p}}\nabla_{\mathrm{p}} P \quad (4.64)$$

which, divided by Δt and integrated in $\Delta \vec{p}$, yields

$$A_{\rm p} - \nabla_{\rm p} D_{\rm pp} = {\rm constant(p)} \qquad (4.65)$$

The functions $A_{\rm p}$ and $D_{\rm pp}$ generally go to zero when $p \to 0$, which implies that the constant on the right-hand side is zero, and therefore

$$A_{\rm p} = \nabla_{\rm p} D_{\rm pp} \qquad (4.66)$$

If we replace this result in equation 4.60, we obtain the scattering equation in the form

$$\frac{\partial f}{\partial t} + \vec{v} \cdot \nabla_{\rm x} f = \nabla_{\rm p} [D_{\rm pp} \nabla_{\rm p} (f)] \qquad (4.67)$$

4.3.2 Scattering in the Angle of Motion in a Medium at Rest

At this point, we would like to go back to a view of the scattering process based on the behavior of single particles. With this aim, let us consider a particle moving with a pitch angle μ in comparison with the unperturbed magnetic field $\vec{B}_0$. Let us also suppose that there is a perturbation $\vec{B}_1$ of the magnetic field, which we take as perpendicular to $\vec{B}_0$. This perturbation is actually responsible for the change of the pitch angle in comparison with the initial one, namely, for causing the particles' scattering. As we well know, the magnetic field does not change the particle's momentum modulus. We saw in chapter 2 that the perturbations of the magnetic field, which propagate parallel to the unperturbed field, can be expressed as Alfvén waves. We shall call a wave's reference frame the one in which there is only the wave's (stationary) magnetic field, and there is therefore no electric field. In this reference frame, the particles' energy does not change, whereas the pitch angle does; in fact,

$$\frac{d\vec{p}_\parallel}{dt} = \frac{q}{c} [\vec{v} \wedge (\vec{B}_0 + \vec{B}_1)]_\parallel = \frac{q}{c} \vec{v}_\perp \wedge \vec{B}_1 \qquad (4.68)$$

Using $p_\parallel = p\mu$ we immediately have

$$\frac{d\mu}{dt} = \frac{q}{cp} [\vec{v}_\perp \wedge \vec{B}_1]_\parallel \qquad (4.69)$$

If we assume that perturbations have a small modulus in comparison with the preexisting field, namely $B_1 \ll B_0$, we can determine the solution of equation 4.69, assuming that the particle's trajectory is close to the unperturbed one, namely, the trajectory the particle would have if only the field B_0 were present. This approach is the basis of the so-called quasi-linear theory of particles' motion in a perturbed field. The unperturbed trajectory of a particle with a Lorentz factor γ is identified when we know the gyration frequency $\Omega = qB_0/(mc\gamma)$ and the modulus v of its velocity. On the other hand, the field B_1 is known when we also know its modulus and the wave number k: $B_1(z) = B_1 \cos(kz + \phi)$ with $z = v\mu t$ the longitudinal coordinate to the unperturbed field B_0. The median value of the deflection squared $\Delta\mu^2$ in an interval of time t can be expressed as follows:

$$\langle \Delta\mu^2 \rangle = \frac{q^2 v^2 (1 - \mu^2) B_1^2}{p^2 c^2} \int_0^t dt_1 \int_0^t dt_2$$
$$\times \cos[(kv\mu - \Omega)t_1 + \phi]\cos[(kv\mu - \Omega)t_2 + \phi] \quad (4.70)$$

where we can easily average on the irrelevant phase ϕ, provided the waves under consideration have a *random* phase.

$$\langle \Delta\mu^2 \rangle = \frac{q^2 v^2 (1 - \mu^2) B_1^2}{2 p^2 c^2} \int_0^t dt_1 \int_0^t dt_2 \cos[(kv\mu - \Omega)(t_1 - t_2)]$$
$$(4.71)$$

If we use the complex representation of the function cosine and remember the properties of Dirac's delta function, we obtain

$$\left\langle \frac{\Delta\mu^2}{\Delta t} \right\rangle = \frac{\pi q^2 v^2 (1 - \mu^2) B_1^2}{p^2 c^2} \delta(kv\mu - \Omega)$$
$$= \frac{\pi q^2 v (1 - \mu^2) B_1^2}{p^2 \mu c^2} \delta\left(k - \frac{\Omega}{v\mu}\right) \quad (4.72)$$

In a realistic situation, instead of a single wave with a fixed wave number $\vec{k}$, there is a series of waves with an energy distribution per wave number. If $W(k)dk$ is the energy contained in the waves with a wave number k and $k+dk$, we can rewrite equation 4.72 in the following way:

$$\left\langle \frac{\Delta \mu^2}{\Delta t} \right\rangle = \frac{\pi}{4}(1-\mu^2)\frac{1}{v\mu}\Omega^2 \int dk \frac{W(k)}{B_0^2/8\pi}\delta(k-\Omega/v\mu)$$

$$= \frac{\pi}{4}(1-\mu^2)\Omega \frac{k_{\text{res}}W(k_{\text{res}})}{B_0^2/8\pi} \qquad (4.73)$$

where $k_{\text{res}} = \Omega/v\mu$ is the wave number resonant with particles having a momentum $p = mv\gamma$. The scattering coefficient in the pitch angle is usually defined as the average value of the scattering angle squared, rather than of its cosine, namely,

$$\nu = \left\langle \frac{\Delta \theta^2}{\Delta t} \right\rangle = \frac{\pi}{4}\Omega \frac{k_{res}W(k_{res})}{B_0^2/8\pi} \qquad (4.74)$$

The scattering coefficient D_{pp}, defined by equation 4.62, in the absence of change in the modulus of $\vec{p}$, can also be written as

$$D_{\text{pp}} = p^2 \left\langle \frac{\Delta \mu^2}{\delta t} \right\rangle = p^2(1-\mu^2)\nu \qquad (4.75)$$

Replacing this result in the transport equation, (eq. 4.67), we easily obtain

$$\frac{\partial f}{\partial t} + \vec{v}\cdot\nabla_{\text{x}}f = \frac{1}{2}\frac{\partial}{\partial \mu'}\left[(1-\mu'^2)\nu\frac{\partial f}{\partial \mu'}\right] \qquad (4.76)$$

where we put an apex on μ to indicate that it is a quantity calculated in the waves' reference frame.

4.3.3 Scattering and Convection in a Medium in Motion

We can now find an equation describing at the same time the scattering in an angle, namely, a spatial scattering, as well as the particles' transport with a fluid that is supposed in motion. With this aim, we must take into account the change of the momentum vector in passing from one reference frame to another. In particular, we shall call $\vec{p}$ the momentum in the laboratory reference frame, where particles have a distribution function $f(\vec{p}, \vec{x}, t)$. In the reference frame in motion with a velocity $\vec{u}$, waves move with a typical velocity, which is the Alfvèn velocity v_A. In cases of interest $v_A \ll u$, and therefore u is also a good approximation of the waves' velocity in the laboratory reference frame. In the waves' frame, as we have often underlined, the electric field vanishes, so that the particles' momentum vector can only change in direction, not in modulus. Let $\vec{p}'$ be the momentum of particles in such reference frame. Since we suppose $u \ll c$ (Newtonian limit) the relation between $\vec{p}$ and $\vec{p}'$ can be written as

$$\vec{p} = \vec{p}' + E'\vec{u} \qquad (4.77)$$

as we can easily see when we write the Lorentz transformation and take its limit $u \ll c$. To order u/c (our stated order of accuracy) the conservation of the number of particles, passing from the laboratory to the waves' reference frame, gives $f(\vec{p}, \vec{x}, t) = f(\vec{p}', \vec{x}, t)$. If we now expand the right-hand side in the small quantity $E'\vec{u}$, we obtain

$$f(\vec{p}, \vec{x}, t) = f(\vec{p}', \vec{x}, t) - E'\vec{u}\nabla_{p'} f(\vec{p}' = \vec{p}, \vec{x}, t) \qquad (4.78)$$

In the case of nonrelativistic motion, the particle's velocity, transformed to the wave's reference frame, becomes $\vec{v}' + \vec{u}$, so that the Vlasov equation becomes

$$\frac{\partial f'}{\partial t} + (\vec{v}' + \vec{u}) \cdot \nabla_x f' - (\vec{v}' + \vec{u}) \cdot \nabla_x (E'\vec{u})\nabla_{p'} f' = -\dot{\vec{p}}' \cdot \nabla_{p'} f' \qquad (4.79)$$

If we neglect once again terms of order $(u/c)^2$, we obtain

$$\frac{\partial f'}{\partial t} + (\vec{v}' + \vec{u}) \cdot \nabla_x f' - (\vec{p}' \cdot \nabla_x)\vec{u}\nabla_{p'} f' = -\dot{\vec{p}}' \cdot \nabla_{p'} f' \qquad (4.80)$$

Let us now introduce a unit vector $\hat{n}(\vec{x}, t)$ oriented along the direction of the local magnetic field, and let us suppose that a set of Cartesian coordinates is placed at the same point, with the z axis oriented along $\hat{n}$ (as in fig. 4.3). Moreover, let μ' be the cosine of the angle between $\hat{n}$ and $\vec{v}'$, so that

$$\vec{v}' \cdot \nabla_x f' = v' \mu' \hat{n} \cdot \nabla_x f' + v'_x \frac{\partial f}{\partial x} + v'_y \frac{\partial f}{\partial y} \qquad (4.81)$$

Physically, the dependence of the distribution function from x and y must be limited to the indirect dependence of μ' from two spatial coordinates, namely $\partial f'/\partial x = (\partial f'/\partial \mu')(d\mu'/dx)$, and likewise for the partial derivative in y (here v' is kept constant). It follows that

$$\frac{\partial \mu'}{\partial x} = \frac{\partial}{\partial x}\left[\frac{\vec{v}' \cdot \hat{n}}{v'}\right] = \frac{\partial}{\partial x}[\sin \theta' \cos \phi' n_x + \sin \theta' \sin \phi' n_y + \cos \theta' n_z]$$

$$= \frac{\partial n_x}{\partial x} \sin \theta' \cos \phi' \qquad (4.82)$$

This relation expresses the fact that the pitch angle is changing because the unit vector $\hat{n}$ is changing its orientation and follows the direction of the local magnetic field. Proceeding in a similar manner, we also obtain that

$$\frac{\partial \mu'}{\partial y} = \frac{\partial n_y}{\partial y} \sin \theta' \sin \phi' \qquad (4.83)$$

As we said above, the angle ϕ' is irrelevant from a physical viewpoint. We can therefore average over ϕ':

$$\langle \vec{v}' \cdot \nabla_x f' \rangle_{\phi'} = v' \mu' \hat{n} \cdot \nabla_x f' + v' \frac{\partial f'}{\partial \mu'} \left\langle (\sin \theta' \cos \phi')^2 \frac{\partial n_x}{\partial x} \right.$$

$$\left. + (\sin \theta' \sin \phi')^2 \frac{\partial n_y}{\partial y} \right\rangle_{\phi'} = v' \mu' \hat{n} \cdot \nabla_x f'$$

$$+ \frac{1}{2} v'(1 - \mu'^2)(\nabla_x \cdot \hat{n}) \frac{\partial f'}{\partial \mu'} \qquad (4.84)$$

The third term of the equation 4.80 must also be averaged over the angle ϕ, so that we have

$$\langle (\vec{p}' \cdot \nabla_x) \vec{u} \nabla_{p'} f' \rangle_{\phi'} = p' \frac{\partial f}{\partial p'} \left\langle \left(\sin\theta' \cos\phi' \frac{\partial}{\partial x} + \sin\theta' \sin\phi' \frac{\partial}{\partial y} \right. \right.$$

$$\left. + \cos\theta' \frac{\partial}{\partial z} \right) \wedge (u_x \sin\theta' \cos\phi'$$

$$\left. + u_y \sin\theta' \sin\phi' + u_z \cos\theta') \right\rangle_{\phi'} \qquad (4.85)$$

Here we have once again explicitly exploited the fact that f' does not depend explicitly on ϕ', namely, $\partial f'/\partial \phi' = 0$. After calculating the average, we obtain

$$\langle (\vec{p}' \cdot \nabla_x) \vec{u} \cdot \nabla_{p'} f' \rangle_{\phi'} = p' \frac{\partial f}{\partial p'} \left[\frac{1}{2}(1 - \mu'^2) \nabla_x \cdot \vec{u} \right.$$

$$\left. + \frac{3\mu'^2 - 1}{2} (\hat{n} \cdot \nabla_x)(\hat{n} \cdot \vec{u}) \right] \qquad (4.86)$$

The results we obtained in equations 4.84 and 4.86 may now be used in the transport equation 4.80. We should first note, however, that the right-hand side in equation. 4.80 must be averaged over the phase ϕ. Since we are in the waves' reference frame, the momentum modulus does not change, and therefore the average of the term due to the Lorentz force (without an electric field) is clearly zero:

$$\langle \dot{\vec{p}} \cdot \nabla_{p'} f' \rangle_{\phi'} = B \left\langle \left[v'_y \frac{\partial f'}{\partial p'_x} - v'_x \frac{\partial f'}{\partial p'_y} \right] \right\rangle_{\phi'}$$

$$= B v' \frac{\partial f}{\partial p'} \langle \sin^2\theta' \cos\phi' - \sin^2\theta' \sin\phi' \rangle_{\phi'} = 0$$

$$(4.87)$$

At this point, we should replace the expressions found in equation 4.80 and remember the above result (eq. 4.76) that expresses the connection between the scattering coefficient in the pitch angle and the transport operator:

$$\frac{\partial f'}{\partial t} + (v'\mu'\hat{n} + \vec{u}) \cdot \nabla_x f'$$

$$- \left[\frac{1 - \mu'^2}{2}(\nabla_x \cdot \vec{u}) + \frac{3\mu'^2 - 1}{2}(\hat{n} \cdot \nabla_x)(\hat{n} \cdot \vec{u}) \right]$$

$$\times p'\frac{\partial f'}{\partial p'} + \frac{1 - \mu'^2}{2}v'(\nabla_x \cdot \hat{n})\frac{\partial f'}{\partial \mu'}$$

$$= \frac{1}{2}\frac{\partial}{\partial \mu'}\left[(1 - \mu'^2)\nu\frac{\partial f}{\partial \mu'}\right] \qquad (4.88)$$

In the case of fluids in nonrelativistic motion, we expect that on spatial scales large compared to the *scattering* length, namely, that which determines the process of scattering in pitch angle, the particles' distribution function is almost isotropic. We can therefore think of considering the "true" distribution function f' as the sum of terms representing growing orders of anisotropy:

$$f' = f'_0 + f'_1 + f'_2 \qquad (4.89)$$

with $f'_k \sim (u/c)^k f'_0$ and $u \ll c$. We also recall that

$$\frac{\partial}{\partial t} \sim u\frac{\partial}{\partial x}$$

The isotropic function to zeroth order obviously satisfies the condition

$$\frac{1}{2}\frac{\partial}{\partial \mu'}\left[(1 - \mu'^2)\nu\frac{\partial f'_0}{\partial \mu'}\right] = 0 \qquad (4.90)$$

The next-order terms are

$$v'\mu'(\hat{n} \cdot \nabla_x)f'_0 = \frac{1}{2}\frac{\partial}{\partial \mu'}\left[(1 - \mu'^2)\nu\frac{\partial f'_1}{\partial \mu'}\right] \qquad (4.91)$$

Integrating the latter equation, we get

$$\int_{-1}^{\mu'} d\mu' \mu' v'(\hat{n} \cdot \nabla_x)f'_0 = \frac{1}{2}(1 - \mu'^2)\nu\frac{\partial f'_1}{\partial \mu'} \qquad (4.92)$$

which implies

$$\frac{\partial f_1'}{\partial \mu'} = -\frac{v'}{\nu}(\hat{n} \cdot \nabla_x) f_0'$$

(4.93)

Finally, at the following order we obtain

$$\frac{\partial f_0'}{\partial t} + (\vec{u} \cdot \nabla_x) f_0' + \mu' v'(\hat{n} \cdot \nabla_x) f_1'$$
$$- \left[\frac{1-\mu'^2}{2} \nabla_x \cdot \vec{u} + \frac{3\mu'^2 - 1}{2} (\hat{n} \cdot \nabla_x)(\hat{n} \cdot \vec{u}) \right] p' \frac{\partial f_0'}{\partial p'}$$
$$+ \frac{1-\mu'^2}{2} v'(\nabla_x \cdot \hat{n}) \frac{\partial f_1'}{\partial \mu'} = \frac{1}{2} \frac{\partial}{\partial \mu'} \left[(1 - \mu'^2)\nu \frac{\partial f_2'}{\partial \mu'} \right]$$

(4.94)

In general, we shall be interested in the distribution function averaged on the pitch angle, so we must calculate the average value of each single term in equation 4.94. Let us note, first of all, that if we use integration by parts, we can write

$$\langle \mu' f_1' \rangle_{\mu'} = \frac{1}{2} \int_{-1}^{1} d\mu' \mu' f_1'$$
$$= \frac{1}{2} \left[\frac{1}{2}(f'(1) - f'(-1)) - \frac{1}{2} \int_{-1}^{1} d\mu' \mu'^2 \frac{\partial f_1'}{\partial \mu'} \right]$$

(4.95)

and, taking advantage of the fact that $\frac{1}{2}(f'(1) - f'(-1)) = \int_{-1}^{1} d\mu' \frac{\partial f_1'}{\partial \mu'}$, we can write

$$\langle \mu' f_1' \rangle_{\mu'} = \int_{-1}^{1} d\mu' \frac{(1-\mu'^2)}{4} \frac{\partial f_1'}{\partial \mu'} = \frac{1}{2} \left\langle (1 - \mu'^2) \frac{\partial f_1'}{\partial \mu'} \right\rangle_{\mu'}$$

(4.96)

if we use the result expressed in equation 4.93:

$$v'\langle \mu' f_1' \rangle_{\mu'} = -\frac{1}{2} \left\langle (1 - \mu'^2) \frac{v'^2}{\nu} \right\rangle_{\mu'} (\hat{n} \cdot \nabla_x) f_0'$$

(4.97)

Now we introduce the scattering coefficient in the form

$$D_{\parallel}(\vec{p}', \vec{x}) = \frac{1}{2} \left\langle (1 - \mu'^2) \frac{v'^2}{\nu} \right\rangle_{\mu'}$$

(4.98)

We can rewrite equation 4.94 averaged on pitch angle in the form

$$\frac{\partial f_0'}{\partial t} + (\vec{u} \cdot \nabla_x) f_0' - (\hat{n} \cdot \nabla_x) D_{\|} (\hat{n} \cdot \nabla_x) f_0' - (\nabla_x \cdot \hat{n}) D_{\|} (\hat{n} \cdot \nabla_x) f_0'$$

$$- \frac{1}{3} (\nabla_x \cdot \vec{u}) p' \frac{\partial f_0'}{\partial p'} = 0, \tag{4.99}$$

or, in a more compact way,

$$\frac{\partial f_0'}{\partial t} + (\vec{u} \cdot \nabla_x) f_0' - \nabla_x \cdot \left[\hat{n} D_{\|} (\hat{n} \cdot \nabla_x) f_0' \right] = \frac{1}{3} (\nabla_x \cdot \vec{u}) p' \frac{\partial f_0'}{\partial p'} \tag{4.100}$$

In a single dimension, with a magnetic field oriented along the fluid's direction of motion, this equation assumes the simpler and more often used form

$$\frac{\partial f_0}{\partial t} + u \frac{\partial f_0}{\partial x} - \frac{\partial}{\partial x} \left[D_{\|} \frac{\partial f_0}{\partial x} \right] = \frac{1}{3} \frac{du}{dx} p' \frac{\partial f_0}{\partial p} \tag{4.101}$$

Here the indexes have been removed because, at the order we are interested in, the distribution function is $f_0' \approx f_0$. The first term of this equation expresses the obvious explicit dependence of the distribution function on time. The second term describes the variation of the distribution function because of the particles' convection with the fluid in motion. On the other hand, the spatial scattering of particles because of fluctuations in the magnetic field are described by the term containing $D_{\|}$. The increase or decrease of the particles' energy, due to adiabatic compression or decompression of the fluid, is taken into account in the term du/dx. The latter term, however, plays an even more important role in the cases in which the velocity profile of the fluid has a discontinuity. This is exactly what happens when shock surfaces are formed. In this case, equation 4.101 correctly describes the so-called first-order Fermi acceleration mechanism.

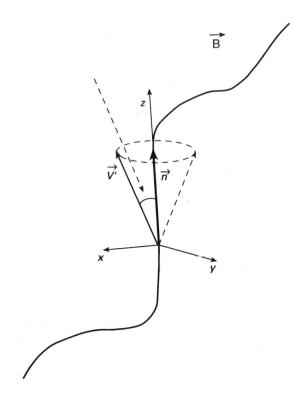

Figure 4.3. Schematic view of a magnetic field line.

4.4   General Discussion

Within the limit in which the total pressure exerted by non-thermal particles is negligible, which is sometimes called the *test particle limit*, we can introduce the problem in general terms for arbitrary scattering law and shock speed (Vietri 2003). The difficulty of this theory consists in the fact that the angular distribution of particles at the shock is not known a priori as in the case of Newtonian shocks (where it is isotropic) but must be determined together with the particles' spectral index.

As a consequence of the theory, the spectrum is always given by a power law, $dN \propto p^{-k}dp$, and the spectral index k

is determined. However, in the relativistic regime, there is no universal limit, in the sense of the Newtonian regime, and the spectral indexes we obtain depend in detail on the scattering properties of the medium and on the fluid velocity, before and after the shock.

In this paragraph, we shall use units such that $c = 1$, and we shall only consider hyperrelativistic particles, because these have momentum $p \approx E$, which simplifies our work. The generalization to Newtonian particles is simple and without interest.

We should once again place ourselves in the shock's reference frame, with the shock located at $z = 0$. Let us assume that the fluid moves from left to right, so that its velocity is always $u > 0$. We shall assume that the shock is planar and stationary, so that for any physical quantity X we have $\partial X/\partial t = \partial X/\partial x = \partial X/\partial y = 0$. In order to describe the particles, we shall use once again their distribution function $f(t, x, y, z, p_x, p_y, p_z)$ which gives the total number of particles contained in a volume $d^6 V = d^3 V d^3 p$ in phase space. Because of the hypotheses of stationarity and symmetry, we must simply have:

$$f = f(z, p, \mu) \qquad (4.102)$$

where p is the momentum modulus, and $\mu \equiv p_z/p$ is the momentum component along the normal to the shock, namely the cosine of the angle between the normal to the shock and the particle's direction of motion.

At this point, we recall that f is a scalar with respect to Lorentz transformations; see problem 2.

4.4.1 An Equation for f

Here we shall derive an equation for the evolution of f. We shall also show that this equation is covariant, albeit not explicitly so. The equation must obviously be of the type

$$\frac{df}{ds} = \left(\frac{\partial f}{\partial s}\right)_{\text{coll}} + \phi_{\text{inj}} \qquad (4.103)$$

where the left-hand side term (often called *convective deriva-tive*) only tells us how the distribution function changes as a consequence of the interaction of particles with external fields:

$$\frac{df}{ds} = \frac{dx^\nu}{ds}\frac{\partial f}{\partial x^\nu} + \frac{dp^\mu}{ds}\frac{\partial f}{\partial p^\mu} \qquad (4.104)$$

Here, as usual, s is the Lorentz invariant $ds^2 = dt^2 - dx^2 - dy^2 - dz^2$.

If in equation 4.103 the right-hand side term were identically zero, we would be in the presence of the Liouville equation, whose solution is well known. Unfortunately, there are two terms that describe two distinct processes: The first term, $(\partial f/\partial t)_{\text{coll}}$, describes the collisions of particles with the irregularities of the magnetic field, which appear in any astrophysical fluid and are due to space inhomogeneity of the magnetic field, and to turbulence (in other words, time inhomogeneity). The second term, ϕ_{inj}, indicates that there might be injection of nonthermal particles at the shock, as discussed earlier on. That is why we assume for ϕ the form

$$\phi_{\text{inj}} = G_{\text{inj}}(p, \mu)\delta(z) \qquad (4.105)$$

where $\delta(z)$ is the Dirac delta.

The two terms on the right-hand side of equation 4.104 are both relativistic invariant, so that we can safely calculate them in different reference frames. For the first term, we can use the shock's reference frame, described above. It follows that $\partial f/\partial t = \partial f/\partial x = \partial f/\partial y = 0$; thus the only term left is $\propto \partial f/\partial z$. However, for reasons that shall soon become clear, we should express the component of the four-velocity of particles in the shock's reference system, dz/ds, in terms of quantities in the fluid's reference system. We easily find that if the particle moves along the given direction μ, with a Lorentz factor γ_{p}, computed in the fluid's reference frame, which, in its turn, has a velocity and a Lorentz factor—with respect to the shock—u, γ, respectively, we have

$$\frac{dz}{ds} = \gamma_{\text{p}}\gamma(u + \mu) \qquad (4.106)$$

Therefore, the first term on the right-hand side in equation 4.104 reduces to

$$\gamma_p \gamma (u + \mu) \frac{\partial f}{\partial z} \tag{4.107}$$

We compute the second term on the right-hand side of equation 4.104 in the fluid's reference frame. In this way, we know that only the magnetic field is present, because the electric field vanishes. Thus we see that the energy of each particle is conserved, and the term containing $dp^0/ds = \gamma dE/dt$ disappears, because $dE/dt = 0$ in the presence of the magnetic field only. The three remaining components can be rewritten as follows (Vietri 2003):

$$\frac{dp^\mu}{ds} \frac{\partial f}{\partial p^\mu} = \frac{dt}{ds} \frac{dp^\mu}{dt} \frac{\partial f}{\partial p^\mu} = -\gamma_p \omega \frac{\partial f}{\partial \phi''} \tag{4.108}$$

Here $\omega = eB/E$ is the Larmor frequency of the particle, and ϕ'' is the azimuthal angle in the local direction of the magnetic field, *not* of the shock's normal, an angle that we shall call ϕ.

The collision term is given, in general terms, by the expression

$$\left(\frac{\partial f}{\partial t} \right)_{coll} = -d(\mu, \phi) f + \int w(\mu, \phi, \mu' \, \phi') f(\mu', \phi') d\mu' d\phi' \tag{4.109}$$

This collision term has several features, which must be illustrated. First, it is linear in the distribution function, so that it *does not* describe the collisions of particles with each other, but against something else, in this case, a background of disturbances to the average magnetic field in the fluid. Furthermore, since these are exclusively perturbations of the magnetic field, particles' energies are not modified by these collisions; indeed, we see that only the variables μ and ϕ are involved, while their energies are not. Finally, there are two terms, a positive one and a negative one, because these collisions cannot change the total number of particles of a given energy, but only reshuffle their directions of motion. Therefore, the number of particles is conserved: integrating the above equation on $d\mu d\phi$, we can easily see that this is obtained by assuming

$$d(\mu, \phi) \equiv \int w(\mu', \phi', \mu, \phi) d\mu' d\phi' \qquad (4.110)$$

Our key coefficient is therefore $w(\mu, \phi, \mu', \phi')$, which represents the probability per unit path length that a particle, initially moving along the direction μ', ϕ', is deflected toward a new direction, μ, ϕ. In general, we have difficulties in predicting the exact shape of w, for which we refer to Schlickheiser (2001).

Now, in equation 4.104, we need $\partial f/\partial s$, which is given by

$$\frac{\partial f}{\partial s} = \frac{dt}{ds}\frac{\partial f}{\partial t} = \gamma_{\rm p}\frac{\partial f}{\partial t} \qquad (4.111)$$

Putting all of this together, we find

$$\gamma(u + \mu)\frac{\partial f}{\partial z} = -d(\mu, \phi)f + \int w(\mu, \phi, \mu', \phi')f(\mu', \phi')d\mu'd\phi'$$

$$+ \omega\frac{\partial f}{\partial \phi''} + G_{\rm inj}(p, \mu)\delta(z) \qquad (4.112)$$

which is the equation for the distribution function we were looking for.

In the following, we shall completely neglect the presence of a magnetic field with a long scale length, so that we may neglect the term proportional to ω. In this case, we see that the system is totally symmetric for rotations around the normal to shock, and therefore that f does not depend on ϕ. In this case, we can simply average on the angle ϕ of the above equation, to obtain

$$\gamma(u+\mu)\frac{\partial f}{\partial z} = -D(\mu)f(\mu) + \int W(\mu, \mu')f(\mu')d\mu' + G_{\rm inj}(p, \mu)\delta(z)$$

$$(4.113)$$

where, obviously,

$$W(\mu, \mu') \equiv \int w(\mu, \phi, \mu', \phi') d\phi' \ , \ D(\mu) \equiv \int W(\mu', \mu) d\mu'$$
$$(4.114)$$

Notice that W *does not* depend on ϕ, because of the assumption of axial symmetry.

As a general rule, we expect that nonthermal particles, which are injected at the shock, possess rather low energies, and that the following process, described by equation 4.113, will push them to high energies, at which the injection term G_{inj} does not appear. We can solve equation 4.113 in its generality, showing that the spectrum of particles is actually the sum of particles' spectra that have crossed the shock $0, 1, 2, \ldots N, \ldots$ times. However, figure 4.2 makes it clear that the total spectrum does not depend on the spectrum of particles that have completed exactly n acceleration cycles; we could also formally demonstrate this statement. From now on, we shall consider the simpler equation

$$\gamma(u + \mu)\frac{\partial f}{\partial z} = -D(\mu)f(\mu) + \int W(\mu, \mu')f(\mu')d\mu' \quad (4.115)$$

which describes the evolution of the phase space density of nonthermal particles at energies vastly exceeding those of injection.

Finally, we should notice that an equation like 4.115 is valid *before* and *after* the shock, but, since the fluids before and after the shock are in relative motion, the quantities we have decided to define in the fluid's reference frame, u, γ (speed and Lorentz factor of the fluid with respect to the shock) and μ, p (direction of motion and momentum of particles) are different for the two fluids. That is why we shall indicate as $u_u, \gamma_u, \mu_u, p_u$ and $u_d, \gamma_d, \mu_d, p_d$ the quantities before and after the shock for the fluid, respectively. We recall that u_u and u_d (therefore also γ_u and γ_d) are linked by Taub conditions at the shock, whereas μ and p are linked by Lorentz transformations, which give

$$p_{\mathrm{u}} = p_{\mathrm{d}}\gamma_{\mathrm{r}}(1 - u_{\mathrm{r}}\mu_{\mathrm{d}}) \; , \;\; \mu_{\mathrm{u}} = \frac{\mu_{\mathrm{d}} - u_{\mathrm{r}}}{1 - u_{\mathrm{r}}\mu_{\mathrm{d}}} \qquad (4.116)$$

Here u_{r} and γ_{r} are the modulus of the relative velocity between the two fluids, and its Lorentz factor.

4.4.2   The Small Pitch Angle Seattering Limit

Before illustrating how to solve equation 4.115, let us first introduce the form it assumes in a case often discussed in the literature, namely, the case in which the deflection of nonthermal particles is very small in every scattering event. This limit is called *small pitch angle scattering* (SPAS).[3]

Esquation 4.115 can describe scattering for arbitrary angles, small and large. However, when the scattering angle is small, we can transform this integral equation in a more accessible differential equation. The procedure with which this is done is typical of a wide range of equations, called Fokker-Planck equations. That is why it is worthwhile to illustrate this passage here in some detail.

In order to start this derivation, it is necessary to discuss a property of the scattering function, W:

$$W(\mu, \mu') = W(\mu', \mu) \qquad (4.117)$$

This property is the mathematical formulation of the *principle of detailed balance*, according to which, in thermal equilibrium, every process is balanced by its time-reverse. This principle is based on the time-reversal symmetry of quantum mechanics and is therefore a basic principle of any mechanics process.

We shall now rewrite equation 4.115 by means of a new distribution of probability, w, thus defined:

$$w(\mu'; \alpha)d\alpha \equiv W(\mu' + \alpha, \mu')d(\mu' + \alpha) \qquad (4.118)$$

---

[3] This paragraph can be left out at a first reading.

w is simply the probability that a particle, initially moving along the direction μ', is deflected toward a new direction $\mu' + \alpha$. Thanks to w, equation 4.115 can be rewritten as

$$\gamma(u + \mu)\frac{\partial f}{\partial z} = \int_{-1}^{+1} d\alpha (w(\mu - \alpha; \alpha)f(\mu - \alpha) - w(\mu; -\alpha)f(\mu))$$

(4.119)

while the equation of detailed balance becomes

$$w(\mu - \alpha; \alpha) = w(\mu; -\alpha) \tag{4.120}$$

Equation 4.119 is absolutely identical to the one given by Landau and Lifshtiz (1981a, section 21, eq. 21.1), so that their derivation of the Fokker-Planck equation immediately applies also to our case. Please note, however, a difference in the sign convention: what we call α is, for them, $-q$. Following their derivation step by step, we find

$$\gamma(u + \mu)\frac{\partial f}{\partial z} \approx \frac{\partial}{\partial \mu}\left(Af + B\frac{\partial f}{\partial \mu}\right) \tag{4.121}$$

with the definitions

$$A \equiv -\int d\alpha \, \alpha w(\mu; \alpha) + \frac{\partial B}{\partial \mu} \tag{4.122}$$

$$B \equiv \frac{1}{2}\int d\alpha \, \alpha^2 w(\mu; \alpha) \tag{4.123}$$

We now show that, as a consequence of detailed balance, $A = 0$. Notice that

$$\frac{\partial}{\partial \mu}w(\mu; \alpha) = \frac{\partial}{\partial \mu}w(\mu'; \mu - \mu') \tag{4.124}$$

where $\mu' = \mu + \alpha$, because of detailed balance, so that

$$\frac{\partial}{\partial \mu}w(\mu; \alpha) = \frac{\partial}{\partial \alpha}w(\mu'; \alpha) \tag{4.125}$$

Let us now introduce this formula into the expression

$$\frac{\partial B}{\partial \mu} = \frac{1}{2}\int d\alpha \, \alpha^2 \frac{\partial}{\partial \mu}w(\mu; \alpha) = \frac{1}{2}\int d\alpha \, \alpha^2 \frac{\partial}{\partial \alpha}w(\mu'; \alpha)$$

(4.126)

which can be integrated by parts:

$$\frac{\partial B}{\partial \mu} = - \int d\alpha \; \alpha \, w(\mu'; \alpha) \tag{4.127}$$

We shall use, once again, detailed balance, as well as the change of variable $x \equiv -\alpha$ in the above integral, thus obtaining

$$\frac{\partial B}{\partial \mu} = \int d\alpha \; \alpha \, w(\mu; \alpha) \tag{4.128}$$

which gives, once introduced into equation 4.122, $A = 0$. With this we obtain, for equation 4.115 in the SPAS limit (also called Fokker-Planck regime),

$$\gamma(u + \mu)\frac{\partial f}{\partial z} = \frac{\partial}{\partial \mu}\left(B\frac{\partial f}{\partial \mu}\right) \tag{4.129}$$

with B given by equation 4.123. The use of detailed balance implies that the scattering integral reduces, in the SPAS limit, to the divergence of a vector, proportional to the gradient of f, exactly as in the classical problem of heat conduction.

In general, the coefficient B is a function of μ: $B = B(\mu)$. However, in an important limiting case, which is often called the *isotropic limit* in the literature, B assumes the simple form

$$B = K_1(1 - \mu^2) \tag{4.130}$$

where K_1 is a constant. The meaning of *isotropic* in this case needs clarification. Normally, we use the expression *process of isotropic scattering* to indicate the case in which the particle direction of motion after the scattering process does not depend on the direction before the event: the final direction is simply random. This definition, obviously, cannot be applied in this case, where the angle of deflection is infinitesimal, so that the initial and final directions motion are almost perfectly correlated! On the other hand, in this case, the term *isotropic* indicates that the coefficient of elementary scattering $w(\theta, \phi, \theta', \phi')$ (eq. 4.112) depends only on the angle Θ

between the two directions,

$$\cos\Theta \equiv \mu\mu' + \sqrt{1-\mu^2}\sqrt{1-\mu'^2}\cos(\phi-\phi') \qquad (4.131)$$

and nothing else. This property can hold also for scattering through a finite (as opposed to infinitesimal) angle, but, when the scattering angles are infinitesimal, it leads to the above equation, as we now show.

The function w may be expanded in a Legendre series of polynomials:

$$w(\theta,\phi,\theta',\phi') = w(\cos\Theta) = \sum_l a_l P_l(\cos\Theta) \qquad (4.132)$$

where the coefficient a_l is given by

$$a_l = \frac{2l+1}{2}\int d\cos\Theta\, w(\cos\Theta)P_l(\cos\Theta) \qquad (4.133)$$

These relations hold for isotropic functions w, but they are otherwise arbitrary, namely, we have not imposed that w be in the SPAS limit. However, when this happens, the coefficients a_l are easy to calculate. Indeed, in the SPAS limit, we can suppose that the function w is $\neq 0$ only in a small interval around $\cos\Theta = 1$. In other words, we take, in the SPAS limit, $w(\cos\Theta) = f((1-\cos\Theta)/\epsilon)$, where $f(x)$ is a function with only one peak, for $x = 0$, $\epsilon \ll 1$, and the normalization condition,

$$\int_{-\infty}^{+\infty} f(x)dx = K \qquad (4.134)$$

for every ϵ. It is then easy to see that

$$\int_{-1}^{+1} \frac{1}{\epsilon} f\left(\frac{1-\cos\Theta}{\epsilon}\right)\cos^n\Theta\, d\cos\Theta = \epsilon^n g_n \qquad (4.135)$$

where $g_n = \mathcal{O}(1)$, so that the integral vanishes in the SPAS limit, $\epsilon \to 0$. It follows that, in the SPAS limit, the coefficients a_l are given (eq. 4.133) by

$$a_l = \frac{2l+1}{2}P_l(0)K \qquad (4.136)$$

In order to calculate $P_l(0)$, we use Laplace's integral representation:

$$P_l(x) = \frac{1}{\pi} \int_0^\pi \left(x + \sqrt{x^2 - 1} \cos \phi \right)^l d\phi \qquad (4.137)$$

thence we can easily see that

$$P_{2n+1}(0) = 0 \qquad (4.138)$$

$$P_{2n}(0) = (-)^n \frac{(2n-1)!!}{(2n)!!} \qquad (4.139)$$

From these we find

$$a_0 = \frac{K}{2}, \quad a_1 = 0, \quad a_2 = -\frac{5K}{4} \dots \qquad (4.140)$$

In order to calculate the integral in equation 4.114, we shall use the well-known identity

$$P_l(\cos \Theta) = P_l(\mu) P_l(\mu') + 2 \sum_{m=1}^{l} \frac{(l-m)!}{(l+m)!} P_l^m(\mu) P_l^m(\mu')$$
$$\cos(m(\phi - \phi')) \qquad (4.141)$$

which immediately gives

$$W(\mu, \mu') = \sum_l a_l P_l(\mu) P_l(\mu') \qquad (4.142)$$

This can now be substituted into equation 4.123 to obtain

$$B = \left(\frac{a_0}{3} - \frac{a_2}{15} \right) - \mu^2 \left(a_0 + \frac{a_2}{5} \right) \qquad (4.143)$$

in which we can now use equation 4.140:

$$B = \frac{K}{4}(1 - \mu^2) \qquad (4.144)$$

which has the form of equation 4.130.

Thus we obtain, for the equation of scattering in the SPAS limit,

$$\gamma(u + \mu)\frac{\partial f}{\partial z} = \frac{\partial}{\partial \mu}\left(D(\mu)(1 - \mu^2)\frac{\partial f}{\partial \mu}\right) \qquad (4.145)$$

with $D(\mu) = $ constant in the isotropic case. This equation approximates the more complex equation 4.115 in the SPAS limit.

Is it always possible to replace equation 4.115 with 4.145? The answer is no, even though the SPAS limit is valid. In fact, we see from equations 4.122 and 4.123 that we have used *average quantities* for the angle of deflection and its variation; indeed, A is the average cosine of the angle of deflection, and B the average of the square cosine. This can always be done when the number of deflections is large, but there is at least one circumstance in which this is wrong, namely, when the shock is hyperrelativistic. In that case, an infinitesimal deflection, $\approx 1/\gamma$, is sufficient, where γ is the Lorentz shock factor, for the particle to be overcome by the shock. In the limit $\gamma \to \infty$, this assures us that only one deflection will be sufficient for it to return to the postshock region, so it is a bad approximation to replace a random deflection with its average value. Therefore, the SPAS approximation requires, in order to be valid, *both* the SPAS regime, *and* the fact that the particle undergoes many deflections before returning to the other side of the shock.

You may reasonably wonder whether the approximation of isotropy (in the sense we have just defined) is physically motivated or not. We know, for example, through simulations or analytic computations (Schlickheiser 2001), that particles are deflected back toward the shock at a distance $\approx r_{\rm L}$, with $r_{\rm L}$ the Larmor radius for the particle in question, and that the most effective waves that may provoke this scattering are exactly Alfvén waves, with a wavelength $\lambda \approx r_{\rm L}$. This circumstance tells us that isotropy is unlikely to be a good approximation. It would have been good if the distance from the shock at which the direction of motion is reversed had been $\gg r_{\rm L}$; instead, we expect anisotropies in the coefficient B, of the order $\mathcal{O}(1)$. Therefore, as a general rule, we expect $B = B(\mu)$, but, at the moment, we still cannot specify a reasonable functional form for this dependence.

4.4.3 Distributions of Probability $P_{\rm u}$ and $P_{\rm d}$

In order to determine the particles' spectral index, we shall not proceed as usual, namely, solving separately equation 4.115 in the fluid before and after the shock, subject to reasonable boundary conditions, and connecting the two solutions.

We shall consider instead two *conditional* distributions of probability, $P_{\rm u}(\mu_{\rm out}, \mu_{\rm in})$ and $P_{\rm d}(\mu_{\rm in}, \mu_{\rm out})$ thus defined. $P_{\rm u}(\mu_{\rm out}, \mu_{\rm in})$ is the conditional probability that a particle, which has entered the region before the shock, moving along a direction that forms an angle $\theta_{\rm in}$ with the normal to the shock, such that $\cos\theta_{\rm in} = \mu_{\rm in}$, gets out of the preshock region toward the postshock one along a direction $\theta_{\rm out}$ such as $\cos\theta_{\rm out} = \mu_{\rm out}$. In the same way, $P_{\rm d}(\mu_{\rm in}, \mu_{\rm out})$ is the conditional probability that a particle, having entered the postshock region along a direction $\mu_{\rm in}$, gets out of it, entering the preshock one, along a direction $\mu_{\rm out}$.

We shall soon give the equations that determine $P_{\rm u}$ and $P_{\rm d}$ from the fluid velocity and the scattering properties of the medium (function W). But, for the moment, we shall assume we know $P_{\rm u}$ and $P_{\rm d}$ and see how they determine the particles' spectrum.

The flux of particles dJ, which cross the shock per unit time from the postshock to the preshock fluid is given, in terms of quantities in the shock reference frame, by

$$dJ = \mu_{\rm s} dN = \mu_{\rm s} f p_{\rm s}^2 dp_{\rm s} d\mu_{\rm s} \qquad (4.146)$$

Here $\mu_{\rm s}$ is the component of velocity (in units of c) of the particle perpendicular to the shock, in the shock's reference system, and dN is the number of particles included in the small volume of phase space, with a momentum p_s and a direction of motion μ_s. Since we have assumed that the system is stationary, in the shock's reference frame, dJ does not depend on time: it is always the same, per unit time. However, we should express everything in terms of quantities in the postshock fluid's reference frame. With a simple change of variable, following a Lorentz transformation (eq. 4.116), we find

$$dJ = \gamma_d(u_d + \mu_d)p_d^2 f d\mu_d dp_d \qquad (4.147)$$

We recall that the velocity of a particle with respect to the shock, $v_s = \mu_s$, is given by

$$v_s = \frac{u_d + \mu_d}{1 + u_d\mu_d} \qquad (4.148)$$

The particles going from the postshock to the preshock region come back, $v_s < 0$, and therefore $\mu_d \leq -u_d$ in order to go back:

$$-1 \leq \mu_{out} \leq -u_d \qquad (4.149)$$

The particles that, instead, cross the shock in the opposite direction, that is, from the preshock to the postshock fluid, must have $v_s > 0$, and therefore $\mu_d > -u_d$:

$$-u_d \leq \mu_{in} \leq 1 \qquad (4.150)$$

We can now find the angular distribution $J_{out}(\mu_{out})$ of the particles crossing the shock toward the preshock fluid, knowing the angular distribution of the particles entering the postshock zone. In fact, from the definition of P_d, we have

$$J_{out}(\mu_{out}) = \int_{-u_d}^{1} d\mu_{in} P(\mu_{in}, \mu_{out}) J_{in}(\mu_{in}) \qquad (4.151)$$

In the preshock fluid, a very similar relation holds, apart from an important difference: when a particle enters the preshock fluid with momentum p_i, it goes back to the postshock fluid with a momentum $p \neq p_i$ (we remind the reader that all momenta are measured in the postshock fluid's reference frame). Let us find the link between p_i and p. The particle entering with momentum p_i and angle μ_{in} has, in the preshock fluid's reference frame, a momentum p' given by equation 4.116:

$$p' = p_i \gamma_r (1 - u_r \mu_{in}) \qquad (4.152)$$

This momentum is conserved, because, in the fluid's reference frame, there is only a magnetic field, which does not change the particles' energies. However, when a particle with momentum p' in the preshock frame reenters the postshock fluid along a direction μ_{out}, it has a new momentum (in the postshock fluid's system), which is again given by equation 4.116 as

$$p' = \gamma_r p (1 - u_r \mu_{out}) \qquad (4.153)$$

If we eliminate p' between the two equations above, we find

$$p = \frac{1 - u_r \mu_{out}}{1 - u_r \mu_{in}} p_i \equiv G p_i \qquad (4.154)$$

where G is the *energy gain* of the single particle.

A comment on G. From equations 4.149 and 4.150, we can easily find that

$$1 \leq G \leq \frac{1 + u_r}{1 - u_r} \qquad (4.155)$$

The particles always gain energy! This is the essence of the Fermi mechanism around the shock.

Therefore, the particles entering with a momentum p_i, exit with momentum p, so that P_u connects particles with slightly different momenta. We easily find that

$$J_{in}(p, \mu_{in}) = \int_{-1}^{-u_d} d\mu_{out} P_u(\mu_{out}, \mu_{in}) \left(\frac{1 - u_r \mu_{in}}{1 - u_r \mu_{out}} \right)^3 J_{out}(p_i, \mu_{out}) \qquad (4.156)$$

and p and p_i obey equation 4.154.

Equations 4.156 and 4.151 are the two equations to be solved.

4.4.4 The Particles' Spectrum

In order to do this, we shall now demonstrate that $J \propto p^{-s}$, where s is a simple number (the spectral index, which we have yet to determine). Let us remove J_{out} between equations 4.156 and 4.151, thus obtaining

$$J_{\text{in}}(p, \mu) = \int_{-1}^{-u_d} d\mu_{\text{out}} P_u(\mu_{\text{out}}, \mu_{\text{in}}) \left(\frac{1 - u_r \mu_{\text{in}}}{1 - u_r \mu_{\text{out}}}\right)^3$$

$$\times \int_{-u_d}^{1} d\xi P_d(\xi, \mu_{\text{out}}) J_{\text{in}}\left(\left(\frac{1 - u_r \mu_{\text{in}}}{1 - u_r \mu_{\text{out}}}\right) p, \xi\right) \quad (4.157)$$

Let us note that this equation is obtained by neglecting the injection term in equation 4.113 and by reducing equation 4.115 to its homogeneous version. Now, let us imagine two problems in which particles are injected with momenta p_0 and $p_0 + \delta p$, respectively, with the solutions $J_{\text{in}}(p, p_0, \mu)$ and $J_{\text{in}}(p, p_0 + \delta p, \mu)$. The difference between these two solutions, since the problem is linear, must be a solution of the associate homogeneous equation 4.157. Therefore, $\partial J_{\text{in}}/\partial p_0$ is a solution of equation 4.157 or, which amounts to the same thing, of equation 4.115. However, for purely dimensional reasons, we must have

$$J_{\text{in}} = \frac{h(p/p_0, \mu)}{p^3} \quad (4.158)$$

In fact, $J_{\text{in}} p^2 dp$ is a number of particles. Thus we have

$$\frac{\partial J_{\text{in}}}{\partial p} = -\frac{3 J_{\text{in}}}{p} + \frac{1}{p^3 p_0} \frac{\partial h(p/p_0, \mu)}{\partial p/p_0}, \quad \frac{\partial J_{\text{in}}}{\partial p_0} = -\frac{1}{p^2 p_0} \frac{\partial h(p/p_0, \mu)}{\partial p/p_0}$$

$$(4.159)$$

and, if we eliminate the derivative of h,

$$p \frac{\partial J_{\text{in}}}{\partial p} = -3 J_{\text{in}} - \frac{1}{p_0} \frac{\partial J_{\text{in}}}{\partial p_0} \quad (4.160)$$

From this we see that $p\partial J_{\text{in}}/\partial p$ is the linear combination of two functions, solutions of the homogeneous equation 4.157 (provided $p \gg p_0$). Therefore, if the homogeneous equation has only one solution, $p\partial J_{\text{in}}/\partial p$ must be a multiple of the only solution, J_{in}:

$$\frac{1}{p}\frac{\partial J_{\text{in}}}{\partial p} = -sJ_{\text{in}} \qquad (4.161)$$

which has the simple solution

$$J_{\text{in}} = \frac{(u_d + \mu)g(\mu)}{p^s} \qquad (4.162)$$

This completes the proof.

Now that we have this simple expression for J_{in}, we can determine g by introducing it into equation 4.157 to obtain

$$(u_{\text{d}} + \mu)g(\mu) = \int_{-u_{\text{d}}}^{1} d\xi Q^{T}(\xi, \mu)(u_{\text{d}} + \mu)g(\mu) \qquad (4.163)$$

$$Q^{T}(\xi, \mu) \equiv \int_{-1}^{-u_{\text{d}}} d\nu P_{\text{u}}(\nu, \mu)P_{\text{d}}(\xi, \nu)\left(\frac{1 - u_{\text{r}}\mu}{1 - u_{\text{r}}\nu}\right)^{3-s} \qquad (4.164)$$

In general, this integral equation *does not* have a solution. The equation with a solution is

$$(u_{\text{d}} + \mu)g(\mu) = \lambda \int_{-u_{\text{d}}}^{1} d\xi Q^{T}(\xi, \mu)(u_{\text{d}} + \mu)g(\mu) \qquad (4.165)$$

which is an equation with eigenvalue λ. Therefore, if we require that equation 4.163 has a solution, we must simply vary the parameter s until the eigenvalue λ assumes the value $\lambda = 1$. This determines the spectral index s, as well as the index of energy distribution, $k = s - 2$.

This equation gives only the flux entering the postshock region. In order to obtain the remaining part, that is, the outgoing flux, we simply introduce the solution we have just found in equation 4.151 to obtain the full solution. At this point, we have to find only the equations determining P_u and P_d.

4.4.5 The equations for P_u and P_d

The P_d

Let us start with P_d, and consider a bunch of particles crossing the shock, all of them along one direction μ_{in}. These ingoing particles have a distribution function f given by

$$f(\mu) = \frac{F_0}{\gamma_d(u_d + \mu)}\delta(\mu - \mu_{in}), \quad u_d + \mu \geq 0 \qquad (4.166)$$

Here, the constant F_0 takes into account the normalization of the particles' flux:

$$F = \int_{\mu \geq -u_d} d\mu \gamma_d(u_d + \mu) = F_0 \qquad (4.167)$$

Therefore, F_0 is the number of particles, per unit time and surface, that cross the shock. By definition, the distribution function for particles *going out* of the postshock zone is given by

$$-(u_d + \mu)f(\mu_{out}) = F_0 P_d(\mu_{in}, \mu_{out}), \quad u_d + \mu \leq 0 \quad (4.168)$$

There is a minus sign because both terms on the right are positive, just like f, too, while $u_d + \mu_{out} < 0$.

In this section, we should have a notation to distinguish the particles moving away from the shock ($u_d + \mu \geq 0$) from the ones going back to the shock, $u_d + \mu \leq 0$. Let us therefore call

$$f_+ = f, \quad u_d + \mu \geq 0$$
$$f_- = f, \quad u_d + \mu \leq 0$$
$$f = f_- + f_- \qquad (4.169)$$

This allows us to rewrite equation 4.115 as

$$\gamma_d(u_d + \mu)\frac{\partial f_-}{\partial z} = -D(\mu)f_- + \int_{-u_d}^{1} d\mu' W(\mu, \mu')f_+(\mu')$$

$$+ \int_{-1}^{-u_d} d\mu' W(\mu, \mu')f_-(\mu') \qquad (4.170)$$

$$\gamma_d(u_d + \mu)\frac{\partial f_+}{\partial z} = -D(\mu)f_+ + \int_{-u_d}^{1} d\mu' W(\mu, \mu')f_+(\mu')$$

$$+ \int_{-1}^{-u_d} d\mu' W(\mu, \mu')f_-(\mu') \qquad (4.171)$$

Now we can use equations 4.166 and 4.167 in the previous ones, computed at $z = 0$, to obtain

$$\gamma_d(u_d + \mu)\frac{\partial f_-}{\partial z}|_0 = \frac{F_0 D(\mu)}{\gamma_d(u_d + \mu)}P_d(\mu_{in}, \mu) + \frac{F_0 W(\mu, \mu_{in})}{\gamma_d(u + \mu_{in})}$$

$$- \int_{-1}^{-u_d} d\mu' \frac{W(\mu, \mu')F_0 P_d(\mu_{in}, \mu')}{\gamma_d(u_d + \mu')}$$

$$\gamma_d(u_d + \mu)\frac{\partial f_+}{\partial z}|_0 = -\frac{F_0 D(\mu)}{\gamma_d(u_d + \mu)}\delta(\mu - \mu_{in}) + \frac{F_0 W(\mu, \mu_{in})}{\gamma_d(u + \mu_{in})}$$

$$- \int_{-1}^{-u_d} d\mu' \frac{W(\mu, \mu')F_0 P_d(\mu_{in}, \mu')}{\gamma_d(u_d + \mu')} \quad (4.172)$$

These equations are useful, because the right-hand sides *do not* depend on f or its derivatives. Therefore, we should simply find another relation between the left-hand side members, to have a relation involving P_d and W only.

In order to find this further relation, let us consider what happens *not* at the shock, but at a distance z from it. The particles going back to the shock are given by

$$\gamma_d(u_d + \mu)f_-(\mu) = -\int_{-u}^{1} f_+(\mu')\gamma_d(u_d + \mu')P_d(\mu', \mu, z)d\mu'$$

$$(4.173)$$

Since we are at a distance z from the shock, the probability to return to the shock, $P_{\rm d}(z)$, as a general rule, will not necessarily be equal to the probability at the shock, $P_{\rm d}$. But, in this specific case, which describes a stationary and infinite situation in the direction z, we *do* have instead $P_{\rm d}(z) = P_{\rm d}(0)$, independently of the value of z. The reason is that the amount of fluid to the right of the surface $z \neq 0$ is infinite, just like the amount of fluid to the right of the surface $z = 0$; therefore, the scattering properties must be the same. In other words, if we add or take away a finite thickness of matter to a semi-infinite sector, the latter remains the same: $P_{\rm d}(z)$ is the same as $P_{\rm d} = P_{\rm d}(0)$.

Therefore, in the preceding equation, $P_{\rm d}$ does not depend on z; now we can take its derivative with respect to z, thus obtaining

$$\gamma_{\rm d}(u_{\rm d} + \mu)\frac{\partial f_-}{\partial z} = -\int_{-u_{\rm d}}^{1} \gamma_{\rm d}(u + \mu')P_{\rm d}(\mu', \mu)\frac{\partial f_+}{\partial z} \quad (4.174)$$

We shall now use equation 4.172 to obtain

$$P_{\rm d}(\mu_{\rm in}, \mu)\left(\frac{D(\mu_{\rm in})}{u - d + \mu_{\rm in}} - \frac{D(\mu)}{u_{\rm d} + \mu}\right)$$

$$= \frac{W(\mu, \mu_{\rm in})}{u_{\rm d} + \mu_{\rm in}} + \int_{-u_{\rm d}}^{1} d\mu' \frac{P_{\rm d}(\mu', \mu)W(\mu', \mu_{\rm in})}{u_{\rm d} + \mu_{\rm in}}$$

$$- \int_{-1}^{-u_{\rm d}} d\mu' \frac{P_{\rm d}(\mu_{\rm in}, \mu')W(\mu, \mu')}{u_{\rm d} + \mu')}$$

$$- \int_{-u_{\rm d}}^{1} d\mu' P_{\rm d}(\mu', \mu)\int_{-1}^{-u_{\rm d}} d\mu'' \frac{W(\mu', \mu'')P_{\rm d}(\mu_{\rm in}, \mu'')}{u_{\rm d} + \mu''} \quad (4.175)$$

Both f and its derivatives have disappeared from this equation, which is therefore an equation for $P_{\rm d}$, since we have assumed all the other terms are known.

The P_u

For the region before the shock, we can obviously proceed in the same way. The incoming flux is

$$f(\mu) = -\frac{F_0}{\gamma_u(u_u + \mu)}\delta(\mu - \mu_{\rm out}), \qquad u_u + \mu_{\rm out} \leq 0 \quad (4.176)$$

whereas the outgoing flux is

$$f(\mu) = \frac{F_0}{\gamma_u(u_u + \mu)}P_u(\mu_{\rm out}, \mu), \qquad u_u + \mu \geq 0 \quad (4.177)$$

In the same way, at a distance z from the shock we find, by using exactly the same principle of symmetry that we have found before,

$$\gamma_u(u_u + \mu)\frac{\partial f_+}{\partial z} = -\int_{-1}^{-u_d} \gamma_u(u_u + \mu')P_u(\mu', \mu)\frac{\partial f_-}{\partial z} \quad (4.178)$$

On the other hand, equation 4.172 is still valid, and we can eliminate f and its derivatives at the shock:

$$P_u(\mu_{\rm out}, \mu)\left(\frac{D(\mu_{\rm out})}{u_u + \mu_{\rm out}} - \frac{D(\mu)}{u_u + \mu}\right)$$

$$= \frac{W(\mu, \mu_{\rm out})}{u_u + \mu_{\rm out}} + \int_{-1}^{-u_u} d\mu' \frac{P_u(\mu', \mu)W(\mu', \mu_{\rm out})}{u_u + \mu_{\rm out}}$$

$$- \int_{-u_u}^{1} d\mu' \frac{P_u(\mu_{\rm out}, \mu')W(\mu, \mu')}{u_u + \mu'} - \int_{-1}^{-u_u} d\mu' P_u(\mu', \mu)$$

$$\times \int_{-u_u}^{1} d\mu'' \frac{W(\mu', \mu'')P_u(\mu_{\rm out}, \mu'')}{u_u + \mu''} \quad (4.179)$$

This is the equation determining P_u.

Comments

The $P_u(\nu', \mu')$ that we have just found is expressed in terms of the angles measured in the preshock fluid's reference system, whereas, in section 4.4.4, we used the same function, which we call $\hat{P}_u(\nu, \mu)$, expressed in terms of angles measured in the postshock fluid's reference system. The relation between the two is obviously

$$\hat{P}_u(\nu, \mu) = P_u(\nu', \mu')\frac{d\mu'}{d\mu} = P_u(\nu'\mu')\left(\frac{1 - u_r^2}{(1 - u_r\mu)^2}\right) \quad (4.180)$$

and the relation between the angles expressed in the two reference systems is given by equation 4.116. Therefore, in section 4.4.4, we must use $\hat{P}_u$, which is determined with the transformation we have just given, as soon as equation 4.179 is solved.

The equations 4.179 and 4.175 are nonlinear integral equations, which can be solved numerically by iteration. As the number of iterations increases, the approximate solution converges toward the only physical solution, namely, the only one satisfying $P < \infty$ and $P \geq 0$.

Therefore, the problem of identification of the spectral index for shocks of arbitrary velocity and arbitrary scattering properties, is solved by determining P_d and P_u from equations 4.175 and 4.179, and then determining the only value of s that satisfies equation 4.163. The spectral index in energy is obviously given by

$$dN = fp^2 dp \propto p^{2-s}dp \propto E^{-k}dE , \quad k = s - 2 \quad (4.181)$$

4.4.6 Results

Since we do not know the deflection function $W(\mu, \mu')$ that is realized in concrete situations, all computations introduced so far have used isotropic scattering functions, in the sense discussed above. We should therefore choose a scattering function that is easy to handle, and study in detail the spectrum of accelerated particles in these conditions, while waiting for a more physically motivated scattering form. Such a *simple* choice consists in taking

$$w(\mu, \mu', \phi, \phi') \equiv \frac{1}{\sigma}\exp\left(-\frac{1 - \cos\Theta}{\sigma}\right) \quad (4.182)$$

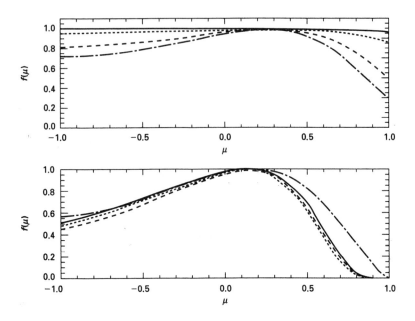

Figure 4.4. Distribution function at the shock, for different values of the parameter $\gamma\beta = 0.04, 0.2, 0.4$, and 0.6 (upper panel), and $\gamma\beta = 1, 2, 4$, and 5 (lower panel). Blasi and Vietri 2005.

where Θ, the angle between the directions of entry and exit is given by

$$\cos\Theta = \mu\mu' + \sqrt{1-\mu^2}\sqrt{1-\mu'^2}\cos(\phi - \phi') \qquad (4.183)$$

Using equation 4.114, we find

$$W(\mu,\mu') = \int w(\mu,\mu',\phi,\phi')$$

$$= \frac{2\pi}{\sigma}\exp\left(-\frac{1-\mu\mu'}{\sigma}\right)I_0\left(\frac{\sqrt{1-\mu^2}\sqrt{1-\mu'^2}}{\sigma}\right)$$

$$(4.184)$$

where $I_0(x)$ is a Bessel function. This equation has the advantage of representing well a system in the SPAS limit (when

$\gamma\beta$	u	u_{d}	s
0.04	0.04	0.01	4
0.2	0.196	0.049	3.99
0.4	0.374	0.094	3.99
0.6	0.51	0.132	3.98
1.	0.707	0.191	4.00
2.	0.894	0.263	4.07
4.	0.97	0.305	4.12
5.	0.98	0.311	4.13

Table 4.1. Special Indices for relativistic shocks

$\sigma \to 0$), and one in the large-angles scattering limit, when $\sigma \to \infty$.

We now concentrate on the SPAS case and find, for $\sigma = 0.01$ as the factor $\gamma\beta$ for the shock changes, the distribution function in figure 4.4. We can see from this figure that when the Lorentz factor of the shock increases, the anisotropy of the distribution function grows: for $\beta \ll 1$, we obviously have $f \approx$ constant, as deduced in our analytic discussion of the Newtonian limit above. However, as γ increases, the function differs significantly from an isotropic one. Unfortunately, this anisotropy depends in detail on the form assumed for W and thus gives no universal results. The spectral slopes calculated for this regime are given in table 4.1.

From this table we see that the spectral slopes diverge from those predicted for the Newtonian regime (eq. 4.22), but despite this the spectral index does not seem to wander too far from the Newtonian value $s \approx 4$. This coincidence is due to the fact that while the average amplification of energy per cycle around the shock increases up to the value $A \approx 2$, the probability to return to the shock decreases to $P \approx 0.4$, thus making the likelihood of many cycles (and thus of strong energy gains) no greater than in the Newtonian limit.

In order to appreciate the differences in the spectral index, we shall also present the spectral index in the case of

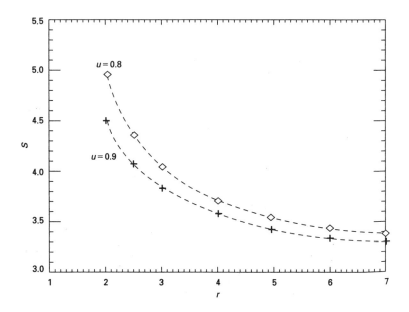

Figure 4.5. Spectral index for $\sigma \gg 1$, as the compression relation changes r. Continued lines are the results of the Ellison
et al., simulations (1990); the dots are the results obtained
by Blasi and Vietri (2005) with the method described above.

large-angles scattering, $\sigma \gg 1$, in figure 4.5. From this we
can easily see that it is very easy to obtain spectra steeper
than $s \approx 4$, when the properties of the scattering function W
vary.

4.5   The Unipolar Inductor

Apart from acceleration around shocks, there is another
mechanism that accelerates high-energy particles, using electric fields. In general, as we have often emphasized, electric
fields are negligible in astrophysics, because of the abundance
of free electric charges, which can shorten out any electric
field due to charge separation. The real exception to this
excellent empirical rule concerns electric fields generated by

the movement of magnetic fields. Indeed, we know from the Lorentz transformations that in this case, an electric field appears, of amplitude (in order of magnitude, for $v \ll c$) $E \approx vB/c$. Short-circuiting this electric field would mean halting the object's motion, but, when the object in question is a huge rotating star, this may simply be impossible for the free charges around the star since they have a mass negligible in comparison with that of the star.

In order to be definite, let us consider the simple case of the *unipolar inductor*: it is a metallic sphere, magnetized so as to have a pure magnetic field $\vec{B}$ when it is not rotating. If this sphere is now made to rotate with angular velocity $\vec{\Omega}$, the electrons on its surface are subject to a net force (i.e., an unbalanced force) given by $e(\vec{\Omega} \wedge \vec{r}) \wedge \vec{B}/c$. This net force produces a separation of charges, which lasts until an excess of electrons is created repelling other electrons; there must therefore arise an electric field $\vec{E}$ that causes the total Lorentz force to vanish. At this point, electrons can only corotate with the star. We thus have, as an equilibrium condition,

$$\vec{E} + \frac{\vec{\Omega} \wedge \vec{r}}{c} \wedge \vec{B} = 0 \tag{4.185}$$

This equation holds anywhere within the sphere, since it has been assumed to be an excellent conductor, and it holds in particular on its surface. The above equation immediately implies that there is a potential difference between points on the surface of the sphere. If we assume that the field $\vec{B}$ is a pure dipole, with the south-north pole axis parallel to the rotation axis of the sphere, we find that the potential difference between pole and equator is given by

$$\phi = - \int_0^\pi E_\theta R d\theta = \frac{B_{\mathrm{p}} \Omega R^2}{2c} \tag{4.186}$$

where R is the sphere's radius, and B_{p} is the magnetic field at the pole. For the dipole field at a distance R_{o} and an arbitrary direction $\hat{n}$, we have assumed

$$\vec{B} = \frac{3\hat{n}(\hat{n} \cdot \vec{M}) - \vec{M}}{R_{\mathrm{o}}^3} \tag{4.187}$$

where the total magnetic momentum $\vec{M} = B_{\mathrm{p}} R^3 \hat{z}$.

If now, with a sliding contact, we close a circuit between pole and equator, this potential difference generates a current that can easily be measured in the laboratory.

A pulsar acts exactly like the unipolar inductor described above: it is a (small, $R \approx 10^6$ cm) star, which is a good conductor (in fact, inside, it is even a superconductor, even though this is irrelevant to our aims), endowed with a very intense magnetic field ($B_{\mathrm{p}} \approx 10^{12} - 10^{14}$ G) and large rotation velocities corresponding to periods $T \gtrsim 0.03$ s. In this case, equation 4.186 predicts a large potential difference: $e\phi \lesssim 3 \times 10^{15}$ to 3×10^{17} eV. However, a more accurate analysis, to be presented in chapter 8, shows that there is a mechanism canceling most of this potential difference, so that the remaining part that can be used for the acceleration of particles is given by equation 8.27:

$$\phi \approx \frac{BR}{2} \left(\frac{\Omega R}{c} \right)^2 \tag{4.188}$$

which is smaller than equation 4.186 by a factor $\Omega R/c \ll 1$. In order of magnitude, we obtain

$$e\phi \lesssim 2 \times 10^{13} - 2 \times 10^{15} \text{ eV} \tag{4.189}$$

Therefore, normal pulsars can accelerate particles up to energies comparable, at most, to a few times 10^{15} eV.

We can easily realize that acceleration cannot take place near the surface of a neutron star. First, here the magnetic field is so strong that the gyration radius of particles has a value of $\approx 10^{-8}$ cm, so that particles can only slide along magnetic lines; but they are free to do so, and thus magnetic lines act as live wires of enormous conductivity. Second, as we shall demonstrate in chapter 8, there must be many free electric charges making the magnetic field lines equipotential, that is, the potential difference between different points on the same magnetic line vanishes, even if the difference of potential among points belonging to different magnetic lines does not. However, since we have just seen that particles have very small gyroradii, they do not perceive this potential difference. This situation can change only for those magnetic field lines that manage to go significantly far away from the star, because the magnetic field weakens, the gyration radius increases, particles may move away from their magnetic field line, and thus they feel the potential difference. Therefore, the acceleration of particles can take place only at large distances from the pulsar.

Please note also that there are two types of magnetic field lines (chapter 8): those that close within a distance c/Ω, where Ω is the angular velocity of the pulsar (the distance c/Ω identifies a locus called a *light cylinder*), and those that cross the light cylinder, which, instead, open up and close only at infinity. The magnetic lines closing within the light cylinder are too strong to allow particles to migrate from one magnetic line to another, so there can be no acceleration within the light cylinder. Instead, all magnetic lines crossing the light cylinder reach infinity, and there the field becomes so weak it allows particles to feel the potential difference. The magnetic lines crossing the light cylinder are those leaving the pulsar within an angle $\Delta\theta^2 \approx \Omega R/c$ of the vertical, while the others close within the cylinder. This explains the introduction of the limiting factor $\Omega R/c$ in equation 4.188.

We shall discuss and further illustrate these ideas in chapter 8.

Can we increase the estimate of equation 4.188? At first sight, it appears we can. Probably, there are stars with magnetic fields of the order of $B \approx 10^{15}$ G, called *magnetars*, and there are (certainly!) pulsars with periods of $T = 1.667$ ms; but, unfortunately, magnetars have very long rotation periods, $T \gtrsim 1$ s, and millisecond pulsars have weak magnetic fields, $B_{\rm p} \approx 10^8$ G. Increasing the estimate of equation 4.188 means requiring at the same time the existence of both stronger magnetic fields and shorter rotation periods. For this reason some physicists have proposed that this estimate can be increased in young pulsars. In fact, it has been suggested that they form with strong magnetic fields (which can, perhaps, decay with time later on) and short rotation periods. Just to play safe, assuming $B_{\rm p} = 10^{14}$ G and $T = 1$ ms, we find

$$e\phi_{\rm max} = 4 \times 10^{18} \text{ eV} . \qquad (4.190)$$

This energy is still low in comparison with the highest-energy cosmic rays observed so far, for which $E \approx 10^{20}$ eV; but, it has been noticed that if instead of considering electrons or protons, we consider iron nuclei, whose atomic number is $Z = 26$, we find

$$Ze\phi_{\rm max} = 10^{20} \text{ eV} \qquad (4.191)$$

in agreement with observations. The presence of wholly ionized iron nuclei around newly formed pulsars is probably realistic, so the acceleration by unipolar inductor is currently considered a realistic possibility for the highest-energy cosmic rays.

4.6   Problems

1. The quantity that scales like a power law is not $dN/dE \propto E^{-p}$, but $f(p) \propto p^{-s}$, for relativistic particles and for Newtonian particles. For relativistic particles, derive from equation 4.1 the link between the spectral indexes s and p. Can you explain why $f \propto p^{-s}$?

2. Show that f is a Lorentz invariant. (Hint: $dN = f d^3V d^3p$; you should note that dN is obviously invariant, and d^3p/E as well).

3. It is possible to derive Bell's result, eq. 4.29, from eq. 4.101, applied to a shock. As a matter of fact, eq. 4.101 gives f everywhere, not just at downstream infinity. Can you see how? (Blandford and Ostriker 1978.)

Chapter 5

Spherical Flows: Accretion and Explosion

Accretion is, with the rotation of celestial bodies, the main source of energy in high-energy astrophysics; ultimately, it is responsible for almost all emission. The fundamental idea is quite simple. During accretion, because of energy conservation, the increase of the potential energy of an element of fluid (in modulus) must be compensated by an increase of kinetic energy. If a part of this kinetic energy becomes internal energy (namely, if temperature increases or if nonthermal particles are efficiently accelerated), the various processes seen in chapter 3 may dissipate this internal energy, and emit copious amounts of radiation.

The prominence of accretion is due to its high efficiency. We shall see in appendix B that a quantity of mass M accreted onto a compact object, such as a neutron star or a black hole, can radiate a quantity of energy equal to $\eta M c^2$, with $\eta \approx 0.1 - 0.4$. In contrast, the burning of the main nuclear fuels, such as H and He, does not reach $\eta \lesssim 0.01$, and even the burning of heavier elements, closer to the peak of the nuclear binding energy, which is represented by Fe, is much less efficient than 0.01, since a large fraction of the energy liberated is lost to neutrinos, which interact very little with matter and are therefore completely lost. This is why Zel'dovich proposed, back in 1964, that quasars are very massive black holes, accreting mass from the surrounding environment; for a given total luminosity, accretion requires less fuel, and this makes the problem of feeding the central engine less severe.

This chapter discusses, first of all, accretion in spherical or quasi-spherical conditions, in the limits in which we can, or cannot, neglect the temperature of the accreting gas. We shall then discuss the importance of the feedback of the source luminosity on the flux, and, finally, why spherical accretion flows onto black holes (but not onto neutron stars) have a low radiative efficiency.

After accretion, we shall also discuss the hydrodynamics of explosions, and we shall focus on two fundamental limits: on one hand, Newtonian motions with a description of the properties of supernovae, and on the other hand, highly relativistic motions and gamma ray bursts.

5.1   Accretion from Cold Matter

Initially, we can consider a star with a mass M that moves in the interstellar medium of constant density ρ. We shall suppose that this is an essentially cold medium, in the sense that we shall assume that the sound speed of the gas is much lower than the star velocity V. Therefore, assuming $T = 0$, the gas motion will be essentially ballistic (namely, only subject to the force of gravity); in cosmology, this is called the *dust approximation.* If we neglect the star's gravity, it gains mass at the rate

$$\dot{M} = \pi R^2 V \rho \qquad (5.1)$$

where V is the star's velocity. On the other hand, if we take into account gravity, we see that the fluid particles' trajectories are curved, and we can imagine that the star accretes all the matter reaching the surface. In this case, fluid particles move according to the law

$$\frac{1}{r} = \frac{GM}{b^2 V^2}(1 - e\cos\theta) \qquad e = \left(1 + \frac{b^2 V^4}{G^2 M^2}\right)^{1/2} \tag{5.2}$$

where b is the impact parameter, and e the orbit's eccentricity. We can see at once that $1/r$ has a maximum for $\cos\theta = -1$, corresponding to the distance of closest approach to the star. We can therefore assume that all fluid elements with a minimum distance shorter than the star radius R are accreted onto the star. The maximum impact parameter for material to be accreted is given by

$$\left(\frac{1}{r}\right)_{\min} = \frac{1}{R} \tag{5.3}$$

Solving this we find the maximum impact parameter $b_{\max}$ that is accreted on the star,

$$b_{\max} = R\left(1 + \frac{2GM}{RV^2}\right)^{1/2} \tag{5.4}$$

which gives us the accretion rate in this situation:

$$\dot{M} = \pi R^2 V \rho \left(1 + \frac{2GM}{RV^2}\right) \tag{5.5}$$

However, this is not yet the most complete expression; in fact, we can see from figure 5.1 that matter not hitting the star actually hits matter coming from the other side of the star, exactly on the axis of motion of the star itself. In this case, it is legitimate to think that the two colliding fluid elements completely dissipate the component of the momentum perpendicular to the axis of motion, while conserving the velocity component along the axis. It may then happen that this residual component of velocity is not larger than the escape velocity at that point; in this case, matter must fall on the star. We should therefore find the impact parameter b_a such that when matter reaches the axis of motion of the star,

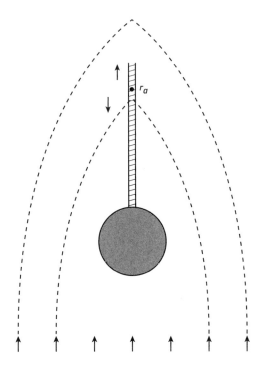

Figure 5.1. Graph illustrating matter accretion for cold gas.

it has a kinetic energy (calculated using only the momentum component along the axis) equal to its potential energy at that point.

The angle corresponding to the direction of arrival has $\cos\theta_a = 1/e$; we can see this from equation 5.2, since the direction of arrival is the one in which $r \to \infty$. The axis of motion is exactly π radians larger than this. It follows that the angle θ_m corresponding to the axis of motion has

$$\cos\theta_m = \cos(\theta_a + \pi) = -\cos\theta_a = -\frac{1}{e} \qquad (5.6)$$

Along this direction, if the impact parameter is b, equation 5.2 tells us that the distance from the star is

$$\frac{1}{r_a} = \frac{2GM}{b^2V^2} \qquad (5.7)$$

The equation of energy conservation yields

$$\frac{v_\theta^2 + v_r^2}{2} - \frac{GM}{r} = \frac{V^2}{2} \tag{5.8}$$

where v_θ and v_r are the two components of the velocity, perpendicular and parallel to the star speed, respectively. The point we are looking for is, by definition, the one in which the total energy, once we have subtracted the kinetic energy $v_\theta^2/2$ corresponding to motion perpendicular to the axis (which, we recall, is dissipated), vanishes:

$$\frac{v_r^2}{2} - \frac{GM}{r} = 0 \tag{5.9}$$

because this is exactly the condition for which the velocity remaining after the dissipation equals the local escape velocity. Comparing the two equations above, we can see that we must have, for this specific orbit,

$$v_\theta = V \tag{5.10}$$

Now we can use the conservation of angular moment

$$bV = r_a v_\theta \tag{5.11}$$

which, compared with the preceding one, gives

$$b = r_a \tag{5.12}$$

This is, in *spatial* terms, not in terms of velocity, the condition that the velocity along the axis equals the escape velocity. Let us now go back to equation 5.7, which, together with the preceding one, gives us what we were looking for:

$$b_a = \frac{2GM}{V^2} \tag{5.13}$$

and the *columnar* accretion rate

$$\dot{M}_{col} = \pi b_a^2 V \rho = 4\pi\rho \frac{G^2 M^2}{V^3} \tag{5.14}$$

We can describe the column of accreting matter more accurately (see problem 1, or see directly its solution in the original article by Bondi and Hoyle [1944]), but this further work is not particularly useful. Indeed, the Bondi-Hoyle accretion is unstable (Cowie 1977), and the structure of the accretion flux is completely different (see section 5.3).

5.2   Accretion from Hot Matter

So far we have neglected the fact that matter has a nonzero temperature and thus resists (through its pressure) compression. Here we shall treat this case by considering a star (with mass M) that is at rest with respect to gas at infinity, where density and pressure have values ρ_∞ and p_∞. In order to simplify matters, we shall assume that the gas is polytropic, with an exponent $1 \leq \gamma \leq 5/3$, and that we are in a stationary situation: $\partial/\partial t = 0$.

The equations we must apply are those of conservation of mass, equation 1.6, and of momentum. Taking into account stationarity and spherical symmetry equation 1.6 yields

$$0 = \frac{\partial \rho}{\partial t} + \nabla \cdot (\rho \vec{v}) = \nabla \cdot (\rho \vec{v}) = \frac{1}{r^2}\frac{\partial}{\partial r}\left(r^2 \rho v_{\mathrm{r}}\right) = 0 \quad (5.15)$$

The integration of the latter equation gives

$$\dot{M} = 4\pi \rho v_{\mathrm{r}} r^2 \tag{5.16}$$

where $\dot{M}$, the mass accreted per unit of time, does not depend on r but is, for the moment, an unknown constant. We shall see that the main result of this problem is to fix it precisely.

We shall not directly use Euler's equation (eq. 1.10), but its integral. Indeed, we know that for polytropic fluids, Euler's equation admits an integral, Bernoulli's theorem. We shall use equation 1.26, where, for the gravitational potential due to the star, we obviously have $\phi = -GM/r$:

$$\frac{v_{\mathrm{r}}^2}{2} + w - \frac{GM}{r} = w_\infty \tag{5.17}$$

where the constant appearing on the right-hand side is the value of the left-hand side evaluated at infinity. The only term that does not vanish there is the specific enthalpy, $w = \gamma p/(\gamma - 1)\rho$.

What we have, mathematically, is a system of algebraic equations (eqs. 5.16 and 5.17) in two unknown quantities, ρ and v_r. In fact, since we have assumed a polytropic fluid, the relation $p \propto \rho^\gamma$ holds, and the pressure is a known function of density. However, this problem contains a subtlety, and we should follow Bondi (1952) to simplify our task. We should first of all take as the unit of velocity the sound speed at infinity $c_s^2 \equiv \gamma p_\infty/\rho_\infty$, as the unit of length the so-called *accretion radius* $r_a \equiv GM/c_s^2$, and as the unit of density ρ_∞. We shall thus take

$$r = r_a \xi ; \qquad \rho = \rho_\infty R(\xi) \tag{5.18}$$

where $R(\xi)$ is obviously a dimensionless function of a quantity ξ, also dimensionless. The accretion rate of mass, which is still unknown, is given, in dimensionless units, by a parameter λ, which is connected to physical units by

$$\dot{M} = 4\pi\lambda \frac{G^2 M^2 \rho_\infty}{c_s^3} \tag{5.19}$$

As the other independent variable, we should *not* take the velocity in the unit of the sound speed at infinity, as appears natural, but instead the local Mach number, namely, the ratio of velocity and local sound speed. We know that the local sound speed is given by $c = \sqrt{\gamma p/\rho}$, which can be rewritten in units of speed at infinity as $c = c_s \sqrt{p\rho_\infty/p_\infty\rho} = c_s(\rho/\rho_\infty)^{(\gamma-1)/2} = c_s R^{(\gamma-1)/2}$. It follows that we define a new dependent variable $\mathcal{M}$ as

$$v_r \equiv \mathcal{M}c = c_s \mathcal{M} R^{(\gamma-1)/2}(\xi) \tag{5.20}$$

Through these definitions, we can rewrite and combine equations 5.16 and 5.17 into a single one, thus removing $R(\xi)$:

$$f(\mathcal{M}) = \lambda^{-s} g(\xi) \tag{5.21}$$

where

$$f(\mathcal{M}) \equiv \frac{1}{2}\mathcal{M}^{4/(\gamma+1)} + \frac{1}{(\gamma-1)\mathcal{M}^s}$$

$$g(\xi) \equiv \left(\frac{\xi^{4(\gamma-1)/(\gamma+1)}}{\gamma-1} + \frac{1}{\xi^{(5-3\gamma)/(\gamma+1)}} \right)$$

$$s \equiv 2\frac{\gamma-1}{\gamma+1} \tag{5.22}$$

We can easily see from this equation that f and g are the sum of two powers of their arguments, one positive and the other negative (provided $1 < \gamma < 5/3$). Therefore, they both certainly have a minimum. It is easy to realize that the two minima are reached at

$$\mathcal{M}_{\mathrm{m}} = 1 , \qquad \xi_{\mathrm{m}} = \frac{5-3\gamma}{4} \tag{5.23}$$

The minimum of $f(\mathcal{M})$ is reached when $\mathcal{M} = 1$, which is the point in which the accretion speed equals that of sound; this point is called the *sonic point*.

Equation 5.21 must be solved: we have to find, for a given value of the radius ξ, the Mach number $\mathcal{M}$ at that point. Physically, we obviously expect a simple relation between the two, so we may wonder which value of $g(\xi)$ corresponds to the value $f(1)$, which is the absolute minimum of f. If $\xi > \xi_{\mathrm{m}}$, this would mean that when we consider smaller values of the radius, we could not satisfy equation 5.21: as ξ decreases, also $g(\xi)$ decreases, because it approaches its minimum, whereas $f(\mathcal{M})$ cannot decrease any further. If we now suppose that $\xi < \xi_{\mathrm{m}}$, we face the same problem when ξ increases: since $g(\xi)$ decreases, whereas f can only increase, equation 5.21 can have no solution.

As a consequence, if we want to find a solution, the two minima, equation 5.23, should exactly match. However, since the minima of f and g do not satisfy equation 5.21, λ must have a value such that

$$f(\mathcal{M}_{\mathrm{m}}) = \lambda^{-s} g(\xi_{\mathrm{m}}) \tag{5.24}$$

namely,

$$\lambda = \left(\frac{1}{2}\right)^{(\gamma+1)/(2(\gamma-1))} \left(\frac{5-3\gamma}{4}\right)^{-(5\gamma-3)/(2(\gamma-1))} \tag{5.25}$$

When we introduce this result into equation 5.19, we see that there is only one accretion rate that is compatible with our hypotheses, namely, fixed conditions at infinity, and a stationary flow. This accretion rate depends on the star mass, on the conditions at infinity (a very hot fluid, with a large sound speed, does not want to be trapped in the star), and even on the equation of state, because the eigenvalue λ depends on γ. This dependence is such that the larger γ (provided $\gamma < 5/3$), the smaller the accretion. This is physically obvious: the larger γ, the larger the amount of compression work stored as internal energy, which keeps the system's sound speed high. This is so true that strictly speaking, the minimum of $g(\xi)$, the sonic point for $\gamma = 5/3$ is reached only at $r = 0$: the gas is desperately resisting compression, so much so that it almost makes it impossible.

The accretion rate in equation 5.19, with λ given by equation 5.25, is called the *Bondi* rate, after the cosmologist Hermann Bondi, who first derived it.

When $r \ll r_a \xi_m$, it is easy to show that equation 5.17 is well approximated by

$$\frac{v^2}{2} \approx \frac{GM}{r} \tag{5.26}$$

Beyond the sonic point, the gas is in free fall, and nothing can stop its fall onto the star. On the other hand, when $r \gg r_a \xi_m$, the kinetic term in equation 5.17 is a small correction: the gas is slowly settling down, almost in hydrostatic equilibrium. The sonic point is the point separating these two regimes: inside, free fall, outside a hydrostatic quasi equilibrium.

Figure 5.2 illustrates different solutions, all subject to the same boundary conditions, but endowed with different accretion rates $\dot{M}$. We see that the solution corresponding

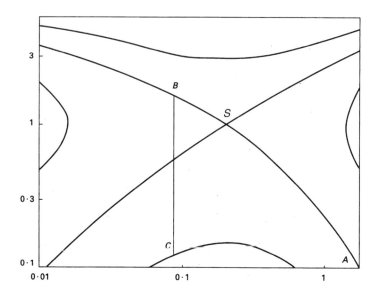

Figure 5.2. The Bondi solution. The point $\mathcal{S}$ indicates the sonic point, and the decreasing solution that crosses it is the physically significant one. The line ABC indicates a possible solution with a shock wave. Adapted from Bondi 1952.

to our physical intuition (flow essentially in a hydrostatic quasi equilibrium at a large radius, and in free fall at radii small relative to Bondi's critical radius) is the unique solution corresponding to Bondi's accretion rate.

When we take into detailed account the gas heating and cooling processes, the results described above do not change, at least qualitatively. However, the mass accretion rate compatible with the hypothesis of stationarity, and the position of the sonic point do change. That is why we generally use as a reference for the starting point of the free fall the accretion radius r_a, which normally implies only a small error; there is a true difference only in the very idealized case in which there is no radiative loss, $\gamma = 5/3$. It is also to be noted that there is a general relativistic treatment of this problem

(Michel 1972) whose conclusions are absolutely similar to those we have just illustrated.

5.2.1 The Critical Point

Bondi's description of spherical accretion (essentially identical to the one presented above) skillfully hides an important detail, namely, that sonic points are singular (a more precise term is *critical*) points of the flow. For teaching purposes, here we shall briefly discuss the Bondi problem in a different form.

We shall use once again the equation of mass conservation in the form of equation 5.16, but, instead of using the Bernoulli theorem (eq. 5.17), we shall directly use Euler's equation (eq. 1.10), where we take, as external force, the gravitational force of the point mass. We shall use the polytropic condition $p = K\rho^\gamma$. In spherical symmetry, for a stationary problem, we find

$$v_r \frac{dv_r}{dr} = -\frac{c_s^2}{\rho} \frac{d\rho}{dr} - \frac{GM}{r^2} \tag{5.27}$$

If we take the logarithmic derivative (the logarithmic derivative of X with respect to R is $d\ln X/dR$) of equation 5.16, we find

$$\frac{1}{\rho} \frac{d\rho}{dr} + \frac{2}{r} + \frac{1}{v_r} \frac{dv_r}{dr} = 0 \tag{5.28}$$

Now we eliminate from this expression the derivative of ρ using the previous equation, thus obtaining

$$\left(v_r^2 - c_s^2\right) \frac{d\ln v_r}{d\ln r} = 2c_s^2 - \frac{GM}{r} \tag{5.29}$$

which is the equation we were looking for. Here we can immediately see the mathematical problem: the coefficient of $d\ln v_r/d\ln r$ vanishes at the sonic point, and, if the right-hand side of the equation does not vanish at the same point, there is a singularity in v_r, which obviously makes no physical sense—why should velocity diverge at the sonic point, at a finite distance from the star? Physically, this is an unacceptable situation. As a consequence, the right-hand side must vanish simultaneously. However, a brief reflection shows that the right-hand side will not, in general, vanish at the same point. What then? The problem contains an eigenvalue, $\dot{M}$, namely, a quantity that is not specified a priori, but that we can vary freely in order to satisfy the condition that the right-hand side vanishes when the left-hand one does.

In fact, the above equation depends on $\dot{M}$: from equation 5.16, we see that ρ is defined in terms of $\dot{M}$, and c_s, in equation 5.29, is given by $c_\infty(\rho/\rho_\infty)^{\gamma-1}$, where c_∞ and ρ_∞, the sound speed and the gas density, are boundary conditions and are therefore known.

This problem appears whenever the flow has a sonic point, namely, wherever the fluid becomes supersonic with continuity, rather than by passing through a shock wave. Its mathematical structure is always the same: the coefficient of $d\ln v_r/d\ln r$ vanishes, while the right-hand side, in general, does not. We must therefore find a free parameter that allows us to avoid the unphysical singularity. Normally, the problem is tackled numerically, because Euler's equation can only rarely be solved in an analytic form. However, when Bernoulli's theorem holds, that is, when the fluid is polytropic, Euler's equation has a first integral, so that we can find the value of the free parameter (in our case, λ or, which is the same, $\dot{M}$) in an analytic way.

What is the physical root of the mathematical problem? If we consider two fluid elements on both sides of the sonic point, they are not in causal contact, since the subsonic one cannot send a signal capable of reaching its supersonic companion because the signal moves at the speed of sound. Therefore, at least a priori, the supersonic fluid might well have thermodynamic properties radically different from the subsonic ones. The condition, which requires the simultaneous vanishing of right- and left-hand sides of the above equation, may be reexpressed also in the following terms: the fluid properties beyond the sonic point are simply the continuation of those of the subsonic fluid.

5.3   The Intermediate Case

The intermediate case is the one in which the star velocity in the medium is of the same order of magnitude as the sound speed. This situation is characterized by a dimensionless number, namely, the Mach number $\mathcal{M} \equiv V/c_s$ given by the *star* velocity with respect to the unperturbed medium in units of the sound speed of the gas. Obviously, we reduce to the case of negligible pressure when $\mathcal{M} \to \infty$, and to the Bondi case for $\mathcal{M} \to 0$. However, the intermediate case has a flow structure that is completely different from both problems we discussed above. If the star has a hard surface, like neutron stars, but *different* from black holes, for $\mathcal{M} \geq 1$ a stationary shock wave must form close to the star. This shock wave deflects the incoming material away from the star's hard surface and is called a *bow shock*. Moreover, astern of the star a stagnation point forms, namely, a point (which is stationary in the star's reference frame) where the gas velocity vanishes; this point has a large pressure, and it is exactly this pressure that deflects away matter that would otherwise enter the star's trail.

Finally, we should notice that the functional dependence of the mass accretion rate from all the relevant physical parameters, equations 5.14 and 5.19, is exactly the same in the two different situations we have treated, whether or not it is possible to neglect the temperature of the accreting matter. This suggests that a reasonable approximation can be found in the intermediate cases in the form

$$\dot{M}_{\mathrm{B}} = 4\pi \frac{G^2 M^2 \rho_\infty}{(V^2 + c_s^2)^{3/2}} \tag{5.30}$$

Numerical simulations confirm the validity of this approximation; see Shima et at. (1985).

5.4   Doubts about the Bondi Accretion Rate

Recently, doubts have been raised about the validity of the Bondi accretion rate, equation 5.30. These doubts arise from the fact that in numerical simulations, the structure of the Bondi accretion flow is very different from that of flows with small—but not strictly zero—amounts of angular momentum and magnetic field. In other words, it seems that the structure of the Bondi flux is *nongeneric*: it changes dramatically as soon as we relinquish the hypothesis of zero angular momentum in the incoming material.

This problem has originated many discussions in the literature, accompanied by not entirely consistent simulations. The situation has not been cleared up yet, but we notice that the most relevant simulations are those by Proga et al. (2003a,b), which start from initial conditions endowed with a small angular momentum and low magnetic fields. They completely neglect any radiative loss. The results of the simulations depend significantly on time: initially, matter close to the rotation axis falls on the compact object, because it has very little angular momentum, and the gravitational force is not balanced. After an initial transient, the situation tends toward a stationary state, which should be described (but it is not) by the Bondi solution.

In this problem, there is a critical amount of angular momentum: it is the angular momentum per mass unit (sometimes called *specific angular momentum*) of the smallest stable circular orbit around a compact object. Here we shall call it $l_{\rm mso}$, and we shall discuss it in appendix B. The initial distribution of specific angular momentum is such that the dynamic importance of rotation is negligible ($GM/R^2 \gg \omega^2 R$), at least initially. Despite this, some matter has angular momentum $>l_{\rm mso}$ and therefore cannot directly fall on the compact object; it must form a thick accretion disk. However, matter near the rotation axis has low angular momentum and may collapse directly onto the compact object. This transient phenomenon goes on for a while until the thick accretion disk initiates a strong wind along the rotation axis, which not only blocks the accretion of further low-angular-momentum matter that has not fallen on the star yet, but also reverses the direction of motion of accreting material. After a while, a stationary situation is reached, with strong fluctuations on several different time scales.

This outflow of matter along the polar axis reduces significantly the accretion rate with respect to Bondi's value. It is a purely magnetohydrodynamic effect; purely hydrodynamic simulations present much weaker winds, as well as mass accretion rates closer to the Bondi one, even though still a bit lower than that. Proga and Begelman (2003b) clearly identify the mechanism amplifying the magnetic field, namely the magnetorotational instability, which we shall discuss in the next chapter. The magnetohydrodynamic generation of the wind is not similar to the models of Blandford (1976), Lovelace (1976), and Blandford and Payne (1982), which all require a large–scale ordered magnetic field, which these simulations do not possess.

Simulations give clear and unambiguous results as regards the flow structure and the reduction of accretion with respect to the Bondi case. However, they are much less precise as far as quantitative estimates are concerned. The accretion rate of these simulations has been parameterized as follows:

$$\dot{M} = \left(\frac{r_{\rm in}}{r_{\rm a}}\right)^p \dot{M}_{\rm B} \qquad (5.31)$$

where r_{in} is an internal radius $\ll r_{\rm a}$, and the spectral index varies in the interval $0 \leq p \leq 1$ (Blandford and Begelman 1999). On the basis of currently available simulations, it is impossible to go any further. As a consequence, we do not exactly know, as yet, the reduction factor of the accretion rate with respect to Bondi's.

5.5   The Eddington Luminosity

One of the fundamental ideas of accretion is that the luminosity produced by the object on which accretion takes place, or in its immediate surroundings, can affect the amount of matter that falls on the compact object. The easiest way to visualize this effect consists of considering a spherical accretion flow like the one we have just described, with matter composed of completely ionized gas. In this case, the main interaction between the photons produced by the compact object and the accreting matter is through scattering. Let us consider a source with photons of energy $\lesssim 100$ keV, and Newtonian accretion velocity, so that scattering can be treated in the Thomson limit.

In order to visualize what happens, imagine riding astride an electron. You see a flux of photons coming toward you, all moving in the radial direction. Scattering has the usual back/front symmetry, equation 3.114; as many photons are scattered in the initial direction, as in the opposite one. However, this was not true for the initial photons, which were moving outward. It follows that in the electron's reference frame, scattered photons have no net momentum along the radial direction, but the prescattering ones do. The difference between the two is the usual *Compton drag force*, which slows down the electron in its radial fall.

If the source has a luminosity L, the momentum carried by photons per unit time is L/c; at a distance r from the source, the momentum per unit time and surface is $L/4\pi r^2 c$. Each electron absorbs a quantity $\sigma_T L/4\pi r^2 c$ of this, and if there are n_e photons per unit of volume, the total (wholly radial!) momentum absorbed per unit time (namely, the force) will be

$$F_r = n_e \sigma_T \frac{L}{4\pi r^2 c} \tag{5.32}$$

For this expression to be correct, the fraction of solid angle covered by the cross sections of all electrons must be small with respect to the total, 4π: otherwise, the photons would all be scattered by the elctrons closest to the compact object, while all the distant ones would be left unaffected. The fraction of sky covered by electrons located between r and $r + r$ is obviously

$$d\Omega_T = \frac{4\pi r^2 n_e(r)\sigma_T dr}{4\pi r^2} \qquad (5.33)$$

whence we can see that the total fraction of sky covered is

$$\Omega_T = \sigma_T \int^\infty n_e(r) dr \equiv \tau_T \qquad (5.34)$$

Indeed, Ω_T is the total optical depth for Thomson scattering, τ_T. As a consequence, this computation is valid only in the limit $\tau_T \ll 1$. We shall soon see how to generalize it in the opposite limit.

This force, exerted on all electrons, must be compared with the gravitational force. If we call $\mu_e m_p$ the mass per electron (we shall soon explain the reason for this unusual definition), we find that the gravitational force on the electrons in a unit volume is

$$F_g = n_e \mu_e m_P \frac{GM}{r^2} \qquad (5.35)$$

Which one is strongest: the repulsive radiative force, or the attractive gravitational force? Comparing the two equations above, we find that both forces scale with the same power of radius, so that if one is larger at a given radius, it is larger everywhere. Equality holds when luminosity reaches a value L_E, called the *Eddington luminosity*:

$$L_E \equiv \frac{4\pi c G M \mu_e m_p}{\sigma_T} = \mu_e 1.5 \times 10^{38} \frac{M}{M_\odot} \text{ erg s}^{-1} \qquad (5.36)$$

When $L > L_E$, the radiation pressure force exceeds gravity. We have written $\mu_e m_p$ for the electron's mass because the Compton drag acts (almost) only on electrons, and the gravitational force (almost) only on protons, but electrons and protons are tightly coupled by the electrostatic force, which prevents them from going their separate ways. The Compton drag and the gravitational force act simultaneously on each electron-proton pair. In this way, it is easy to see that the electron behaves as if it had a mass $m_e + m_p$, so that the mass per electron is $\mu_e = (m_e + m_p)/m_p \approx 1$.

In a sense, you have already seen the Eddington luminosity in action in elementary courses of stellar evolution: in fact, there are no stars with masses $\gtrsim 100\ M_\odot$. The luminosity of main sequence stars is approximately given by

$$L \approx L_\odot \left(\frac{M}{M_\odot} \right)^{3.5} \tag{5.37}$$

This exceeds the Eddington luminosity for $M \gtrsim 70\ M_\odot$. In this case, the radiation pressure on surface layers of the star exceeds the gravitational force: luminosity destroys the star. Thus, stars more massive than $\approx 70 - 100\ M_\odot$ cannot exist because they exceed the Eddington luminosity.

Another consequence of the existence of the Eddington luminosity is that if the luminosity is produced by accretion, there will be an upper limit (L_E) to the luminosity of stationary (i.e., not explosive) sources. Indeed, if $L \gg L_E$, the radiation pressure would halt the accretion flow, which is the very fuel that generates the luminosity (according to ways that we shall see in chapter 6). For example, the galactic X-ray sources satisfy $L \lesssim L_E$. On the other hand, sources such as supernovae have $L \gg L_E$, which is acceptable because we do know that supernovae are the products of the (near) total disintegration of a massive star; the source is not stationary.

As shown in appendix B, the efficiency η with which radiation is generated from a flow of mass rate $\dot{M}$, defined as

$$L \equiv \eta \dot{M} c^2 \tag{5.38}$$

is estimated to be in the interval $0.06 \leq \eta \leq 0.42$. If we assume a cautious value, $\eta = 0.1$, we can estimate the quantity of mass per unit time $\dot{M}_E$ that must be accreted on a compact source to produce the Eddington luminosity:

$$\dot{M}_{\mathrm{E}} \equiv \frac{L_{\mathrm{E}}}{\eta c^2} = 2.5 \times 10^{-8} \frac{M}{M_\odot} M_\odot \frac{0.1}{\eta} \, \mathrm{yr}^{-1} \tag{5.39}$$

At this point, we have an obvious question: If I increase $\dot{M}$ well beyond $\dot{M}_{\mathrm{E}}$, why may I not violate the Eddington limit? In this case, I could have $\tau_{\mathrm{T}} \gg 1$, thus contradicting one of the hypotheses according to which we have derived the Eddington luminosity.

The answer is discussed here, *in spherical symmetry*, because there are configurations of disklike shape that violate this limit (Abramowicz, Calvani, and Nobili 1980, chapter 7). First, we have cheated, because the efficiency given by η can be reached when we have disk accretion. There is no certainty that this high efficiency applies also to spherical symmetry. In particular, we shall soon show that $\eta \ll 0.1$ in spherical symmetry. However, even if we neglect this fact, we have, second, that even when a large luminosity is generated, the optical depth is so huge that photons cannot reach infinity; they are de facto trapped. What happens is that $\dot{M} \gg \dot{M}_{\mathrm{E}}$, but from this it does not follow that $L \gg L_{\mathrm{E}}$.

The heart of the matter is that when $\dot{M}$ is very large, the optical depth for scattering may become very large too, so that photons cannot get out. In order to be definite, we shall consider the region $r \ll r_{\mathrm{a}}$, in which the gas is in free fall. So we have

$$v_{\mathrm{r}} \approx \left(\frac{2GM}{r} \right)^{1/2} \tag{5.40}$$

and, from the mass flux conservation, equation 5.16,

$$\rho = \frac{\dot{M}}{4\pi r^2 v_{\mathrm{r}}} \approx \frac{\dot{M}}{4\pi r^{3/2}\sqrt{2GM}} \tag{5.41}$$

The total optical depth for Thomson scattering from a point at a distance r to infinity is given by

$$\tau_{\mathrm{T}} = \int_{\mathrm{r}}^{\infty} \frac{\rho \sigma_{\mathrm{T}}}{\mu_{\mathrm{e}} m_{\mathrm{p}}} dr \tag{5.42}$$

We should write it in the following form:

$$\tau_T = \left(\frac{2GM}{c^2 r}\right)^{1/2} \frac{\dot{M}}{\dot{M}_E} \tag{5.43}$$

The first term in parenthesis represents the radius in Schwarzschild units, raised to the $-1/2$ power. Obviously, we expect that most luminosity is produced in the innermost zones of the accretion flow, namely, where r is a few times the Schwarzschild radius. Therefore, in those zones, as soon as $\dot{M} \gtrsim \dot{M}_E$, the flux becomes opaque.

The consequences of this fact can be seen as follows. When the optical depth is large, photons cannot escape and are advected by the flow toward the compact object and none of these photons will ever go back to the observer. The reason for this phenomenon is the following. The time necessary for a fluid element to fall on the compact object is obviously given, in order of magnitude, by $t_a \approx r/v_r = \sqrt{r^3/2GM}$. In order to escape from a region of dimension r and optical depth $\tau_T \gg 1$, it takes a photon a time $t_d \approx r\tau_T/c$. The condition for the photon to remain trapped is $t_d \gtrsim t_a$, which can now be rewritten as

$$\tau_T \gtrsim \left(\frac{rc^2}{2GM}\right)^{1/2} \tag{5.44}$$

Comparing the two equations above, we can see that the photons are advected by the matter flow inside the black hole when they are in the region

$$r \lesssim \frac{2GM}{c^2} \frac{\dot{M}}{\dot{M}_E} \tag{5.45}$$

Therefore, the necessary condition for the photons' advection is

$$\dot{M} > \dot{M}_E \tag{5.46}$$

Obviously, most of the luminosity is produced at small radii, where the temperature is largest. Therefore, when $\dot{M}$ grows, as soon as the above condition is realized, the luminosity reaching the observer at infinity stops growing. So, $L < L_E$.

5.6 The Efficiency of Spherical Accretion

Since we have just derived the density of spherical accretion, we can now calculate its total luminosity, in an approximate manner. In order to do this, we specify that we are considering the accretion on a black hole. This implies that there can nowhere be a stationary shock wave. The reason is that since the black hole has no real surface (the horizon of events is not a physical surface, but only the locus of points separating the region that can send us signals, from the one that cannot do it), there can nowhere be the large amount of matter, almost at rest, producing the pressure necessary to block the accretion flow. Therefore, the solution will be everywhere the Bondi solution (apart from general-relativistic corrections, which are, however, negligible; see later on).

In this model, there are no nonthermal populations of electrons, because, obviously, there are no shock waves to accelerate them. This implies that neither the inverse Compton nor the synchrotron processes are possible radiative mechanisms. There is only Bremsstrahlung left (eq. 3.41), which we integrate in an approximate way on frequency, assuming the Gaunt factor to be constant with frequency. We thus obtain

$$\epsilon_{\rm br} = 1.4 \times 10^{-27} n_e n_i \bar{g}_{\rm br} T^{1/2} \text{ erg s}^{-1} \text{ cm}^{-3} \qquad (5.47)$$

Let us consider a flow with $\gamma = 4/3$, for which $T \propto \rho^{1/3}$. On the other hand, from equation 5.41, we find $\rho \propto r^{-3/2}$, so that $T \propto r^{-1/2}$. We obtain, scaling with the value of the Schwarzschild radius, $R_{\rm s} \equiv 2GM/c^2$:

$$T = T_{\rm o} \left(\frac{R_{\rm s}}{r} \right)^{1/2} \qquad (5.48)$$

In order to identify $T_{\rm o}$, we argue as follows: the sound speed, $\sqrt{\gamma kT/m_{\rm p}}$, must equal the accretion velocity, equation 5.40, at the accretion radius, $r_{\rm a} = 2GM/c_\infty^2$, where we now indicate with c_∞ the sound speed at infinity. Taking $T_\infty = 10^4$ K, an absolutely standard value, we find

$$T_{\rm o} \approx \left(\frac{m_{\rm p}c^2 T_\infty}{\gamma k}\right)^{1/2} \approx 2 \times 10^8 \text{ K} \qquad (5.49)$$

The luminosity is

$$L = 4\pi \int_{R_{\rm s}}^\infty \epsilon_{\rm br} r^2 dr \qquad (5.50)$$

Putting together all the preceding equations, we find

$$\frac{L}{L_{\rm E}} \approx 10^{-9} \left(\frac{\dot{M}}{\dot{M_{\rm E}}}\right)^2 \qquad (5.51)$$

in agreement with a more accurate computation by Shapiro (1973), who, however, uses $\gamma = 5/3$. Even for this value of γ, he obtains a transonic solution because radiative losses are explicitly included, which explains the difference with respect to Bondi's purely adiabatic solution. The computation for accretion on a black hole with galactic dimensions $M \gtrsim 10^8 M_\odot$ gives a slightly less extreme result: $L/L_{\rm E} \approx 10^{-4}$.

This shows that the efficiency of spherical accretion is really very small; most potential energy liberated remains in the form of bulk kinetic energy, rather than of internal energy, and as such it cannot be radiated away.

The conclusion would be different if a stationary shock wave could form around the black hole, since it is around shocks that bulk kinetic energy is transformed into internal kinetic energy. Obviously, the problem is that the black hole does not have a hard surface, and thus the necessary counter-pressure cannot be created in order to block the momentum of accreting matter. Attempts to suggest something of this kind have always been received with skepticism.

In the case of neutron stars, the stationary shock wave can instead form because the star has a hard surface and the accreting matter is halted by the surface, transforms its kinetic energy into internal energy, and its pressure increases greatly and generates a shockwave halting the accretion flow. It is easy to see from figure 5.2 how a solution of this kind is obtained: the solution is everywhere of the Bondi type, up to a point B inside the sonic radius (why?); at that point, the shock forces the flow to make a transition to one of the subsonic solutions, at small radii: the point C. In this case, the fluid has been heated to high temperatures and can emit copiously. This case, however, is of more academic than practical interest: neutron stars rarely have spherical accretion flows and very often have such intense magnetic fields that their dynamic effect cannot be neglected (see chapter 7). The only real exception seems to consist of old and isolated neutron stars accreting from the interstellar medium; in this case, the magnetic field has probably decayed to dynamically negligible values.

5.7   Explosive Motions

There is, in astrophysics, a large number of phenomena that lead stars and galaxies to lose mass, and consequently energy, in the form of wind. In elementary courses on astronomy, one becomes acquainted with stellar winds, planetary nebulae, galactic fountains, and many different types of jets. These phenomena take place in stationary conditions and have a relatively low energy, so they lie outside the scope of this book. On the other hand, in the universe there are also bona fide explosions, such as supernovae (SNe) and gamma ray bursts (GRB), which deeply alter the nature of the object that generates them. In this section, we shall study the hydrodynamic properties of these explosions, leaving out the discussion of the related radiative processes, which, unfortunately, are less well known. Other phenomena, such as novae, dwarf novae, and type I and II bursts from neutron stars, possess some features similar to SNe, but they will not be discussed separately.

The fundamental feature distinguishing SNe from GRBs is the expansion speed: SNe are Newtonian, with $V \lesssim 60,000$ km s^{-1}, whereas GRBs are hyperrelativistic, with

Lorentz factors $\gamma \gtrsim 100$. Hence we shall present the two models separately.

5.7.1 Supernovae

A SN explosion (surely of type II, and probably also of type Ib and Ic, but not of type Ia) takes place when the core of a massive star has exhausted its nuclear fuel and can no longer resist its own gravitational attraction. The star core collapses very quickly,[1] until it forms a proto–neutron star (PNS), while the outer layers of the star, which collapse more slowly, collide against the PNS surface, which halts their fall and reverses their direction of motion. This generates a shock wave moving outward, together with matter immediately after the shock (i.e., closer to the center of the star), which already moves in a supersonic way in comparison with the outer parts of the star. Therefore, so to speak, SN explosions are supersonic at their very outset, and the Bondi solutions for expanding motions, described above, are not applicable because they do not possess a shock wave, and matter passes with continuity from the subsonic to the supersonic phase.

Please note that at the time of writing, while there is agreement that this scenario correctly describes a SN explosion, no simulation has ever succeeded in producing an explosion: the shock expansion stalls, and the explosion fizzles away (Mezzacappa 2005). Probably, some fundamental detail is still missing; perhaps the missing detail is also at the root of the other well-known problem of PNS formation, namely, that these explosions must be really asymmetric to generate the very large speeds observed in so many radio pulsars; the largest peculiar velocities of known pulsars are ≈ 1200 km s^{-1}. For this phenomenon too, there is no explanation as yet.

Models for the expansion of the SN into the circum-stellar environment neglect this incompleteness, assuming that an

---

[1] In the absence of pressure, the characteristic collapse time is $\approx (G\rho)^{-1/2}$, so that the denser the layers, the quicker their collapse.

explosion of suitable energy *does* take place. Since the duration of the explosion is very short compared with typical hydrodynamic expansion times (tens of seconds vs. days) and the star where the explosion takes place is small with respect to observable dimensions (10^{11} cm vs. 10^{16} cm in order of magnitude), it is conventional to assume an instantaneous release of energy in a negligibly small volume; spherical symmetry is also usually assumed, for lack of a good model for the asymmetry of the explosion, as mentioned above. We speak therefore of pointlike, instantaneous explosions. Let us consider then an explosion releasing a mass M_{ej} endowed with a kinetic energy E; this explosion takes place within an interstellar medium of density ρ. The hydrodynamics of SN explosions may be divided into three distinct stages: free expansion, the adiabatic (or Sedov) phase, and the snowplough phase.

At least initially, until the total mass of the interstellar medium that has been swept up, M_s is much lower than the ejected mass M_{ej}, the expansion must proceed freely, with a velocity V given by $E \approx M_{ej}V^2/2$. In these instants, a shock wave must be formed, because, obviously, for typical stellar parameters, the expansion velocity $V \approx 10,000$ km s^{-1} is much larger than the sound speed in the interstellar medium: for typical temperatures, $T \approx 10^4$ K, $c_s \approx 10$ km s^{-1}. The shock wave is obviously collisionless, as discussed in chapter 1; it is followed by the matter that has crossed the shock wave, and this, in its turn, is followed by matter ejected by the SN. These two components must obviously be in pressure equilibrium, but there is no reason why they should have identical thermodynamic quantities. Therefore, the surface of separation is a contact discontinuity that as discussed in chapter 2, is absolutely unstable. As a consequence, the material that passed through the shock, and the ejected material, will give origin to strong turbulent motions, which mix them up.

The phase of free expansion ends when the mass of interstellar medium that has been swept up becomes comparable to the mass ejected by the SN, namely, when

$$\frac{4\pi}{3}\rho R_{\rm S}^3 = M_{\rm ej} \tag{5.52}$$

where $R_{\rm S}$ is the value of the shock radius $R_{\rm s}$ when the equality holds. For typical values, $M_{\rm ej} = 10\ M_\odot$, and $\rho = nm_{\rm p} = 1 \times 10^{-24}$ g, we find $R_{\rm s} = 4$ pc, and $t_{\rm S} = R_{\rm S}/V = 300$ yr. Very slow SNe, like the Crab ($V = 900$ km s^{-1}), have not left the phase of free expansion yet, despite an age of 1000 yr. In this new phase, the inertia of matter swept up by the SN cannot be neglected, as in the phase of free expansion. A further circumstance helps us: the newly shocked gas is so hot that it cannot radiate its internal energy quickly enough.

In order to see this, consider the temperature of electrons, assuming that in the shock wave they do not exchange energy with the protons; as a consequence, their temperature after the shock is given, approximately, by $kT_{\rm e} \approx m_{\rm e}V^2/2$, namely, $T \approx 10^7$ K. For these temperatures and densities 4ρ (appropriate for a strong, adiabatic shock wave in a cold medium), the cooling time is given by (eq. 3.43) $t_{\rm c} \approx 10^6$ yr, which is a much longer time than the age of the SN remnant, $t_{\rm S}$, in this phase. It follows that we can idealize the SN expansion, in this new phase, as adiabatic: the total energy is conserved. We have already seen in chapter 1 that when the total mass is dominated by the mass swept up by the shock wave, the adiabatic solution is given by the Sedov solution. Therefore, as time passes and $t \gg t_{\rm S}$, the expansion approaches the Sedov solution already studied. We know that in this limit we have:

$$R_{\rm s} = \left(\frac{2.02Et^2}{\rho}\right)^{1/5} = 0.26\,{\rm pc}\left(\frac{E}{4\times 10^{50}\ {\rm erg}}\left(\frac{t}{1\ {\rm yr}}\right)^2 \frac{1\ {\rm cm}^{-3}}{n}\right)^{1/5}$$

$$\tag{5.53}$$

$$T_{\rm s} = \frac{3}{100k}\left(\frac{2.02Et^{-3}}{\rho}\right)^{2/5} = 1.5\times 10^{11}\ {\rm K}\left(\left(\frac{1\ {\rm yr}}{t}\right)^3 \frac{1\ {\rm cm}^{-3}}{n}\right)^{2/5}$$

$$\tag{5.54}$$

We should notice that the temperature behind the shock decreases as time passes: the shock slows down with time ($V_s \propto t^{-3/5}$), and therefore the temperature at which the material is heated immediately after the shock decreases. On the other hand, if we look at figure 1.2, we can clearly see that most mass is confined in a thin layer; in fact, half of the total mass is to be found in a thin layer $0.06R_s$ wide. Exact computations (Chevalier 1974) confirm this analysis.

Notice also that according to figure 1.2, there is a zone of low density and high temperature at the center of the SN remnant (SNR), so that the pressure at the center of the cloud is finite. This pressure is responsible for an apparent paradox. Indeed, in the Sedov expansion, kinetic energy is conserved, whereas the mass increases; this implies that the total momentum increases with time. The reason is simply that the central zones, which have a nonzero pressure, push the thin corona ($\approx 0.06R_s$), which contains most of the mass, and increase its momentum.

There is another phenomenon that we briefly describe here, which is due to the fact that the SNR has not always been expanding according to Sedov's law. In the transition from the phase of free expansion to the Sedov phase, the material starts slowing down. This happens through obvious hydrodynamic processes: a wave propagates back from the shock toward the center of the explosion and slows down the expansion of the material. Notice that the material behind the shock moves subsonically with respect to the shock (see chapter 1), so that the slowdown can be mediated by a wave with a noninfinitesimal amplitude. As we already know, waves with finite amplitude are likely to steepen into a shock wave. We call this a *reverse shock*. The SNR structure is thus composed: the central region is in free expansion, then the *reverse shock* separates the material in free expansion from the one that is already moving according to the Sedov solution (i.e., slowing down), and finally, after the shock, the unperturbed matter that has not been reached by the explosion yet. Notice that while the *reverse shock* moves backward, in the sense that it propagates toward the innermost layers of mass, it is not obvious that it moves geometrically toward the center of the cloud; indeed, it is, at least partly, dragged by the overall motion of ejected matter, with a velocity that can be obtained from numerical solutions.

The Sedov solution becomes inaccurate when radiative losses become important, and the solution loses its adiabatic character. This happens when the gas temperature decreases below $\approx 10^6$ K, because the most abundant *metal* ions, $C, N,$ and O, start acquiring bound electrons and can thus emit radiation in the prohibited and semiprohibited lines described in chapter 3. Using the formulae given above, the temperature decreases below 10^6 K after a time $t_{SP} \approx 2 \times 10^4$ yr; at that point, electrons have the same temperature as protons, because the equipartition time, equation 2.100, has become short enough, ≈ 500 yr. The radius reached at time t_{SP} is about $R_{SP} \approx 15$ pc.

In order to describe the expansion in this new phase, we make the following remark. When radiative losses become important, the SNR expansion can no longer follow the Sedov solution, which assumes $E = $ constant, but it must still conserve linear momentum. The evolution of the shock radius can then be derived from the law of momentum conservation:

$$MV_s = \frac{4\pi}{3} R_s^3 \rho V_s = P_0 \qquad (5.55)$$

which immediately gives

$$R_s \propto t^{1/4} , \quad V_s \propto t^{-3/4} \qquad (5.56)$$

Almost all identified SNR are far from reaching this phase, also because it is highly idealized; indeed, we completely neglect in this approximation the pressure of the gas inside the SNR cavity, which, on the other hand, keeps pushing and increasing the total momentum, just as happens during the Sedov phase. Obviously, sooner or later, even this very hot material will have to cool down, but, because of the very low densities and very high temperatures inside the cavity, this will happen much more slowly than for the material immediately after the shock. Numerical simulations (Mansfield and Salpeter 1974) show that after the Sedov phase, in which $R_s \propto t^{2/5}$, and before the radiative phase, in which $R_s \propto t^{1/4}$, there is a long phase in which $R_s \propto t^{0.31}$ between the two, exactly due to the pressure of the hot central zones.

This third and last phase of the expansion is often called *snowplough* because most matter is accumulated in a very thin layer immediately after the shock, as we now show. In the radiative phase, the cooling time of the material t_c is by definition short with respect to the hydrodynamic expansion time R_s/V_s; as a consequence, although the shock increases the temperature of the swept-up material, T returns very quickly toward the value it had before being run over by the shock. This being the case, we can imagine that the cooling layer of length $\approx V_s t_c \ll R_s$ is contained within the shock width, so that we can apply Rankine-Hugoniot conditions to the whole layer, even to the zone in which the cooling takes place, as shown in figure 5.3.

Notice that since the Rankine-Hugoniot conditions require only that neither mass nor momentum be created within the layer, they must obviously be satisfied. Therefore we have

$$\rho_1 u_1 = \rho_2 u_2 \tag{5.57}$$

$$p_1 + \rho_1 u_1^2 = p_2 + \rho_2 u_2^2 \tag{5.58}$$

where, once again, we have chosen as the reference frame the one in which the shock is at rest. We have deliberately omitted the Rankine-Hugoniot condition that assures energy conservation; the reason is that we have applied these conditions to a layer that includes the cooling layer, from which energy is irretrievably lost, thus energy conservation does not apply. To solve the problem, we must, however, provide a third equation that describes this radiative energy loss. But, to keep things easy, we shall simply assume that the radiative

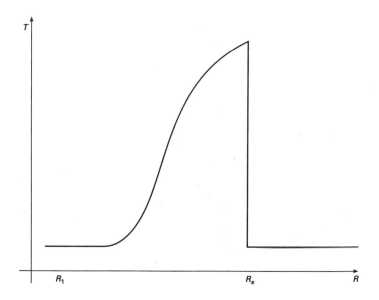

Figure 5.3. Figure illustrating the isothermal shock. The shock is at R_s, but we can apply the conditions of Rankine-Hugoniot between the surfaces $R = R_s$ and $R = R_1$, which is particularly useful because the fluid is isothermal between these two surfaces and has therefore the simple equation of state $p = c_s^2 \rho$, with c_s^2 a constant.

losses are so fast that the gas returns to its initial temperature, $T_2 = T_1$, in which case the equation of state is evidently

$$p = \rho c_s^2 \tag{5.59}$$

with $c_s^2 = kT_1/m_p$ a constant. From these three equations, we easily find that

$$\frac{\rho_2}{\rho_1} = \frac{u_1^2}{c_s^2} = \mathcal{M}^2 \gg 1 \tag{5.60}$$

where the inequality comes from the fact that we are assuming the shock to be strong. It follows that for these shocks, called *isothermal* for obvious reasons, the temperature soon returns to the initial value, but the compression factor is huge, $=\mathcal{M}^2 \gg 1$. At the same time,

$$\frac{u_2}{u_1} = \frac{\rho_1}{\rho_2} = \frac{1}{\mathcal{M}^2} \ll 1 \tag{5.61}$$

Now remember that u_2 is the velocity with respect to the shock; therefore, the fact that it tends to zero implies that the cold material tends to move together with the shock; specifically, it tends to form a layer of dense, cold material, which closely follows the shock wave.

It is now easy to see that most soft X-ray radiation, which is so typical of the SNR, is emitted close to the transition from the Sedov phase to the snowplough phase. In fact, during the Sedov phase the material emits very little energy. Moreover, the radiation (which is obviously thermal) is emitted at temperatures $\gtrsim 10^6$ K, which is the temperature of the transition. On the other hand, during the snowplough phase, because of the slowing down of the shock wave, the material tends to radiate at lower temperatures, in the UV. We get the peak of the soft X-ray emission (≈ 1 keV) just around the transition.

The details of this phenomenon, which we have described here, are confirmed by numerical computations (Chevalier 1974; Mansfield and Salpeter 1974). The most questionable assumption introduced so far is spherical symmetry. The main reason seems to be the lack of uniformity of the medium in which the SN propagates, which is understandable, since massive stars giving origin to SNe lose huge quantities of mass through winds. The presence of instabilities during the Sedov phase has also been suggested; in this case, in fact, a shell of dense matter (the material ejected by the explosion) pushes on a lighter one, namely, the interstellar material that passed through the shock. However, this instability is realized only for adiabatic indices $\gamma \lesssim 1.2$ (Ryu and Vishniac 1987), therefore it is absent for $\gamma = 5/3$. The same kind of instability might instead be present during the isothermal phase of the shock evolution; the linear analysis of the phenomenon (Vishniac and Ryu 1989) shows in fact that this is the case.

This instability leads to the formation of small dense clouds called *clumps* (see the numerical simulations by Chevalier and Theys 1975). Moreover, possible magnetic effects will tend to break the spherical symmetry of the clouds, stretching them in the direction parallel to the field.

5.7.2 Gamma Ray Bursts

The case of SNe, which we have just discussed, typically has a value of the parameter

$$\eta \equiv \frac{E}{Mc^2} \ll 1 \tag{5.62}$$

which simply implies that since the fraction of the total energy E in the form of kinetic energy has a value of <1, the following expansion is Newtonian: $V \ll c$. However, it is interesting to consider also what happens in the opposite limit, namely, when

$$\eta \gg 1 \tag{5.63}$$

Explosions of this kind are called *fireballs*. These models naturally apply to GRBs, phenomena in which observations can be understood only if matter emitting photons moves with Lorentz factors $\gamma \gg 1$.

Before discussing these phenomena, I wish to pause to explain why we need relativistic motions to explain GRBs. There are many different possible explanations, but we choose one that allows us to discuss a seldom used but very useful constrain on transient sources originally attributed to Cavallo and Rees (1978). GRBs are short-lived phenomena at cosmological distances, with typical luminosities $L \approx 10^{51}$ erg s^{-1} (neglecting beaming), which last from 0.01 s to hundreds of seconds. Independent of their duration, they turn on suddenly, on time scales of at most 1 s, and sometimes even much less. This implies rates of variation for their luminosities of the order

$$\frac{\Delta L}{\Delta t} \gtrsim 10^{51} \text{ erg s}^{-2} \tag{5.64}$$

This violates the Cavallo-Rees limit on $\Delta L/\Delta t$, which we now derive. A source of typical size R, containing n baryons and electrons per unit volume, which varies on a time scale Δt with average luminosity ΔL in this small interval, must release a total amount of energy $\Delta L \, \Delta t$ given by

$$\Delta L \, \Delta t = q\frac{4\pi}{3}R^3 n m_{\mathrm p}c^2 \tag{5.65}$$

Here q is the energy extraction efficiency, that is, the ratio between the energy emitted by an average electron and the proton rest energy, $m_{\mathrm p}c^2$. It is clear that under Newtonian conditions, $q \ll 1$; $q = 1$ would imply that protons evaporate completely, which would be possible only in the presence of as many antiprotons, which clearly violates the assumption of Newtonian conditions. On the other hand, the time for photons to leave the source is

$$\Delta t \gtrsim \frac{R}{c}(1 + \tau) \tag{5.66}$$

where $\tau = nR\sigma_{\mathrm T}$ is the Thomson scattering optical depth, and $\sigma_{\mathrm T}$ the usual Thomson cross section; the reason why we did not set an equality but used $\gtrsim$ instead in the previous expression is that other processes may contribute to the overall opacity, besides Thomson scattering, so that the right-hand side is a lower limit to the escape time.

We can now eliminate R, the source size, between the last two expressions and the definition of τ, we find

$$\Delta t \gtrsim \frac{3}{8\pi}\frac{\sigma_{\mathrm T}}{m_{\mathrm p}c^4}\frac{\Delta L}{q}\frac{(1 + \tau)^2}{\tau} \tag{5.67}$$

If we now consider the above expression as a function of τ, we see that it has an *absolute minimum* for $\tau = 1$, so that we obtain in the end the Cavallo-Rees limit:

$$\frac{\Delta L}{\Delta t} \lesssim q\frac{2\pi}{3}\frac{m_{\mathrm p}c^4}{\sigma_{\mathrm T}} = 2q \times 10^{42} \ \mathrm{erg \ s^{-2}} \tag{5.68}$$

which constrains the rate of luminosity variation in any Newtonian source. It is a remarkable limit in that it does not depend upon unobservable quantities like n, R, or τ, but only on directly observable quantities $\Delta L, \Delta t$, and atomic constants. As an example of its application, SNe reach their peak luminosities, $\approx 3 \times 10^{42}$ erg s^{-1}, in about a day or so, which is perfectly consistent with the limit above and with our knowledge that SNe are Newtonian, for physically realistic values of q.

But observations of GRBs show that they violate the Cavallo-Rees limit by 8 orders of magnitude or more; and since our only assumption is that of Newtonian motion, it is this that GRBs must violate. To show how this affects the Cavallo-Rees limit, we have to follow two tracks. On one hand, it is possible that in the reference frame of the material producing the GRB, the Cavallo-Rees limit holds, but this reference frame moves with Lorentz factor $\gamma \gg 1$ with respect to the observer. On the other hand, perhaps *internal* motions in the reference frame of the material responsible for the emission are relativistic, in which case we have no reason to expect $q \ll 1$, and we can instead have $q \gg 1$ because the electrons manage to radiate the part of the proton mass due to its motion, $(\gamma - 1)m_{\mathrm{p}}c^2$, and this may exceed $m_{\mathrm{p}}c^2$ by a large amount.

We shall not push this argument further, leaving it to one of the exercises, at the end of this chapter and proceed instead to the discussion of the fireball model for GRBs.

The Initial Phases

The initial phases in the evolution of a relativistic fluid ($\eta \gg 1$) are more interesting than those of a Newtonian fluid, exposed above.

Since $\eta \gg 1$, we expect expansion at a velocity near that of light. As a consequence, first, there are no gravitational wells deep enough to confine a plasma with these properties, so we shall consider a free plasma; second, we must use relativistic hydrodynamics to describe these phenomena.

Before writing the equations of relativistic hydrodynamics in spherical symmetry, we should discuss whether we can consider a system composed of baryons, electrons (and maybe positrons), and photons as a fluid. We recall that for this to be possible, the mean free path for interactions among the various components must be small with respect to the dimensions of the system. Let us therefore consider a system with as small a radius as possible: $r_0 = 10$ km, corresponding to the radius of a neutron star, or to some Schwarzschild radii of a black hole with a stellar mass. Since the total energies released in the GRBs are of the order of a fraction of the binding energy of a neutron star, $E \gtrsim 10^{51}$ erg, it seems unlikely that these may be released on a much smaller scale. If there were no photons, the energy per particle would be of the order of $Em_p/M = \eta m_p c^2 \gg 1$ GeV; electrons are hyperrelativistic and therefore able to produce pairs through collisions with protons. Once pairs are present, they tend to form photons through processes like $e^+ + e^- \rightarrow \gamma\gamma$, and the energy density in photons quickly tends to thermal equilibrium. With $E = 10^{51}$ erg released in a sphere of radius $r_0 = 10$ km, photons have a temperature $T \approx 1$ MeV and can therefore produce electron-positron pairs. We know from statistical mechanics that in thermal equilibrium, at temperature T with $kT \lesssim m_e c^2$, there will be a pair density:

$$n_\pm = \frac{\sqrt{2}}{\pi^{3/2}} \left(\frac{m_e c}{\hbar}\right)^3 \left(\frac{kT}{m_e c^2}\right)^{3/2} e^{-m_e c^2/kT}$$

$$= 4.4 \times 10^{30} \left(\frac{kT}{m_e c^2}\right)^{3/2} e^{-m_e c^2/kT} \text{ cm}^{-3} \quad (5.69)$$

At energies of this order of magnitude, the main coupling process for electrons and photons is Compton scattering, which, in this regime, has a cross section $\approx \sigma_T$. As a consequence, the optical depth for inverse Compton is

$$\tau_{IC} \approx n_\pm \sigma_T r_0 \approx 3 \times 10^{12} \gg 1 \quad (5.70)$$

It follows that photons and pairs are tightly coupled in the initial phases. On the other hand, protons and electrons are also tightly bound by the usual Coulomb coupling. Therefore, the whole photons-couples-baryons ensemble may be considered as one fluid. Moreover, since photons are trapped inside r_0, we can assume the evolution to be essentially adiabatic, since only photons emitted within a very thin area, close to the surface, can escape (how thin? about r_0/τ; why?).

The fact that $\tau \gg 1$ implies that photons, pairs, and baryons thermalize; that is, they reach the same temperature, given by $kT \approx m_e c^2$. The energy density in photons is given by aT^4, whereas the one in pairs by $n_\pm kT$. It is easy to see that $aT^4 \gg n_\pm kT$, and that the contribution of baryons is also negligible; therefore, photons dominate energy density, and (*a fortiori*), total pressure. Plasma has therefore the equation of state appropriate to the photon gas, $p = e/3$.

We can now take the equations of relativistic hydrodynamics, 1.102 and 1.103, and specialize them to the case of a spherically symmetric explosion, in which case the only components of the four-velocity different from zero are the timelike and radial ones:

$$u^i = (\gamma, \gamma v, 0, 0) \tag{5.71}$$

From now on, we shall use units for which $c = 1$. Here v is the radial velocity of the fluid, and $\gamma^2 = 1/(1 - v^2)$. Also, as discussed above, $p = e/3$. There are obviously only two independent components in equation 1.102, which, together with the equation for the conservation of baryon number, give

$$\frac{\partial}{\partial t}\left((nm_p + 4e/3)\gamma^2 v\right) + \frac{1}{r^2}\frac{\partial}{\partial r}\left(r^2(nm_p + 4e/3)\gamma^2 v^2\right) = -\frac{1}{3}\frac{\partial e}{\partial r} \tag{5.72}$$

$$\frac{\partial}{\partial t}(e^{3/4}\gamma) + \frac{1}{r^2}\frac{\partial}{\partial r}(r^2 e^{3/4}\gamma v) = 0 \tag{5.73}$$

$$\frac{\partial}{\partial t}(n\gamma) + \frac{1}{r^2}\frac{\partial}{\partial r}(r^2 n\gamma v) = 0 \tag{5.74}$$

We should now use, as independent variables, not r, t, but instead $r, s \equiv t - r$, in terms of which this system becomes

$$\frac{1}{r^2}\frac{\partial}{\partial r}\left(r^2(nm_{\rm p} + 4e/3)\gamma^2 v^2\right)$$

$$= -\frac{\partial}{\partial s}\left((nm_{\rm p} + 4e/3)\frac{v}{1+v}\right) + \frac{1}{3}\left(\frac{\partial e}{\partial s} - \frac{\partial e}{\partial r}\right) \quad (5.75)$$

$$\frac{1}{r^2}\frac{\partial}{\partial r}(r^2 e^{3/4}\gamma v) = -\frac{\partial}{\partial s}\left(\frac{e^{3/4}}{\gamma(1+v)}\right) \quad (5.76)$$

$$\frac{1}{r^2}\frac{\partial}{\partial r}(r^2 n\gamma v) = -\frac{\partial}{\partial s}\left(\frac{n}{\gamma(1+v)}\right) \quad (5.77)$$

We are interested in an approximate solution of this system; we expect in fact, as we show below, that the system evolves toward a relativistic expansion, $\gamma \gg 1$, and we are thus interested in a solution that need be accurate only to the most significant terms in $1/\gamma$. Simple physical considerations show that the evolution of this system leads to a relativistic expansion, $\gamma \gg 1$. We neglected gravity, so that no force opposes expansion; we have neglected radiative losses, so that the total energy is conserved; lastly, $E/Mc^2 = \eta \gg 1$: the system tends to a relativistic expansion. When this is the case, we see that all terms on the right-hand side in the above system are of order $\mathcal{O}(1/\gamma^2)$ smaller than those on the left-hand side; therefore, as soon as the expansion has reached $\gamma - 1 \gtrsim \mathcal{O}(1)$, that is, as soon as the expansion starts to be relativistic, we can neglect the terms on the right-hand side. It follows that

$$r^2(nm_{\rm p} + 4e/3)\gamma^2 = \text{constant} \quad (5.78)$$

$$r^2 e^{3/4}\gamma = \text{constant} \quad (5.79)$$

$$r^2 n\gamma = \text{constant} \quad (5.80)$$

where, of course, we have used $v \approx 1$ because $\gamma \gg 1$.

Since we have assumed $\eta = E/Mc^2 \gg 1$, we must also have (at least initially) $e \gg nm_{\rm p}$; in this case, we immediately find from the above equation

$$\gamma \propto r, \quad n \propto r^{-3}, \quad e \propto r^{-4} \quad (5.81)$$

Notice also that since e is mainly due to the photons' contribution, for which $e = aT^4$, we also find

$$T \propto r^{-1} \propto \gamma^{-1} \tag{5.82}$$

This is important, because the apparent temperature of the fluid, which takes into account the fact that the fluid is moving toward the observer, is not just T, but $T\gamma$, and this is a constant; during this phase, $T\gamma = T_0$, the initial temperature, which never changes.

The system's evolution can also be seen this way: it accelerates, the Lorentz factor grows linearly with the distance from the explosion's site, and the densities (of baryons and of energy) are diluted by the expansion. Notice that in this phase, the fluid's kinetic energy grows, whereas the internal one decreases; the system is simply converting internal energy into kinetic energy, thanks to the PdV work done by radiation pressure.

However, since $n \propto r^{-3}$ and $e \propto r^{-4}$, we cannot assume $e \gg nm_{\mathrm{p}}$ indefinitely; a time comes when $nm_{\mathrm{p}} \gg e$. When this happens, the expansion follows a new law, simply given by

$$\gamma = \text{constant}, \quad n \propto r^{-2}, \quad e \propto r^{-8/3} \tag{5.83}$$

In this phase, matter is expanding freely, with a constant Lorentz factor. This phase is analogous to that of free expansion in SNe and is sometimes called *coasting*. Please note that now $T \propto r^{-2/3}$, so that the apparent temperature detected by an observer, $T\gamma$, decreases during this phase of the expansion.

Now we have a reasonable question: when does the transition occur? It would seem that this must take place not later than the time when $\gamma Mc^2 = E$, namely, when $\gamma = \eta$, because at that point, all the radiative energy has been converted into kinetic energy. This happens at a radius r_{a} such as

$$\gamma \approx \frac{r}{r_0} = \eta \Rightarrow r_{\mathrm{a}} = \eta r_0 \tag{5.84}$$

But this is not completely correct; in fact, at a certain moment, $\tau < 1$, and photons are free to escape from the fireball. As this happens, photons stop exerting their pressure on the fluid, which must necessarily stop accelerating. (Actually, photons can even push matter when $\tau < 1$, because of Compton drag, but this effect will be discussed later.)

In order to determine when $\tau = 1$ takes place, we must distinguish two cases: if the density of leptons is dominated by pairs, or by electrons associated with the baryons present in the fireball. Let us assume, for the moment, that the density of leptons is dominated by pairs (eq. 5.69). The optical depth is

$$\tau = \sigma_T n_\pm r \tag{5.85}$$

however, since $n_\pm$ depends exponentially from T, we can see that the time when $\tau = 1$ depends only logarithmically (and thus very weakly!) from all parameters. The temperature T_c at which $\tau = 1$ is approximately given by

$$kT_c \approx 15 \text{ keV} \tag{5.86}$$

which is the temperature at which pairs disappear. From now on, the fluid can no longer accelerate, because the main contribution to the expansion is missing, namely, the radiation pressure: photons are free to expand without collisions. If we assume that the evolution of temperature and Lorentz factor are still given by equation 5.81,

$$T = T_0 \frac{r_0}{r}, \quad \gamma = 1 \times \frac{r}{r_0} \tag{5.87}$$

We see that $T = T_c$ is reached at a radius $r_c = r_0(T_0/T_c)$, where the largest possible Lorentz factor, under these conditions, is reached, $\Gamma_c = T_0/T_c$, given by

$$\Gamma_c = 2400 \left(\frac{E}{10^{51} \text{ erg}} \right)^{1/4} \left(\frac{10^6 \text{ cm}}{r_0} \right)^{3/4} \tag{5.88}$$

A plasma made only of pairs cannot be accelerated to larger Lorentz factors.

However, if there are baryons, with their electrons, plasma may be accelerated to a larger Lorentz factor, because, if electrons associated to baryons dominate the density of leptons at the time when $T = T_c$, the fireball becomes optically thin later; plasma remains optically thick, holds its photons, and keeps accelerating. Since $n \propto r^{-3}$, the optical depth, due to electrons associated to protons, is given by

$$\tau = n_e \sigma_T r = n_p \sigma_T r = n_0 \frac{r_0^3}{r^2} \sigma_T = \frac{3M}{m_p 4\pi r_0^3} \sigma_T \frac{r_0^3}{r^2} \qquad (5.89)$$

and, using the definition of η, the radius r_t at which $\tau = 1$ becomes

$$r_t = \left(\frac{3E\sigma_T}{4\pi m_p c^2 \eta} \right)^{1/2} \qquad (5.90)$$

For this radius to be relevant, we must have $r_t > r_c$, because otherwise the contribution of electrons associated to baryons becomes negligible before the contribution due to pairs disappears. This condition may also be written as

$$\eta < \eta_c \equiv \frac{3E\sigma_T}{4\pi m_p c^2 \Gamma_c^2} = 6.3 \times 10^9 \left(\frac{E}{10^{51} \text{ erg}} \right)^{1/2} \left(\frac{10^6 \text{ cm}}{r_0} \right)^{1/2} \qquad (5.91)$$

Therefore, if $\eta > \eta_c$, we apply the limit Γ_c, due to the pairs only. But if $\eta < \eta_c$, the fluid is still optically thick when $T = T_c$, and keeps accelerating. This acceleration process continues to a radius r_a, or r_t, according to which comes first; in any case, in fact, acceleration must cease. If r_a is reached first, this happens because the total available energy has been exhausted, whereas if r_t is reached first, this happens because the boost due to radiation pressure is missing. The condition $r_t < r_a$ may also be written as

$$\eta > \eta_t \equiv \left(\frac{3E\sigma_T}{4\pi m_p c^2 r_0^2} \right)^{1/3} = 3.3 \times 10^5 \left(\frac{E}{10^{51} \text{ erg}} \right)^{1/3} \left(\frac{10^6 \text{ cm}}{r_0} \right)^{2/3} \qquad (5.92)$$

In this case, the maximum Lorentz factor is given by $\gamma = r_t/r_0$. If, vice versa, $\eta < \eta_t$, the acceleration goes on until the radius r_a, where the whole initial radiative energy has been converted into kinetic energy; in this case, matter is accelerated to a maximum Lorentz factor given by $\gamma = \eta$.

Which is then the highest Lorentz factor up to which matter may be pushed in this way? If $\eta < \eta_t$, $\gamma = \eta$ *grows* as η grows; if we add matter, the fireball goes more quickly, because it can hold photons for a longer time. On the other hand, if $\eta > \eta_t$, the limit of the Lorentz factor $\gamma = r_t/r_0$ diminishes as η grows. There is, therefore, a maximum possible value of γ, which is given by the condition $r_t = r_a$, because, in this case, photons are free to escape just when they have completed their boosting task. The condition $r_t = r_a$ is obviously equivalent to $\eta = \eta_t$, and the maximum Lorentz factor that can be reached, Γ_m, is also given by $\Gamma_m = \eta_t$.

Summing up, for $\eta < \eta_t$, the coasting phase is reached at r_a, when all the initial internal energy has been converted into kinetic energy, in which case the Lorentz factor in the coasting phase is given by $\gamma = \eta$. For $\eta > \eta_t$, the fireball becomes transparent to photons before consuming all the initial internal energy, thus reaching a Lorentz factor, in the coasting phase, of r_t/r_0. However, if $\eta > \eta_c$, the opacity due to the pairs prevails on the one due to the electrons associated to baryons, and a maximum Lorentz factor of Γ_c is reached, which cannot be reduced by the increase of η.

Notice also that if $r_t < r_a$, that is, $\eta > \eta_t$, only a fraction $r_t/(\eta r_0)$ of the total energy has been converted into kinetic energy; therefore, the remaining part, $1 - r_t/(\eta r_0)$, is in the form of photons, which are released at the moment in which the fireball has reached the radius r_t. This energy is obviously thermalized at the temperature $\gamma T = T_0 \approx 1$ MeV; however, since GRBs are never observed to be preceded by thermal components, it follows that probably in nature we can see only cases with $r_t > r_a$, that is, $\eta < \eta_t$. Although we do not understand why this happens, we are delighted to note that

these cases possess a maximum efficiency, where all the initial
internal energy is converted into kinetic energy.

We have so far assumed, in a simplistic manner, that photons cease pushing matter when $\tau < 1$. In fact, this is not
true, because of Compton drag. This effect leads all mod-
els, with any value of η, provided $\eta > \eta_t$, to expand, in the
coasting phase, with a Lorentz factor $\gamma = \eta_t$. The reader not
interested in these details, which are not relevant for realistic
models, can move to the next section.

We have seen in chapter 3 that an electron with a velocity
v with respect to the reference frame in which photons have
a thermal (and therefore isotropic) distribution is subject to
a force given by equation 3.141:

$$m_e \frac{dv}{dt} = -\frac{4}{3}\sigma_T e \frac{v}{c} \qquad (5.93)$$

This force tends to reduce the peculiar velocity of the electron with respect to the reference frame in which photons
are distributed in an isotropic manner, on the time scale
$t_C \approx m_e c/(\sigma_T e)$. However, if we also have protons, we must
take into account that, though they are not directly subject
to the Compton drag, all the same, they are strongly coupled
to electrons by the Coulomb force, so that the braking force
is exerted on the whole electron-proton system. Therefore,
the damping of the peculiar velocity of electrons and protons
does not take place on the time scale t_C, but on t_{fC} given
by

$$t_{fC} = \frac{(m_p + m_e)c}{\sigma_T e} \qquad (5.94)$$

which is longer than t_C by the factor $(m_e + m_p)/m_e \approx$
m_p/m_e.

Let us now assume that $\tau < 1$. Photons tend to expand
freely at a velocity that is necessarily larger than that of
matter. Electrons are left behind and tend to acquire a peculiar velocity v, which we can take as Newtonian ($v \ll c$);
this peculiar velocity can, however, be damped, provided the

time scale on which this occurs is sufficiently short. In the matter's reference frame, what we now see is an isotropic ball of photons, which expands freely (because $\tau < 1$), and some electrons that are not confined and are therefore pushed by pressure to expand freely; but since electrons have mass, they are left behind by photons. The width of the zone in which electrons are found is, in the comoving reference frame, δr, and their expansion time $t_e \approx \delta r/c$. As a consequence, the Compton drag lowers the peculiar velocities of electrons, provided $t_{fC} < t_e$. When this happens, the matter's expansion continues as if the contribution of the radiation pressure had never been missing.

The reason why this happens deserves an explanation. Indeed, the condition $\tau < 1$ implies that the average photon will no longer suffer collisions. However, there are many more photons than electrons; notice that the ratio between the number of photons $n_\gamma \approx aT^4/kT$ and of electrons $n_e = n$, does not depend on the radius and can therefore be calculated at the initial moment. We find

$$\frac{n_\gamma}{n_e} = \frac{Em_p}{kTM} = \eta \frac{m_p c^2}{kT} \approx \eta \frac{m_p}{m_e} \gg 1 \qquad (5.95)$$

Therefore, even though only a small fraction of photons collides with an electron, there are so many photons per electron that the ensuing number of collisions is sufficiently large to allow electrons to keep up with photons.

We can therefore show (Meszaros, Laguna, and Rees 1993) that the acceleration due to photons continues until $\gamma = \Gamma_c$, independently from η, provided $\eta > \eta_t$.

The Next Phases

The more the ejected matter expands in the interstellar medium, the more interstellar matter is swept up by the shock; obviously, when the matter swept up becomes dynamically significant, the expansion cannot continue expansion at constant Lorentz factor. When discussing SNe, we have seen that the deceleration of the shock starts when the amount of matter swept up, M_s, equals the amount of matter ejected, M_e, but in the relativistic case this is not correct. In order to understand what happens, it proves convenient to place one-self in the reference frame of the ejected matter, which moves toward the observer with Lorentz factor γ. In this reference frame, matter is by definition at rest but receives a momentum from the matter swept up by the shock, of the order of $\gamma M_s v \approx \gamma M_s$, which, shared with all the ejected matter, generates a recoil four-velocity $\gamma M_s/(M_s + M_e)$. Obviously, this is small as long as $\gamma M_s \ll M_e$, but we can no longer neglect the slowdown after the time when

$$M_s \approx \frac{M_e}{\gamma} \tag{5.96}$$

which is a much weaker condition than $M_e \approx M_s$. In other words, the shock in the relativistic case leaves the free expansion phase much earlier than in the Newtonian case.

Notice that when the above condition is satisfied, the system has as much energy in the ejected matter, M_e, as in matter swept up from the interstellar medium, M_s. The energy in the ejected matter is almost exclusively kinetic, because the deceleration radius is much larger than the pairs' recombination radius, where $kT \approx 15$ keV; the gas is therefore cold, and $E_e \approx M_e\gamma$. On the other hand, we have seen that the energy density of matter crossing a relativistic shock (eq. 1.113) is given by $\approx \gamma^2 m_p$ per particle. Therefore, $E_s \approx M_s\gamma^2 = M_e\gamma$; the two quantities of total energy are almost equal.

Since, from now on, the total energy is all concentrated in the swept-up matter, for which $M_s \approx \rho R_s^3$ (where ρ is the rest mass density of the interstellar medium, and R_s the shock's instantaneous radius), the total energy becomes $E \approx \rho R_s^3 \gamma^2$, and, if we assume once again that the system does not lose energy appreciably through radiation or otherwise, as in the case of SNe, from the constancy of E we find

$$\gamma \propto (E/\rho)^{1/2} R_s^{-3/2} \tag{5.97}$$

From now on, the dynamic importance of ejected matter decreases until it disappears, so we may apply the asymptotic solution by Blandford and McKee, studied in chapter 1. In this solution, which is valid only for Lorentz factors $\gg 1$, for the shock (Γ), and for all the matter distributed after the shock (γ), there is a self-similarity variable given by

$$\chi \equiv 1 + 8\Gamma^2 \frac{R_s - r}{R_s} \tag{5.98}$$

The quantities after the shock are given by

$$e(r,t) = 2\Gamma^2 e_1 r_1(\xi) \quad \gamma^2(r,t) = \frac{1}{2}\Gamma^2 r_2(\xi) \quad n(r,t) = \sqrt{8}\Gamma n_1 r_3(\xi) \tag{5.99}$$

and the functions r_1, r_2, and r_3 are given by

$$r_1(\chi) = \chi^{-17/12} \quad r_2(\chi) = \chi^{-1} \quad r_3(\chi) = \chi^{-7/4} \tag{5.100}$$

Finally, the dependence of the shock's Lorentz factor, Γ, is given by the law of total energy conservation, equation 1.134:

$$E = \frac{8\pi}{17}\rho_0 c^5 t^3 \Gamma^2 \tag{5.101}$$

In this case too, the shock's slowdown is accompanied by the formation of a *reverse shock*, whose properties are somewhat complicated; for a further discussion of this point, see work by Sari and Piran (1995).

In slowing down, the shock becomes Newtonian, and the evolution must tend to the Sedov solution, because the fluid is adiabatic and Newtonian. There is, however, a subtlety. So far, we have always used, for the equation of state, the relation $p \propto \rho^{5/3}$, suitable for an isentropic, Newtonian fluid. However, the gas behind the shock, when the latter becomes Newtonian, has passed through a relativistic shock, so that its equation of state is different, and probably much closer to $p \propto \rho^{4/3}$, suitable to an isentropic fluid, which is also relativistically hot. The transition from one equation of state to another will take place later on in the evolution of the system.

The Sedov solution for an arbitrary γ has been discussed in chapter 1; it presents no novelties. Notice that the

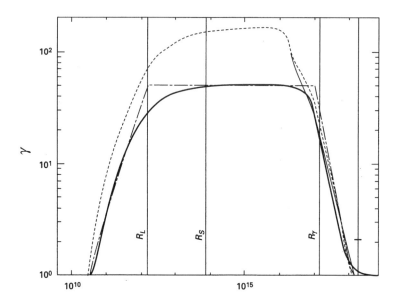

Figure 5.4. Temporal dependence of the Lorentz factor for a fireball. From Kobayashi, Piran, and Sari 1999.

constant q in the relation

$$R_{\rm s}(t) = q \left(\frac{Et^2}{\rho} \right)^{1/5} \qquad (5.102)$$

is $q = 0.99$ for $\gamma = 4/3$.

A numerical solution for the shock's Lorentz factor as a function of time is illustrated in figure 5.4. Here one can easily distinguish the three regimes, namely, acceleration, coasting, and the self-similar solution by Blandford-McKee. For further discussions, especially on the emissivity of the fluid during the different hydrodynamic phases, see the fine article by Piran (2005).

5.8   Problems

1. This is a difficult problem. In the treatment of columnar accretion, we have simply assumed that all fluid particles with a radial velocity lower than the escape velocity fall onto the star. A better treatment of the problem consists in assuming that there is a column of material, subject to the gravitational force but also to the continuous injection of momentum in a radial direction. Neglecting the gas temperature, can you describe the conditions in the column and determine the accretion rate? For the solution, see Bondi and Hoyle (1944).

2. There is also an Eddington luminosity for neutrinos. In order to determine it, you might look up the cross sections for neutrino scattering off electrons, but there is another, simpler way. In fact, the observations of neutrinos from the SN 1987A have shown neutrinos of ≈ 10 MeV, arriving at Earth over a time interval of about 10 s. The neutrinos are produced inside a proto-neutron star of radius $R \approx 10$ km and mass $M \approx 1.4$ $M_\odot$. Knowing that the neutron star emits almost all its binding energy in the form of neutrinos, explain why SNe explode.

3. Let us consider a black hole accreting mass with a fixed efficiency η, generating a luminosity equal to a given fraction ϵ of its Eddington luminosity. Show that the mass grows exponentially and that the time scale on which it grows is exclusively determined by atomic constants and is of order 10^8 yr.

4. The computation of radiative efficiency in spherical symmetry has shown that it varies by about 5 orders of magnitude when we pass from galactic black holes to stellar-mass black holes. Which part of the argument is not invariant through scale transformations?

5. How is the Cavallo-Rees limit (eq. 5.68) modified in the presence of relativistic motions? Do GRBs violate or satisfy this constraint?

6. We have shown that in GRBs, when $\eta < \eta_t$, all the internal energy is converted into kinetic energy. How does this agree with the second principle of thermodynamics, according to which it is impossible to realize a transformation whose only consequence is the complete transformation of heat into work?

Chapter 6

Disk Accretion I

In almost all astrophysical situations, matter has too much angular momentum to fall in a mostly radial direction on a compact object. It is possible to define, in a quantitative way, when the specific angular momentum (i.e., per mass unit) angular is excessive. In appendix B, we show that in general relativity, there is a last stable circular orbit, with the minimum specific angular momentum j_{min} compatible with a circular orbit; it is of the order of $j_{min} = qGM/c$, where the dimensionless parameter q depends on the rotation of the compact object, but, in any case, is $\mathcal{O}(1)$. Therefore, all fluid elements with $j \gtrsim j_{min}$ cannot directly fall on the dead star.

Let us provide two quantitative examples. If the black hole has a galactic mass ($M \gtrsim 10^8 M_\odot$) like those sitting at the center of active galactic nuclei (AGNs), the accreted material comes from the whole galaxy; it will therefore have typical distances of the order of 1 kpc and rotation speed ≈ 100 km s^{-1}. In this case $j \approx 3 \times 10^{28}$ cm^2 s^{-1}, whereas $j_{min} \approx 4 \times 10^{23}$ cm^2 s$^{-1} \ll j$. If, on the other hand, we consider a compact star of approximately solar mass $M \approx 1$ M$_\odot$, in a binary system in which the companion, at a distance $\approx 10^{11}$ cm, transfers mass on the compact star, the orbital velocity is 300 km s^{-1}, so that $j \approx 3 \times 10^{18}$ cm^2 s^{-1}, whereas $j_{min} \approx 4 \times 10^{15}$ cm^2 s$^{-1} \ll j$.

It thus seems reasonable that accreting mass settles into a configuration flattened by rotation. At least initially, the flattening need not be extreme, but, as time passes and gas radiates away its internal energy, it must settle down to a *thin* disk configuration. The reason is simply that while we know processes that carry away the gas energy, we know of no such process capable of ridding the gas of its angular momentum; matter is subject, in fact, only to gravity, which is a central force.

Thus matter must eventually end in a disk perpendicular to the rotation axis. We also know something else about this disk before making any computation: we know that it will be relatively cold, in the sense that the sound speed $c_s \ll v_K$, the orbital velocity, which, in the innermost zones is of the order of the speed of light. It is easy to see this. We know that the disk must have a luminosity comparable, at most, to the Eddington luminosity, L_E, and that most of this luminosity comes from the innermost zones, with a radius $r_S \approx 2GM/c^2$. Therefore, if we assume that the disk emits like a blackbody, we find

$$L = 2\pi r_S^2 \sigma T^4 \approx L_E \qquad (6.1)$$

and, solving for T, we find

$$T = \left(\frac{L_E}{2\pi r_S^2 \sigma} \right)^{1/4} \approx 5 \times 10^7 \text{ K} \left(\frac{M}{M_\odot} \right)^{-1/4} \qquad (6.2)$$

This equation shows that at most, the sound speed is of the order of 10^8 cm s$^{-1} \ll c$, proving that the disk is cold. We also see from this relationship that the larger the compact object, the lower the disk's central temperature.

The condition that $c_s \ll \sqrt{GM/r} \equiv v_K$ is also the condition for the disk to be thin, namely, that its thickness H in the direction z is $H \ll r$. In its turn, this implies that the disk is in an almost perfect Keplerian rotation, $\omega^2 \approx GM/r^3$. These two important points will be demonstrated in section 6.5, but we anticipate them here because of their importance.

In this chapter, we shall study the structure and the properties of thin accretion discs and derive the properties of stationary solutions, discussing also their stability and why they lie in the equatorial plane of the compact object around which they rotate. Further developments of the theory will be treated in the next chapter.

6.1 Qualitative Introduction

Our task consists in investigating the structure and the emissivity of thin cold disks. First of all, the circular orbit, for fixed angular momentum, is the one with least energy. Therefore, material will dissipate first its internal energy while conserving its angular momentum, because of the lack of noncentral forces. Thus matter cannot collapse toward the compact object; as long as the gas has angular momentum, it cannot be accreted.

This difficulty arises from the obvious fact that the gravitational force, the only one we have considered so far, is central, and therefore conserves angular momentum. It follows that disk accretion requires a mechanism changing the fluid's angular momentum. The whole difficulty of disk accretion consists in this, namely, in identifying a force that may change the angular momentum.

So far, we have been careful in speaking only of *change* of angular momentum. The reason is that since the forces that give origin to changes of angular momentum are internal, they cannot change the fluid's total angular momentum, but only force its redistribution. The classical case is friction. Let us consider two rings of matter, almost in centrifugal equilibrium, such that their angular rotation speed ω_K is given by

$$\omega_K^2 r = \frac{GM}{r^2} \tag{6.3}$$

Therefore, the inner ring rotates a little faster than the outer one; the two rings seem to be sliding one on the other. Thus there will be friction forces directed along $\hat{e}_\phi$. The force on the inner ring, $\vec{f}_1$, and the one on the outer ring, $\vec{f}_2$, satisfy $\vec{f}_1 = -\vec{f}_2$ and have the same point of application. Therefore, the total angular momentum does not change; the same quantity of total angular momentum that is *lost* by the inner ring is *gained* by the outer ring. We see therefore that in a disk, angular momentum tends to move outward. At this point, it is probably worthwhile to solve problem 1.

Let us call $\mathcal{G}(r)$ the total torque that all the matter external to the radius r exerts on the matter inside the same radius; in order for the internal matter to fall on the star, it must lose angular momentum, so we must have

$$\mathcal{G}(r) < 0 \qquad (6.4)$$

A ring centered in r, with an infinitesimal thickness dr exerts a torque on the internal matter, which we call $\mathcal{G}(r - dr/2)$; there is a similar torque exerted by the external ring on our ring, $\mathcal{G}(r + dr/2)$. The total torque per unit of thickness (dr) of the disk, exerted on our ring is obviously $\partial\mathcal{G}/\partial r$. It follows that if $\partial\mathcal{G}/\partial r < 0$, the ring suffers a net loss of angular momentum and must spiral toward the center. Therefore, the necessary condition for the disk evolution is

$$\frac{\partial\mathcal{G}}{\partial r} < 0 \qquad (6.5)$$

Both of these conditions must be satisfied for matter to be accreted; we will verify later on that this does occur. For the moment, however, we notice that in this process the internal zones lose angular momentum in favor of the external ones: in the disk, there is a flux of angular momentum per unit time, toward the outside.

The physical image of the accretion from the disk is thus the following: mass flows inside, whereas angular momentum is removed and flows outside. We can now study figure 6.1, which shows the surface density distribution as a function of the radius at different times, in a special accretion disk, which admits an analytical solution; this perfectly illustrates the qualitative evolution of the disk.

We have not so far specified the nature of the force between the two rings, which is now well understood. It is a magnetohydrodynamic effect, as it had been supposed since the original paper describing the theory of thin accretion

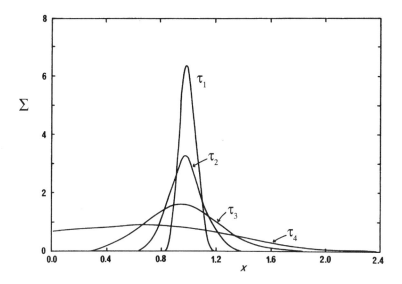

Figure 6.1. Example of an accretion disk with an analytical solution, describing the mass surface distribution at different times. Pringle 1981.

disks by Shakura and Sunyaev (1973). However, since they could not specify the solution in detail, they invented a very useful parametrization of the problem, which is called the α *prescription* for accretion disks.

We shall now proceed this way. First, we shall derive the equations for the accretion disk, without specifying the exact form of the torque. Then, we shall infer some important relations that hold whatever specific form the torque may take. We shall then give the solution for the disk structure under the α *prescription*. Finally, we shall specify the magnetohydrodynamic effect that gives origin to the torque and show how this effect naturally leads to the α prescription for the disks.

6.2 Fundamental Equations

We approximate the disk as infinitely thin and therefore characterized by a mass surface density Σ. Later we shall derive the disk's true thickness, and we shall justify this approximation. The first equation is obviously the mass conservation one,

$$\frac{\partial \Sigma}{\partial t} = -\nabla \cdot (\vec{v}\Sigma) \qquad (6.6)$$

which can be rewritten, taking into account the disk's axial symmetry (the disk therefore has $\partial/\partial\phi = 0$, with obvious meaning), and using cylindrical coordinates, as

$$\frac{\partial \Sigma}{\partial t} + \frac{1}{r}\frac{\partial}{\partial r}(v_r r \Sigma) = 0 \qquad (6.7)$$

In axial symmetry, the equation for momentum conservation becomes an equation for the conservation of the angular momentum along the rotation axis. We could easily derive this equation from those of hydrodynamics in cylindrical coordinates, but we prefer to follow a more intuitive approach.

In order to write this equation, we recall that if $L_z = r^2 \omega \Sigma$ is the density of angular momentum along the axis z per unit area, the equation for its absolute conservation (namely, in the hypothesis that it is neither created nor destroyed) would be

$$\frac{\partial L_z}{\partial t} + \nabla \cdot (\vec{v} L_z) = 0 \qquad (6.8)$$

in agreement with our discussion of the equations of conservation. If we use once again cylindrical coordinates, as well as axial symmetry, this simplifies to

$$\frac{\partial L_z}{\partial t} + \frac{1}{r}\frac{\partial}{\partial r}(v_r r L_z) = 0 \qquad (6.9)$$

However, this equation is *wrong*, because it assumes that the angular momentum within a certain volume may change only as a consequence of inflows or outflows, just like mass or electric charge. But angular momentum can also change because of torques. We have just seen, in the previous section, that the torque on a small ring is $\partial \mathcal{G}/\partial r$, and therefore, on a unit surface the torque is $1/(2\pi r)\,\partial\mathcal{G}/\partial r$. Thus, the correct equation is

$$\frac{\partial L_z}{\partial t} + \frac{1}{r}\frac{\partial}{\partial r}(v_r r L_z) = \frac{1}{2\pi r}\frac{\partial \mathcal{G}}{\partial r} \qquad (6.10)$$

In the stationary case, $\partial/\partial t = 0$, equations 6.7 and 6.10 may easily be integrated. We find

$$\dot{M} = -2\pi v_r r \Sigma \qquad (6.11)$$

where the sign has been chosen so as to make $\dot{M}$ the accretion rate on the compact object; moreover,

$$F_{\rm L} = -r^2 \dot{M}\omega - \mathcal{G}(r) \qquad (6.12)$$

$F_{\rm L}$ is a constant of integration and represents a conserved quantity, exactly like $\dot{M}$. What does it represent? Just like $\dot{M}$ is the mass flux, $F_{\rm L}$ is the flux of angular momentum. This equation tells us that the total flux of angular momentum is given by the sum of two terms. On one side (the first term), there is a flux due to the transport of matter, which is obviously negative; while matter accretes on the compact object, it advects some angular momentum.[1] On the other side, there is a positive flux component due to the torque between disk elements; it removes angular momentum from the inner regions and carries it toward the outer ones. Since we are considering a stationary situation, this flux can only be a constant, as a function of radius and of time, because, in a stationary situation, there can be no accumulation of a physical quantity, except at the origin or at infinity.

What is the value of $F_{\rm L}$? The best way to calculate it is to consider the innermost point of the disk, $r_{\rm m}$. Here the total net flux of angular momentum, $F_{\rm L}$, must end up on the compact object, and there can be no angular momentum flowing

---

[1] This is an *advection* term, namely, a term in which any physical quantity (in this case angular momentum) is carried by the fluid in its motion. The difference between convection and advection is that although there is a transport of a physical quantity in both, in the former there is no net transport of matter, whereas there is in the latter.

outward ($\mathcal{G}(r_\mathrm{m}) = 0$) simply because, by assumption, there is no inner matter from which removal can take place. If the matter falling beyond r_m advected all its angular momentum, then we would have $F_\mathrm{L} = \dot{M}(GMr)^{1/2}$. Let us assume instead that matter manages to advect only a fraction β of its angular momentum. Then we would have

$$F_\mathrm{L} = \beta\dot{M}(GMr)^{1/2} \tag{6.13}$$

Using this expression, equation 6.12 is rewritten as

$$\dot{M}r^2\omega R(r) = -\mathcal{G}(r) \tag{6.14}$$

where the reduction factor $R(r)$ is given by

$$R(r) \equiv 1 - \beta\frac{r_\mathrm{m}^2\omega_\mathrm{m}}{r^2\omega} \tag{6.15}$$

Therefore we see that at large radii $r \gg r_\mathrm{m}$, the two terms, $\mathcal{G}$ and the advection one, balance almost perfectly: matter gets as close as allowed by the removal of the angular momentum.

6.3   Special Relations

Let us consider once again two rings, exerting a reciprocal torque N. The equations of motion for the two rings are

$$\frac{dJ_1}{dt} = -N , \qquad \frac{dJ_2}{dt} = N \tag{6.16}$$

The kinetic energy of the first ring is $\omega_1 J_1/2$, that of the second one is $\omega_2 J_2/2$, so that the variation of kinetic energy is

$$\frac{d}{dt}\left(\frac{\omega_1 J_1 + \omega_2 J_2}{2}\right) = -N(\omega_1 - \omega_2) \approx dr\frac{d\omega}{dr}N < 0 \tag{6.17}$$

In other words, the total energy has decreased. Where did the missing energy, $-drd\omega/drN$, end up? The only possibility, obviously, is the disk's internal energy; therefore, internal torques cause *heating* of the disk. The heating rate, per unit time and surface, is easy to derive from the above:

$$Q = \frac{\mathcal{G}}{2\pi r}\frac{d\omega}{dr} \tag{6.18}$$

In the next paragraph, once we derive equations 6.36 and 6.37, we shall be able to check whether this heating rate is the same as that given by equation 1.52, integrated on the disc's thickness. Now, if we use equation 6.14, we can rewrite it as

$$Q = \frac{\dot{M}\omega^2}{2\pi}\left|\frac{d\ln\omega}{d\ln r}\right|R(r) \tag{6.19}$$

Now, the disk we have in mind is in equilibrium, with the force of gravity balancing the centrifugal force. Neglecting the disk's self-gravity, we have

$$\omega_K^2 \equiv \frac{v_K^2}{r^2} = \frac{GM}{r^3} \tag{6.20}$$

and the local dissipation rate is

$$Q = \frac{3GM\dot{M}}{4\pi r^3}R(r) \tag{6.21}$$

Why do we like these relations? Because they hold no matter what the properties of the internal torque $\mathcal{G}$ are, which we have not specified yet. If we now assume that all the energy dissipated in the internal energy is radiated away, integrating on the whole disk from r_{m} to ∞ we obtain the disk total luminosity as

$$L = \left(\frac{3}{2} - \beta\right)\frac{GM\dot{M}}{r_{\mathrm{m}}} \tag{6.22}$$

For $\beta = 1$, the numerical coefficient is $1/2$. Why is there a $1/2$ factor? Because circular orbits have a kinetic energy, purely due to the angular momentum, which is half their potential energy. If we consider a fluid element starting from infinity with zero total energy, once it arrives at radius r, it cannot have radiated away its whole potential energy at that radius because it has a kinetic energy equal to half the potential one; therefore, its specific total energy (i.e., per mass unit), at the radius r, is $-GM/2r$. The difference between this and the initial energy (≈ 0, because matter is injected into the disk at large distances) is what can be radiated by matter on the disk.

However, the equation above has a paradoxical aspect: when $\beta < 1$, the energy liberated per unit accreted mass *exceeds* $GM/(2r_{\rm m})$, which is the whole binding energy available to us. How can it be? It is because *not* all the kinetic energy of the last orbit falls on the compact object, just an amount $\beta GM/r_{\rm m}$. The remaining kinetic energy, $(1 - \beta)GM/r_{\rm m}$, which is not accreted, must be transformed into heating and thus radiation; the total is $(3/2 - \beta)GM/r_{\rm m}$, as we found above.

We now pause to comment on the physical meaning of β. We have seen that this quantity parameterizes the fraction of angular momentum of the last circular orbit belonging to the disk, which ends up on the compact object. Why not simply take $\beta = 1$? Because we do not know exactly when and how matter on the last circular orbit will fall on the star. While it falls, can it rid itself of some angular momentum? If so, $\beta < 1$, if not, $\beta = 1$. Therefore the quantity β parameterizes our ignorance of what happens in the last instants in which matter spirals down, before being actually swallowed. Problem 2 invites you to think further about this point.

Once we know Q, we can determine the temperature and spectrum of the radiation emitted by the disk, under the assumption that it is in local thermal equilibrium (often called LTE). This assumption is not trivial, namely, it implies that all the heat that has been produced is dissipated locally; we shall see further on, in discussing advection-dominated accretion flows (ADAFs) (section 7.3) the difficulties inherent in this assumption and what happens when it is abandoned. Now, assuming that all the heat is dissipated locally, and taking into account the fact that the disk has two sides, we find

$$2\sigma T^4 = Q \tag{6.23}$$

where $\sigma = ac/4$ is the Stefan-Boltzmann constant. In particular, we find

$$T = \left(\frac{3GM\dot{M}}{8\pi\sigma} \frac{R(r)}{r^3} \right)^{1/4} \tag{6.24}$$

In order of magnitude,

$$T = 8 \times 10^7 \text{ K} \left(\frac{0.1}{\eta} \right)^{1/4} \left(\frac{L}{L_E} \right)^{1/4} \left(\frac{10^{38} \text{ erg s}^{-1}}{L_E} \right)^{1/4} \left(\frac{R(x)}{x^3} \right)^{1/4} \tag{6.25}$$

where

$$x \equiv \frac{r}{r_s} = \frac{rc^2}{2GM} \tag{6.26}$$

is the radius in units of the Schwarzschild radius. For black holes with galactic dimensions, the Eddington luminosity is much larger, and we immediately obtain for the temperature

$$T = 8 \times 10^5 \text{ K} \left(\frac{0.1}{\eta} \right)^{1/4} \left(\frac{L}{L_E} \right)^{1/4} \left(\frac{10^{46} \text{ erg s}^{-1}}{L_E} \right)^{1/4} \left(\frac{R(x)}{x^3} \right)^{1/4} \tag{6.27}$$

Thence we see at once that accretion disks in AGNs emit mainly in the UV, while those around black holes with stellar dimensions reach their peak in the soft X-ray band.

We should notice that because of the presence of the factor $R(x)$, the temperature is always below the value 10^6 K for AGNs, and 10^8 K for compact objects with a stellar mass, see problem 3.

Once we know the temperature (having therefore assumed thermal equilibrium), we also know the spectrum of the disk. In fact, since, at least locally, the spectrum is that of a blackbody, we shall simply superpose the spectrum of many blackbodies. Thus we have, calling ϵ the photon's energy,

$$L_\epsilon = 4\pi r_s^2 \int x dx \frac{2\epsilon^3}{h^3 c^2} \frac{1}{\exp(\epsilon/kT(x)) - 1} \tag{6.28}$$

In order to calculate this integral, we take $T = T_c x^{-3/4}$ (thus neglecting $R(x)$) and define a new integration variable, $\xi \equiv \epsilon/kT(x)$, thus obtaining

$$L_\epsilon = \frac{32\pi}{3} \frac{r_s^2}{h^3 c^2} \epsilon^{1/3} \int_{\xi_{\min}}^{\infty} d\xi \frac{\xi^{5/3}}{e^\xi - 1} \tag{6.29}$$

where

$$\xi_{\min} \equiv \frac{\epsilon}{kT_c} \tag{6.30}$$

The integral has the following asymptotic form:

$$\int_{\xi_{\min}}^{\infty} d\xi \frac{\xi^{5/3}}{e^\xi - 1} \approx \begin{cases} \text{constant} & \xi_{\min} \ll 1 \\ \xi_{\min}^{5/3} e^{-\xi_{\min}} & \xi_{\min} \gg 1 \end{cases} \tag{6.31}$$

We thus see that as long as we consider photons with energy $\epsilon \lesssim kT_c$, the spectrum has the form

$$L_\epsilon \propto \epsilon^{1/3} \tag{6.32}$$

which is a typical signature of accretion disks. On the other hand, when $\epsilon \gg kT_c$, we are looking at the extreme tail of the exponential distribution, and the spectrum takes on the form

$$L_\epsilon \propto \epsilon^2 e^{-\epsilon/kT} \tag{6.33}$$

Let us make two final comments. Equation 6.22 explains why the accretion disk has such a large radiative efficiency. We have seen that in spherical accretion, the total energy of matter remains essentially zero, because the radiative terms are small; basically, all the potential energy released is transformed into the fluid's kinetic energy, not radiation. In disk accretion, on the other hand, one-half of the potential energy released is radiated away, while the other half is required by the conservation of angular momentum. It follows that the total radiative efficiency is given by

$$\eta \equiv \frac{L}{\dot{M}c^2} = \frac{GM}{2r_{\rm m}c^2} = \frac{r_{\rm s}}{2r_{\rm m}} \tag{6.34}$$

where, of course, we have neglected all relativistic effects. Since the radius of the marginally stable orbit is $3r_{\rm s}$, the efficiency thus calculated is $1/6$, whereas an exact computation (appendix B), which includes relativistic effects, gives $\eta = 0.06$ for the nonrotating black hole, and up to $\eta = 0.42$ for the fastest-rotating black hole.

We should, however, notice that not all the internal energy Q need end up in radiation. On the one hand, the disk may generate a wind and use the power Q, or a part of it, to accelerate it. Alternatively, the gas may not succeed in cooling quickly, in which case it heats up while falling on the compact object; the energy released is stored in the gas internal energy. In this case, we talk of advection-dominated accretion flows (ADAFs). We shall soon discuss these possibilities; however, let us note that if just one of them were realized in nature, the disk would radiate less than the amount given by equation 6.22.

6.4   The α Prescription

In order to make further progress, we must specify the properties of the torque $\mathcal{G}$. To this aim, let us follow the classical argument and consider the possibility that it is due to friction. The treatment of friction in hydrodynamics is very simple. There is friction when the fluid has a large velocity gradient: when we rub our hands, the velocity changes significantly along the axis perpendicular to the palms. We can therefore assume that friction depends on the terms $\partial v_i / \partial x_j$, where indexes refer to Cartesian components.

We have seen that in the presence of viscosity, Euler equations are replaced by equation 1.38, where the viscous stress tensor V is given by

$$V_{ij} = \rho \nu \left(\frac{\partial v_i}{\partial x_j} + \frac{\partial v_j}{\partial x_i} - \frac{2}{3} \delta_{ij} \nabla \cdot \vec{v} \right) \qquad (6.35)$$

where the coefficient ν is called *kinematic shear viscosity*, and δ_{ij} is the usual Kronecker delta. (Why have we neglected the second viscosity η?) We easily realize that in axial symmetry, the only component V different from zero is

$$V_{\phi r} = \rho \nu \left(\frac{dv_\phi}{dr} - \frac{v_\phi}{r} \right) = \rho \nu r \frac{d\omega}{dr} \qquad (6.36)$$

Since the quantity $\mathcal{G}$ is the torque between two rings, in order to obtain $\mathcal{G}$ from $V_{\phi r}$ we must integrate on the whole surface separating the two rings:

$$\mathcal{G} = \int r d\phi \int dz \, r V_{\phi r} = 2\pi r^3 \nu \Sigma \frac{d\omega}{dr} \qquad (6.37)$$

This is the expression for the torque due to a viscous force. When the disk is in Keplerian rotation (eq. 6.20), since $d\omega/dr < 0$, we see that $\mathcal{G} < 0$, exactly as needed to carry angular momentum outward.

If we now use the above relation in equations 6.7 and 6.10, using once again stationarity, we find a very useful relation:

$$\frac{d \ln \omega}{d \ln r} = \frac{r v_{\rm r} R(r)}{\nu} \qquad (6.38)$$

whence, for the Keplerian disk,

$$v_{\rm r} = -\frac{3\nu}{2 r R(r)} \qquad (6.39)$$

Later on, we shall need this equation in a slightly different form; if we recall equation 6.11, the above equation becomes

$$\nu \Sigma = \frac{\dot{M}}{3\pi} R(r) \qquad (6.40)$$

If we now put this into equation 6.37 and remember that for a keplerian disk, $\omega \propto r^{-3/2}$, we see that $\partial \mathcal{G}/\partial r < 0$, just like anticipated.

We are now getting to the heart of the matter. Indeed, if we take the coefficients of viscosity for normal materials, we can easily demonstrate that we shall get ridiculously small accretion velocities and, consequently, accretion rates (Shakura and Sunyaev 1973). It follows that the viscosity of accretion disks cannot be the classic atomic one. Shakura and Sunyaev immediately suggested that the nature of viscosity had a magnetohydrodynamic origin but, without a theory, were forced to suggest a parametrization of the problem. They noticed that the stress tensor R_{ij} for an ideal fluid is $R_{ij} = -p\delta_{ij}$, so that in order of magnitude, they postulated that the stress tensor in equation 6.35 is of the same order of magnitude as p. In this way, the coefficient of shear is

$$\nu \approx \frac{p}{\rho dv_{\mathrm{K}}/dr} \approx \frac{c_{\mathrm{s}}^2}{\omega} \qquad (6.41)$$

Thus they defined a parameter α in this way:

$$\nu \equiv \frac{\alpha c_{\mathrm{s}}^2}{\omega} \qquad (6.42)$$

which is the parametrization we shall use from now on. Of course, the parameter α is not specified by the theory; they assumed $\alpha \lesssim 0.1$.

The physical motivation for Shakura and Sunyaev was that the disk might be turbulent. It is a well-known fact that turbulence gives rise to a viscosity (often called *anomalous*) much larger than the usual atomic one. One simple way to see this follows. In the presence of turbulence, there are perturbations of maximum size L and velocity V; these tend to mix up the fluid and to make it more homogeneous, albeit only in a time-averaged sense. However, this effect is similar to that of the stress tensor V_{ij}, because this too tends to smooth the fluid by flattening out velocity gradients. We may then expect turbulence to lead to an effective viscosity of order $\nu \approx LV$. For the maximum velocity, we must take $V \approx c_\mathrm{s}$, because supersonic turbulence generates shock waves and dissipates quickly. As maximum size, we can take the thickness of the disk $L \approx H$, which, as we shall soon show, is $H \approx c_\mathrm{s} r / v_\mathrm{K}$.[2] Putting together these two equalities, we find the above equation.

One of course may wonder why α should be a constant across the whole accretion disk. There is no strong theoretical argument in favor of this constancy; at most, one may invoke the fact that adimensional quantities, in the theory of turbulence, are often constant because of the property of scale-invariance of turbulence. We must thus regard the constancy of α as an educated guess and judge its suitability from the comparison of observations with theoretical predictions.

The problem of this argument is that nobody has ever been able to show that the disk is actually turbulent (in hydrodynamics!). In the literature, various possibilities have been studied—convection (Stone and Balbus 1996), viscosity of photons (Loeb and Laor 1992), self-gravity of the disk (Paczynski 1978)—but none of these is fully convincing. That is why the α parametrization has seemed, at times, a bit optimistic.

6.5   Equations for the Structure of Disks

Equations 6.7 and 6.10 do not contain the gas temperature[3]; our prescription for the disks, equations 6.37 and 6.42, depends on this quantity through c_s. We must therefore determine the disk temperature T, and this is usually done (just think of stellar interiors) by considering energy transport. In other words, we must require that the disk temperature

---

[2]The disk scale height H will be derived without using the α prescription, so there is no circularity in the definition.

[3]Equation 6.24 obviously gives the disc's photospheric temperature, not the temperature inside the disk.

be determined by the equilibrium between the energy produced locally and that lost. The necessary equation is that of radiative transport of heat, since, so far, it has never been demonstrated that the disks are convectively unstable. The heat flux along the z direction is then given by (see section 3.1.1) equation 3.35:

$$F_R(z) = -\frac{ac}{3\alpha_R} \frac{\partial T^4}{\partial z} \qquad (6.43)$$

where α_R is the average coefficient of absorption, called the *Rosseland* coefficient, which gives origin to an optical depth of the disk,

$$\tau = \alpha_R H \qquad (6.44)$$

and H is the thickness (still to be determined) of the disk in the z direction. This procedure is valid only if $\tau \gg 1$, that is, when local thermal equilibrium holds. For the moment, we assume that this condition is satisfied; we shall check it *a posteriori*. In stationary conditions, the whole flux that reaches the surface of the disk, and is lost, must be generated by the heating processes, in particular by equation 6.21:

$$F_R(H) - F_R(0) = \frac{Q}{2} \qquad (6.45)$$

The flux $F_R(0) = 0$ because of symmetry. Replacing derivatives with finite differences, we find that

$$F_R(H) \approx \frac{acT_c^4}{3\tau} \approx \frac{Q}{2} \qquad (6.46)$$

Comparing this equation with equation 6.24, we see that T_c exceeds the photospheric temperature by a modest factor $\tau^{1/4}$, which is $\mathcal{O}(1)$; in other words, the disk is almost isothermal in the z direction.

We must now calculate the thickness H of the disk. The appropriate way to do this consists of considering equation 1.38 in the z direction. We shall write it in a conservative form to illustrate the use of this method, but we shall obtain simple equations. We find

$$\frac{\partial}{\partial t}(\rho v_z) = -\frac{\partial}{\partial x_k}(R_{zk} + V_{zk}) - \rho\frac{\partial\phi}{\partial z} \qquad (6.47)$$

Here, R is, as usual, the Reynolds stress tensor, equation 1.35, whereas V is the viscous one, equation 6.35; the above equation simplifies greatly, for various reasons. First of all, we want a stationary solution; therefore the left-hand side is identically zero. For the same reason, we must have $v_z = 0$; all components of $V_{zi} = 0$ vanish, and the relevant components of the Reynolds tensor simplify: $R_{zi} = p\delta_{zi}$. This leaves

$$\frac{\partial p}{\partial z} = -\rho\frac{\partial\phi}{\partial z} \qquad (6.48)$$

We immediately recognize this as the equation of hydrostatic equilibrium.

We are studying thin disks, and thus we need the gradient of the gravitational potential for small values of z only:

$$-\frac{\partial\phi}{\partial z} = \frac{\partial}{\partial z}\frac{GM}{(r^2 + z^2)^{1/2}} \approx -\frac{GMz}{r^3} \qquad (6.49)$$

where M is only the mass of the compact object, not of the disk, and r is the radius in cylindrical coordinates. We have just shown that the disk is almost isothermal, so that we can take $T(r, z) \approx T(r)$, constant in the vertical direction. In this case, the equation of hydrostatic equilibrium becomes

$$\frac{\partial p}{\partial z} = \frac{kT}{m_p}\frac{\partial\rho}{\partial z} = -\rho\frac{GMz}{r^3} \qquad (6.50)$$

which has the following, very simple, solution:

$$\rho = \rho(r)\sqrt{2/\pi}\exp\left(-z^2/2H^2\right) \qquad (6.51)$$

$$H \equiv \frac{c_s}{v_K}r \qquad (6.52)$$

where we have used the sound speed in the isothermal case, $c_s^2 = p/\rho$. Therefore, the condition that the disk be thin, $H \ll r$, corresponds to requiring

$$c_s \ll v_K \qquad (6.53)$$

or, in other words, the orbital motion of matter in the disk is highly supersonic.

Finally, let us justify the fact that the motion in the disk plane is in almost perfect centrifugal equilibrium. We can take the equation of momentum conservation in the radial direction, which gives us

$$\frac{\partial}{\partial t}(\rho v_r) = -\nabla \cdot (R_{rk} + V_{rk}) - \rho \frac{\partial \phi}{\partial r} \tag{6.54}$$

In order to simplify this equation, let us recall that the left-hand side can be neglected, because we are considering a stationary case. Moreover, on the disk plane, all quantities depend only on r, and the viscous stress tensor disappears. The Reynolds tensor in cylindrical coordinates is easy to calculate:

$$\frac{1}{2}\frac{\partial v_r^2}{\partial r} - \frac{v_\phi^2}{r} = -\frac{1}{\rho}\frac{\partial p}{\partial r} - \frac{GM}{r^2} \tag{6.55}$$

Because of equation 6.53, we can neglect the pressure gradient compared with that of the gravitational potential. Besides, from equations 6.39 and 6.42, we find

$$v_r \approx \frac{\nu}{r} \approx \alpha c_s \frac{H}{r} \ll c_s \tag{6.56}$$

therefore, the first term in equation 6.55 is also smaller than the pressure gradient. As a consequence, the equation reduces to

$$\frac{v_\phi^2}{r} \approx \frac{GM}{r^2} = \frac{v_K^2}{r} \tag{6.57}$$

With a great precision, the orbital velocity, *for thin disks*, is the Keplerian velocity. Please note that we have found the following ordering of velocities:

$$v_\phi \approx v_K \gg c_s \gg v_r \tag{6.58}$$

We can now build a model for the internal structure of the disk in the following *approximate* way. We can take

$$\rho = \frac{\Sigma}{2H}, \quad H = \frac{c_s}{v_K}r \tag{6.59}$$

and the sound speed must be calculated from the equation of state, including radiation pressure:

$$p = \frac{\rho k T_c}{\mu_m m_p} + \frac{a}{3} T_c^4 \tag{6.60}$$

Here T_c is the temperature at the center of the disk, and μ_m the average particle mass, in units of the proton mass, for a solar composition of *metals*, $\mu_m = 0.615$. The above approximation essentially consists of taking the disk structure as almost isothermal, and replacing the derivatives with finite differences.

We thus find the following eight equations for the eight unknown quantities ρ, Σ, H, c_s, p, T_c, τ, and ν, whose solution will be given in terms of the known parameters M, $\dot{M}$, r, α, and $v_K = \sqrt{GM/r}$:

$$\rho = \frac{\Sigma}{2H}$$

$$H = \frac{c_s}{v_K} r$$

$$c_s^2 = \frac{p}{\rho}$$

$$p = \frac{\rho k T_c}{\mu_m m_p} + \frac{a T_c^4}{3}$$

$$\frac{a c T_c^4}{3\tau} = \frac{3 G M \dot{M}}{8\pi r^3} R(r)$$

$$\tau = \alpha_R (T_c, \rho) H$$

$$\nu = \alpha \frac{c_s^2 r}{v_K}$$

$$\nu\Sigma = \frac{\dot{M}}{3\pi} R(r) \tag{6.61}$$

6.6 The Standard Solution

The system of the equations in equation 6.61 is algebraic and can be easily solved (even if this requires a good amount of work) as soon as we identify a suitable form for α_R. The Rosseland coefficient has a good analytical approximation in the form of the so-called Kramers' law:

$$\alpha_R = 6.6 \times 10^{22} \rho^2 T_c^{-7/2} \text{ cm}^{-1} \qquad (6.62)$$

which is valid at low temperatures. At higher temperatures, the average Rosseland coefficient of absorption is dominated by scattering,[4] in which case we find

$$\alpha_T = \sigma_T \frac{\rho}{m_p} = 0.4\rho \text{ cm}^{-1} \qquad (6.63)$$

Let us now consider a zone in which the Rosseland coefficient is given by Kramers' law, equation 6.62, while the pressure is dominated by the gas. For a simpler notation, we define

$$m \equiv \frac{M}{M_\odot}, \qquad \dot{m} \equiv \frac{\dot{M}}{\dot{M}_E}, \qquad x \equiv \frac{r}{r_s} \qquad (6.64)$$

Thus we find

$$\Sigma = 9.2 \times 10^4 \text{ g cm}^{-2} \alpha^{-4/5} \dot{m}^{7/10} m^{1/5} \eta^{-7/10} x^{-3/4} R(x)^{7/10}$$
$$H = 2.1 \times 10^3 \text{ cm } \alpha^{-1/10} \dot{m}^{3/20} m^{9/10} \eta^{-3/20} x^{9/8} R(x)^{3/20}$$
$$\rho = 43.5 \text{ g cm}^{-3} \alpha^{-7/10} \dot{m}^{11/20} m^{-7/10} \eta^{-11/20} x^{-15/8} R^{11/20}$$
$$T_c = 8.1 \times 10^7 \text{ }^\circ K \text{ } \alpha^{-1/5} \dot{m}^{3/10} m^{-1/4} \eta^{-3/10} x^{-3/4} R^{3/10}$$
$$\tau = 58 \alpha^{-4/5} \dot{m}^{1/5} m^{1/5} \eta^{-1/5} R^{1/5}$$
$$v_r = 8.5 \times 10^5 \text{ cm s}^{-1} \alpha^{4/5} \dot{m}^{3/10} m^{-1/5} \eta^{-3/10} x^{-1/4} R(x)^{-7/10}$$
$$(6.65)$$

We can now check, post facto, our assumptions. First of all, the radial flow is highly subsonic: $\mathcal{M} \lesssim 0.01$. Second, the disk is effectively thin: $H/r \approx 0.01 x^{1/8}$, for each reasonable

---

[4]It may seem inconsistent to say that the average absorption is dominated by scattering, but, since the average Rosseland coefficient has been defined in this way for a long time, this slight incoherence is still in use.

value of the accretion rate, $\dot{m} \lesssim 1$; H/r exceeds unity for values of x unrealistically large, $x \approx 10^{16}$. Then, the optical depth $\tau \gg 1$, which justifies the use of equation 3.35. Moreover, all quantities depend only weakly (fortunately for us) on the parameter α, so that the numerical values we have just obtained are reasonably realistic; at least in part, this is what makes the α prescription so useful. Lastly, we remark that we have neglected the disk self-gravity; in problem 4, we show that this is perfectly justified.

The total mass in the disk is not large at all. We easily find that in units of the collapsed object, the disk mass is

$$\frac{M_D}{M} = \frac{2\pi}{M} \int r\Sigma dr = 1\times10^{-10}\alpha^{-4/5}\dot{m}^{7/10}m^{6/5}\eta^{-7/10} \quad (6.66)$$

whence we see that it is always negligibly small, for black holes of stellar origin ($m \approx 1$) and for galactic ones ($m \lesssim 10^8$) (but, in this case, only marginally!). As upper limit for the integral we have used $x_{\text{out}} = 10^6$.

We have assumed that opacity is given by Kramers's law, and that pressure is dominated by the gas. It can be easily shown that the first condition is more restrictive than the second one. In fact, Kramers' absorption coefficient is given by $\alpha_R = \tau/H$, and this must be $\alpha_R > \sigma_T\rho/m_p$, which may be solved with respect to x, thus obtaining $x > x_{\text{mo}}$, where

$$x_{\text{mo}} = 5400\dot{m}^{2/3}m^{-1/3}\eta^{-2/3}R^{2/3} \quad (6.67)$$

while we can check that the ratio between the radiation and the gas pressures is

$$\frac{p_R}{p_g} = 0.3\alpha^{1/10}\dot{m}^{7/10}m^{-1/20}\eta^{-7/10}x^{-3/8}R(x)^{7/20} \quad (6.68)$$

which is certainly $\ll 1$ in the region defined by equation 6.67. This implies that the solution we have just found, equation 6.65, is valid in the external region, with $x > x_{\text{mo}}$.

For the next internal region, we shall take as coefficient of absorption Thomson's, equation 6.63, and, still assuming

$p_R \ll p_g$, we find a new solution (Treves, Maraschi, and Abramowicz 1988):

$$\Sigma = 7.08 \times 10^4 \text{ g cm}^{-2} \alpha^{-4/5} \dot{m}^{3/5} m^{1/5} \eta^{-3/5} x^{-3/5} R(x)^{3/5}$$

$$H = 3.5 \times 10^3 \text{ cm } \alpha^{-1/10} \dot{m}^{1/5} m^{11/10} \eta^{-1/5} x R(x)^{1/5}$$

$$\rho = 10.3 \text{ g cm}^{-3} \alpha^{-7/10} \dot{m}^{2/5} m^{-7/10} \eta^{-2/5} x^{-33/20} R^{2/5}$$

$$T_c = 3.5 \times 10^8 \text{ K } \alpha^{-1/5} \dot{m}^{2/5} m^{-1/5} \eta^{-2/5} x^{-9/10} R^{2/5}$$

$$\tau = 3.1 \times 10^3 \alpha^{-4/5} \dot{m}^{3/5} m^{1/5} \eta^{-3/5} x^{-3/5} R^{3/5}$$

$$v_r = 1.1 \times 10^6 \text{ cm s}^{-1} \alpha^{4/5} \dot{m}^{2/5} m^{-1/5} \eta^{-2/5} x^{-2/5} R(x)^{-3/5}$$

$$(6.69)$$

All comments made above about the outermost solution also apply to this intermediate solution; in other words, the solution is consistent with our assumptions. However, we can see that $p_R = p_g$ at the radius x_{mi},

$$x_{\text{mi}} = 787.6 \alpha^{2/21} \dot{m}^{16/21} m^{2/21} R(x)^{16/21} \qquad (6.70)$$

where our assumption $p_R \ll p_g$ is violated. If we neglect the gas pressure with respect to the radiation pressure, we find

$$\Sigma = 4.2 \text{ g cm}^{-2} \alpha^{-1} \dot{m}^{-1} \eta x^{3/2} R(x)^{-1}$$

$$H = 2.1 \times 10^5 \text{ cm } \dot{m} m \eta^{-1} x R(x)$$

$$\rho = 9.5 \times 10^{-6} \text{ g cm}^{-3} \alpha^{-1} \dot{m}^{-2} m^{-1} \eta^2 x^{3/2} R^{-2}$$

$$T_c = 3.1 \times 10^7 \text{ K } \alpha^{-1/4} m^{-1/4} x^{-3/8}$$

$$\tau = 1.7 \alpha^{-1} \dot{m}^{-1} x^{3/2} \eta R^{-1}$$

$$v_r = 1.5 \times 10^{10} \text{ cm s}^{-1} \alpha \dot{m} \eta^{-1} x^{-5/2} R(x) \qquad (6.71)$$

As you can see in this case, the inequality $\tau \gg 1$, which is necessary in order to use equation 3.35, is *not* satisfied. In this case, the exact configuration of the disk is unknown. It has been supposed, for example, that it may not be thin (Thorne and Price 1975), or that it may contain a two-phase gas (Krolik 1998), with cold, dense clouds in pressure equilibrium with a tenuous, hot medium.

The various zones also have conventional names. The innermost area is called the *A zone*; here electron scattering and radiation dominate opacity and pressure, respectively. The intermediate area is called the *B zone* and is characterized by electron scattering and gas pressure. The external region is called the *C zone*, with Kramers' opacity and the gas pressure.

The alert reader will have noticed that the transition radii are different, depending on whether we use the solution of the external zone or that of the internal one. This is not so important; the true transition radius will be located somewhere between the two radii.

6.7   The Origin of Torque

We have seen in chapter 2 (eq. 2.21) that in magnetohydrodynamics, the fluid's stress tensor includes a part due to the magnetic field. In fact, the stress tensor of the electromagnetic field, called the *Maxwell tensor*, is given by

$$M_{ij} = \frac{-1}{4\pi} \left(E_i E_j + B_i B_j - \frac{\delta_{ij}}{2}(E^2 + B^2) \right) \qquad (6.72)$$

In this tensor, the terms depending on the electric field E are generally neglected, because, in the limit of infinite conductivity, the electric field is given by

$$\vec{E} = -\frac{\vec{v}}{c} \wedge \vec{B} \qquad (6.73)$$

therefore, the part of M_{ij} depending on E is quadratic in v/c, and, in agreement with our assumption—in magnetohydrodynamics—that we neglect the terms of the order $(v/c)^2$ or higher, they may be dropped.

The Maxwell tensor is perfectly capable of transporting angular momentum. Using cylindrical coordinates, the component $rM_{r\phi} = rM_{\phi r}$ represents the flux of momentum along $\hat{e}_\phi$ times r (which is obviously the momentum conjugate to the angle ϕ, and thus the component of the angular

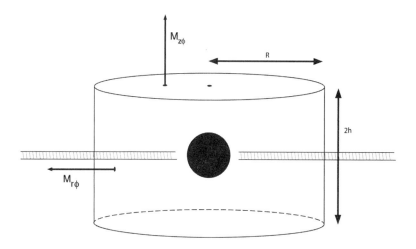

Figure 6.2. The components of Maxwell's stress tensors responsible for the flow of angular momentum across different surfaces.

momentum along the z axis) through a surface that has as a normal the direction $\hat{e}_r$, and thus the angular momentum lost by the matter contained *within* the radius r. This angular momentum obviously ends up in the matter at $> r$. In order to see all this, it proves convenient to consider a cylinder with a radius r and a height z (fig. 6.2), built around the disk, and consider the flux of the ϕ component of the angular momentum crossing the lateral surface of the cylinder or its bases. In this problem, $M_{z\phi}$ may be neglected (why?).

Since we have defined $\mathcal{G}$ as the torque exerted on the internal ring, we have

$$\mathcal{G} = -\int d\phi\, r \int dz\, r M_{r\phi} = \frac{r^2}{2} \int dz\, B_r B_\phi \qquad (6.74)$$

In computing the integral, we have assumed that the product $B_\phi B_r$ is independent of the angle ϕ, that is, the two components are ordered and perfectly correlated. However, in the case of isotropic hydromagnetic turbulence, this is not true; in general, the two components are uncorrelated, and their average value over the surface in question is zero. As a consequence, in the presence of purely hydrodynamic turbulence, this effect produces no loss of angular momentum, as was clear to Shakura and Sunyaev. This is a very serious problem for hydrodynamic turbulence, even assuming that it can somehow be generated.

For a theory to be successful, the product $B_r B_\phi$ must have the same (negative, as eq. 6.4 reminds us) sign along all the curves $r = $ constant; since the initial magnetic field is not necessarily large, we must find an instability (so that the field may grow to appreciable levels), which is independent of ϕ. This instability exists in magnetohydrodynamics and was already known to Velikhov (1959) and to Chandrasekhar (1981), who, however, never thought of applying it to this context. On the other hand, Balbus and Hawley (1991) appreciated its importance for accretion disks. That is why this instability is sometimes called the Balbus-Hawley-Velikhov-Chandrasekhar (BHVC) or *magnetorotational* instability.

We should not be surprised about the presence of a magnetic field in the accretion disk, since it has at least three different plausible origins. First of all, the accreting material comes either from the interstellar medium or from a companion star, and, in any case, this is magnetized material. In the second place, many authors (see, e.g., Torkelsson and Brandenburg 1994) have postulated the existence of a dynamo in the disk itself. Finally, for the accretion on magnetized neutron stars, the disk possesses the magnetic field of the star itself, which, imprisoned at large distances, is later frozen in the accretion flow and dragged toward the star.

It is easy to understand how the instability works (Balbus and Hawley 1991). Let us consider a magnetic field line, directed along the rotation axis, at a distance r_o from the center, and imagine that the part of the line immersed in the disk is perturbed in the form of an S, as in figure 6.3 Because the magnetic field is frozen in the material in which it is placed, the central point of the S, which is at r_o,

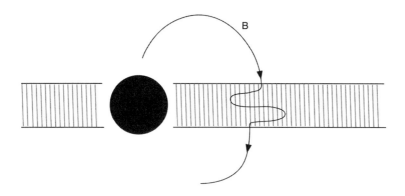

Figure 6.3. How the BHVC instability works.

keeps rotating at the unperturbed frequency ω_o, whereas the belly and hump are in zones that rotate more slowly and more quickly than r_o, respectively. As a consequence, the magnetic line is stretched in a direction having both radial and tangential components. Stretching a magnetic field line requires energy, because the S figure becomes more and more extreme, while the magnetic tension tries to shorten it. Where does the energy come from? The obvious answer is that it comes from the differential rotation. The fluid element at $r_o - dr$ is subject to a force (due to the field tension) whose component along $\hat{e}_\phi$ slows it down; it loses angular momentum and therefore kinetic energy. The fluid element at $r_o + dr$ is subject to an accelerating force and gains angular momentum. Thus the relative angular speeds of the two fluid elements are reduced, and differential rotation is dampened.

At the beginning of section 6.3, we saw that kinetic energy decreases; here, it is used to stretch the magnetic field lines. We easily realize that after a while, as the magnetic field lines are wrapped around the center, their number per unit surface increases, and this means that the magnetic field amplitude grows. The energy is supplied by differential rotation, also called *shear*. Indeed, this mechanism accelerates the slower (external) parts and slows down the faster (internal) ones; it stops working only when the disk rotates as a solid body.

The instability growth rate, for perturbations of the form $e^{ikz}e^{ist}$ in a cold disk, is

$$\frac{s^2}{\omega^2} = q^2 + \frac{1 \pm \sqrt{1 + 16q^2}}{2}, \qquad q \equiv \frac{kv_{\rm A}}{\omega} \qquad (6.75)$$

where $v_{\rm A}$ is the Alfvén velocity. Of course, the solution with the minus sign gives $s^2 < 0$, and thus instability, provided $q \leq \sqrt{3}$, whereas the other solution is always oscillatory and therefore irrelevant to our aims. The maximum growth rate (the maximum of $-s^2$) is reached for $q = \sqrt{15/16}$, where $s = \pm 3i\omega/4$. For $q \geq \sqrt{3}$, we have stability.

Obviously, the minimum value of k is $k_{\rm min} \approx 1/H$, and there will be instability, provided $v_{\rm A} \lesssim \sqrt{3}\omega H \rightarrow v_{\rm A} \lesssim c_{\rm s}$. The instability saturates when the magnetic field becomes too large, because the work necessary to extend the lines of force becomes comparable to the internal energy of the gas.

The heart of this argument is that the instability grows on the time scale $1/\omega$, which is the shortest in the disk. Besides, because of the above-mentioned way in which the magnetic field line is stretched, we see that $B_{\rm r} < 0$ when $B_\phi > 0$, and vice versa; as a consequence, the product $B_{\rm r}B_\phi < 0$ for each value of ϕ, exactly as required.

Many numerical simulations of these phenomena have been carried out (see Balbus 2003 for an exhaustive review), with results that are consistent with the ideas described above, derived from the linear regime. Instability grows on a dynamic time scale, and the importance of the Maxwell term with respect to the Reynolds one is estimated by means of a parameter α that represents an average ratio between the two, just like Shakura and Sunyaev (1973) suggested. The nonlinear regime of instability is reached very quickly and leads to values $\alpha \approx 0.1$. We should notice that these values are not strictly constant in time, even when they oscillate around a well-defined average value. Moreover, when saturation is reached, the values of α seem to be reasonably independent from the initial conditions, even if simulations with slightly different initial conditions seem to produce similar but not identical α values, even though no obvious correlation with any initial quantity has been identified yet.

It should be remarked that oscillating values of α may lead to oscillating values of $\dot{M}$, the accretion rate, and thus of the luminosity of the compact object, which, given the pervasive presence of oscillations in the X-ray luminosity of all compact objects known so far, may lead, hopefully, to a quantitative explanation of at least some part of the observed oscillations.

The last point worth noting is that while a mechanism based on viscosity is intrinsically dissipative, nothing in the BHVC instability implies dissipation. In turbulence, wave coupling transfers the power generated by the large-scale instability to larger and larger wave numbers, until the dissipative terms (in the case of magnetohydrodynamics, the resistive terms) become important. This may be the case here, too, so that the estimates presented in section 6.3 are after all correct. However, strictly speaking, there is no proof that dissipation is *local*, in other words, that the energy produced is not removed and dissipated elsewhere by the global matter flow.

6.8 Disk Stability

Unfortunately, disk stability is an argument concerning which no progress has been made since Pringle's review (1981), which therefore we shall follow.

6.8.1 Time Scales

In perfect analogy with the physics of stellar interiors, we can define characteristic time scales on which dynamic and thermal equilibria are established in a disk. As far as radial equilibrium is concerned, the time scale on which the dynamic equilibrium is established, which is also absolutely the shortest time scale, is the orbital one:

$$t_{\text{din}} \approx \frac{r}{v_{\text{K}}} \approx \frac{1}{\omega} \tag{6.76}$$

However, there is also the time scale on which deviations from hydrostatic equilibrium in z are restored. Since pressure restores equilibrium and sends signals at speed c_{s}, we have

$$t_z \approx \frac{H}{c_{\text{s}}} \approx \frac{r}{v_{\text{K}}} = t_{\text{din}} \tag{6.77}$$

where we have used equation 6.52; therefore, the disk has only one dynamic time scale, independent of the direction.

The thermal time scale is obviously the one on which the disk would dissipate all its internal energy if heat generation were suddenly stopped. Since, in equilibrium, the heat-loss rate is equal to the gain rate, which is given by equations 6.18, 6.37, and 6.42, we have

$$t_{\text{term}} = \frac{\Sigma c_{\text{s}}^2}{Q} = \frac{\Sigma c_{\text{s}}^2}{\nu \Sigma \omega^2 (d \ln \omega / d \ln r)^2} \approx \frac{c_{\text{s}}^2}{\nu \omega^2} \approx \frac{1}{\alpha \omega} \tag{6.78}$$

which is obviously longer than t_{din}.

Finally, we should calculate the viscous time, which is the time scale on which the matter accretes onto the compact object. Using equations 6.39 and 6.42 yields

$$t_{\text{vis}} = \frac{r}{v_{\text{r}}} \approx \frac{r^2}{\nu} \approx \frac{r^2 \omega}{\alpha c_{\text{s}}^2} \approx \frac{1}{\alpha} \frac{r^2}{H^2} \frac{1}{\omega} \tag{6.79}$$

written to make the following order explicit:

$$t_{\text{vis}} \gg t_{\text{term}} \gtrsim t_{\text{din}} \tag{6.80}$$

6.8.2 Instability

Disk stability has been treated in general terms by Piran (1978), who determined dispersion relations for all modes, provided they can be treated in the Wenzel, Kramer, Jeffreys, and Billouin (WKBJ) approximation. In this approximation, we assume that the radial dependence of modes is given by e^{ikr}, and the dispersion relation is calculated to most significant order in $1/k$, namely, for modes tightly wrapped around the center. This treatment is extremely useful, but here, it is worthwhile to introduce a simplified but more intuitive treatment of the problem (Pringle 1981).

Accretion disks are certainly stable on time scales comparable to the dynamic one. This is an interesting result; for example, it applies neither to thick accretion disks (see below), nor to thin ones, made of stars (Toomre 1964).[5] Instead they may be unstable on thermal or viscous time scales. The usefulness of the discussion of time scales is that it suggests how to make useful approximations which simplify computations.

Let us consider thermal instability. In order to do this, let us consider first of all an idealized situation in which the medium is at a given temperature $T_\circ$ and density $\rho_\circ$. The temperature is given by the usual equation in which energy gains Γ (per unit volume and time) balance the losses Λ (also per unit volume and time)[6]:

$$\Gamma(T_\circ, \rho_\circ) - \Lambda(T_\circ, \rho_\circ) \equiv \mathcal{L} = 0 \qquad (6.81)$$

Let us now consider a fluid element with a temperature excess δT, but in density equilibrium with the medium, that is, density $\rho_\circ$. In general, this fluid element is not in thermal equilibrium because $\delta \mathcal{L} \neq 0$, and thus its entropy changes, according to the second principle of thermodynamics:

$$T d(\delta S) = dt \delta \mathcal{L} \qquad (6.82)$$

However, since the transformation takes place at constant density, we also know that

$$T d(\delta S) = C_\rho d(\delta T) \qquad (6.83)$$

---

[5] In comparison with these latter, the difference is that thin disks are not self-gravitating.

[6] Therefore, these definitions differ by a factor ρ from those of chapters 1 and 3.

where C_ρ is, as usual, the specific heat at constant volume. This equation, combined with the one above, gives

$$\frac{d(\delta T)}{dt} = \frac{1}{C_\rho} \left(\frac{\partial \mathcal{L}}{\partial T}\right)_\rho (\delta T) \qquad (6.84)$$

Thence we see that if the coefficient of δT on the right-hand side is positive, the equation has an exponentially growing solution. The disk is then *unstable* when

$$\left(\frac{\partial \mathcal{L}}{\partial T}\right)_\rho > 0 \qquad (6.85)$$

which may be rewritten, also using equation 6.81, as

$$\left(\frac{d\ln\Gamma}{d\ln T}\right)_\rho > \left(\frac{d\ln\Lambda}{d\ln T}\right)_\rho \qquad (6.86)$$

The physical meaning of this equation is rather obvious: if a small bubble is produced, which is warmer than the medium, $\delta T > 0$, the above equation tells that it grows even hotter and will thus never return to the equilibrium temperature $T_\circ$.

This discussion, derived from Field (1965), shows that there are thermally unstable equilibria. It can be easily extended to the case in which the quantity kept constant is P instead of ρ, or any other thermodynamic quantity.

How do we apply the above to disks? The surface density Σ, in the disks, is not necessarily the equilibrium one. In fact, matter is removed from the disk by viscosity, on the viscous time scale, which is the longest of all. Therefore, thermal evolution takes place with density $\Sigma = constant$, simply because the time to make it change is too long. This is the way in which the knowledge of time scales simplifies our analysis of stability. However, if thermal evolution takes place with a constant Σ, we can immediately write the criterion of thermal instability for the disk:

$$\left(\frac{d\ln\Gamma}{d\ln T_c}\right)_\Sigma > \left(\frac{d\ln\Lambda}{d\ln T_c}\right)_\Sigma \qquad (6.87)$$

where we have used $T_{\rm c}$ instead of T because disks are characterized by the temperature $T_{\rm c}$, exactly measured on the plane $z = 0$. The cooling term is easily calculated because the disk emits like a blackbody, so that $\Lambda = 2\sigma T^4$. If we remember that the photospheric temperature T and $T_{\rm c}$ are the same, apart from the small correction $\tau^{1/4}$, we find

$$\frac{d\ln \Lambda}{d\ln T_{\rm c}} = 4 \tag{6.88}$$

The interesting case occurs when pressure is dominated by radiation. In that case (see problem 5),

$$\left(\frac{d\ln Q}{d\ln T_{\rm c}}\right)_\Sigma = 8 \tag{6.89}$$

Therefore, the A region, where pressure is mostly due to radiation, is thermally unstable (Pringle 1976).

Viscous instability (Lightman and Eardley 1974) can now be treated by assuming that the disk is not only in dynamic, but also in thermal equilibrium. In these conditions, there may be an excess of surface density $\delta\Sigma$, and we would like to know whether or not it grows with time, when we consider viscous transport in a radial direction.

First of all, we must derive an equation describing how $\delta\Sigma$ evolves with time. If we combine equations 6.7, 6.10, and 6.37, we find

$$\frac{\partial\Sigma}{\partial t} = -\frac{1}{r}\frac{\partial}{\partial r}\left(\frac{\frac{\partial}{\partial r}(r^3\nu\Sigma\omega')}{\frac{\partial}{\partial r}(r^2\omega)}\right) \tag{6.90}$$

where $\omega' = d\omega/dr$. In order to see whether the system is stable or not, let us consider a small perturbation of Σ, close to the time-independent solution:

$$\Sigma = \Sigma_\circ + \delta\Sigma \tag{6.91}$$

where $\Sigma_\circ$ is independent of time. When we put this in the above equation, and, at the same time, we specialize to a

Keplerian disk ($\omega \propto r^{-3/2}$), we obtain

$$\frac{\partial}{\partial t}\delta\Sigma = \frac{3}{r}\frac{\partial}{\partial r}\left(r^{1/2}\frac{\partial}{\partial r}(r^{1/2}\delta(\nu\Sigma)))\right) \qquad (6.92)$$

Now it is worthwhile to remember that we have assumed thermal equilibrium, so the power radiated must equal the power dissipated:

$$2\sigma T^4 = 2\pi r^3 \nu\Sigma \left(\frac{\partial\omega}{\partial r}\right)^2 \qquad (6.93)$$

This equation tells us that T, and therefore also T_c is a function of $\nu\Sigma$. On the other hand, from the definition

$$\nu = \alpha H c_s \qquad (6.94)$$

we see that ν is a function of T_c and therefore of the product $\nu\Sigma$. We conclude that ν is only a function of Σ:

$$\nu\Sigma = f(\Sigma, r) \qquad (6.95)$$

If we now perturb this equation, we find

$$\delta(\nu\Sigma) = \frac{\partial}{\partial\Sigma}(\nu\Sigma)\delta\Sigma \qquad (6.96)$$

where the coefficient must be computed for the zero-order solution and, as a consequence, is independent of time. When we introduce this into equation 6.92, we find our final relation:

$$\frac{\partial\delta(\nu\Sigma)}{\partial t} = \frac{\partial(\nu\Sigma)}{\partial\Sigma}\frac{3}{r}\frac{\partial}{\partial r}\left(r^{1/2}\frac{\partial}{\partial r}(r^{1/2}\delta(\nu\Sigma)))\right) \qquad (6.97)$$

We immediately realize that this is nothing but a heat diffusion equation, with an unusual diffusion coefficient, $\partial(\nu\Sigma)/\partial\Sigma$. If the scattering coefficient is >0, we get the usual signs, and the equation describes the diffusive damping of the quantity $\nu\Sigma$. However, if the coefficient is <0, we have a *negative* diffusion coefficient, which carries heat to the hottest regions. Obviously, we have instability. Therefore, the condition of instability is

$$\frac{\partial(\nu\Sigma)}{\partial\Sigma} < 0 \tag{6.98}$$

Normally, this condition is not satisfied, but we can once again show that in the A zone, where the radiation pressure dominates the total pressure,

$$\nu\Sigma \propto \frac{1}{\Sigma} \tag{6.99}$$

therefore, once again, it is unstable.

All these results were known at the end of 1970s, and nothing new has been added since. This leaves us with a problem: are the instabilities we have identified real, or are they due to some incorrect assumption? Obviously, since we have assumed the α prescription, we are tormented by this doubt, which increases when we consider that as shown by Lightman and Eardley, if we assume that $\nu = \alpha c_s H$, but we interpret c_s as the *gas* sound speed, thus excluding the radiation pressure even where it dominates (i.e., by assuming $c_s^2 = p_g/\rho$), all instabilities disappear.

One of the possibilities proposed is that the region A of the disk, in analogy with the thermally unstable interstellar medium, may undergo a transition to a *two-phase* equilibrium, with matter distributed either in very cold, dense bubbles, or in a tenuous, hot medium, in pressure equilibrium with the cold phase. Please note that the ensuing hot medium need not be confined to a thin disk any longer, exactly because of its large temperature. Although attractive, this solution has not been proved inevitable yet.

6.9   Lense-Thirring Precession

One of the implicit assumptions of the above paragraphs is that the rotation axis of the disk and that of the compact object are the same. In principle, the two axes might also be distinct; in a binary system, the orbital angular momentum and the two spin momenta (of the donor star and of the compact object) are not necessarily collinear, so that matter torn from the donor star may lie on a plane different from the equatorial plane of the compact object. In the same way, in a galaxy, the gas on an accretion disk around the central black hole may have a specific angular momentum distinct from that of the black hole. If the gas comes from the central regions of the galaxy (the bulge) it may have angular momentum oriented along nearly every direction. Alternatively, matter may come from a galaxy that has just been cannibalized, whose orbital angular momentum is absolutely not correlated with that of the black hole.

For these reasons, it would seem at first sight plausible that the rotation axes of disk and compact object be distinct, but an effect discussed for the first time in 1975 (Bardeen and Petterson) assures us that close enough to the compact object, the disk will lie on the equatorial plane of the compact object. We shall present here a revised version of their argument, from Papaloizou and Pringle (1983). Please note that this is, however, a simplified argument; the complete hydrodynamic treatment is also discussed by Papaloizou and Pringle (1983).

The effect is produced by a combination of two basic effects: the disk viscosity, and the Lense-Thirring (LT) effect of general relativity (see appendix B). In order to be definite, let us consider the case in which the compact object is a black hole. The outer metric of a rotating neutron star differs from that of a rotating black hole, but the same qualitative conclusions apply.

The LT effect consists in the precession of a particle in a circular orbit around a rotating black hole, when the orbital plane is inclined with respect to the black hole's equatorial plane. At large distances from the hole, the particle's angular momentum is subject to the torque

$$\frac{d\vec{L}}{dt} = \vec{\omega}_{\mathrm{LT}} \wedge \vec{L} \qquad (6.100)$$

where the LT angular frequency $\vec{\omega}_{LT}$ is aligned with the angular momentum of the black hole (which we can take as the z axis) and is

$$\vec{\omega}_{LT} = \frac{2GMa}{c^2 r^3}\hat{e}_z \qquad (6.101)$$

Here a is the specific angular momentum of the black hole, for which, as we know, $0 \le a \le GM/c$.

It may seem unusual that a general-relativistic effect, even if small, may play such an important role. The explanation lies in the slowness of the radial accretion; during the time it takes to remove the angular momentum to infinity, each matter ring has the time to precede by a significant amount. In fact, the total precession angle β is given by

$$\beta = \frac{2GMa}{c^2}\int_r^\infty \frac{dr}{v_r r^3}$$

$$= 0.5 \times 10^5 \alpha^{-4/5}\frac{ac}{GM}\frac{M}{3M_\odot}\left(\frac{\dot{M}}{10^{17}\text{gs}^{-1}}\right)^{2/5}\left(\frac{rc^2}{GM}\right)^{-8/5}$$

$$\qquad (6.102)$$

where we have used equation 6.71.

The precession frequency ω_{LT} depends strongly on the radius r, so that we expect a strong differential precession. Even though initially placed on one plane, the disk is not confined to this plane but will be grossly warped (see fig. 6.4). However, as we have already noted in discussing equation 6.35, the presence of speed gradients induces friction forces in the disk, which oppose differential variation (i.e., with the radius) of the frequencies of motion. As a consequence, the motion of the disk is forced by the LT effect and is damped by friction.

In order to compute the disk evolution, we divide it into circular rings of infinitesimal width dr. Each ring is free to rotate around its own center or to be tilted with respect to any axis, but it cannot otherwise expand, contract, or be

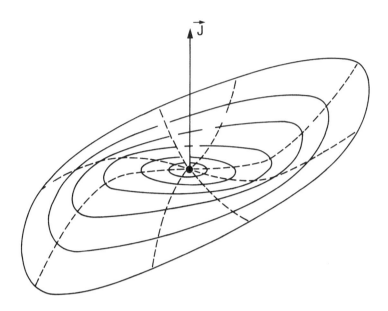

Figure 6.4. Appearance of an accretion disk warped by differential precession due to LT forcing. From Thorne, Price, and McDonald 1986.

distorted. The more correct treatment that allows distortions of the rings from the circular form can be found described by Papaloizou and Pringle (1983). Each ring is characterized by a unit vector in the direction of its angular momentum; in Cartesian coordinates, $\hat{l} = (l_x, l_y, 1)$, with $l_x, l_y \ll 1$. In other words, we introduce here our major approximation: that the inclination angle of the ring plane with respect to the hole's equatorial plane be small. We now proceed to linearize the problem with respect to the small quantities l_x, l_y.

We can easily determine the equation for the conservation of the angular momentum of the small ring by modifying suitably the equation of pure conservation and using equation 6.35.

We use cylindrical coordinates r, z, ϕ. We want to generalize equation 6.10 to include the LT torque and the friction due to the warped form, which the disk has now assumed. We obviously get

$$\frac{\partial L_z \hat{l}}{\partial t} + \frac{1}{r}\frac{\partial}{\partial r}\left(v_r r L_z \hat{l}\right) = \frac{1}{r}\frac{\partial}{\partial r}\left(r^3 \nu \Sigma \hat{l}\frac{d\omega}{dr}\right) + \vec{\omega}_{\mathrm{LT}} \wedge \hat{l} + \frac{1}{r}\frac{\partial}{\partial r}\left(r\vec{T}\right)$$

$$(6.103)$$

where we have used equations 6.37 and 6.100. The term $\vec{T}$ is that part of the viscous torque due to the fact that the instantaneous axis of rotation of infinitesimal rings is not constant in the disk.

Note that we have assumed that the relevant velocity is still along the direction r. In general, this is not correct because the plane of each ring is inclined, and therefore the accretion velocity is directed along the radial direction of the inclined plane, which is *not* just the direction r; it would be so only for rings in the plane $z = 0$. However, the component of $\vec{v}$ along the z and ϕ directions is of the first order in the angle of inclination, which means that if we included this effect, we would be considering an effect of the second order in $l_x, l_y \ll 1$, thus violating the development to the first order. As a consequence, the above equation is approximate but correct to the first order in l_x, l_y.

We can *guess* the form of $\vec{T}$ without much effort; in fact, when all the rings of a disk rotate around the same axis, we must have $\vec{T} = 0$, and therefore $\vec{T} \propto \nabla \hat{l}$. Also, we must take into account the fact that $\hat{l}$ can depend neither on ϕ, because of the axial symmetry of the rings, nor on z, because we have assumed the disk is infinitely thin. Therefore $\vec{T} \propto \partial \hat{l}/\partial r$. Now the coefficient of proportionality can be obtained from dimensional analysis:

$$\vec{T} = \nu_2 \Sigma r^2 \omega \frac{\partial \hat{l}}{\partial r} \qquad (6.104)$$

Note also that we have *not* assumed that the two coefficients of viscosity, ν and ν_2, are identical, for there is no obvious reason for this to be so. Therefore, the equation determining the inclination of the disk at each radius becomes

$$\frac{\partial(L_z\hat{l})}{\partial t} + \frac{1}{r}\frac{\partial}{\partial r}\left(v_r r L_z \hat{l}\right) = \frac{1}{r}\frac{\partial}{\partial r}\left(r^3\nu\Sigma\hat{l}\frac{d\omega}{dr}\right) + \vec{\omega}_{LT} \wedge \hat{l}$$

$$+ \frac{1}{r}\frac{\partial}{\partial r}\left(\nu_2\Sigma r^3\omega\frac{\partial\hat{l}}{\partial r}\right) \tag{6.105}$$

This equation can now be simplified by using the equation for mass conservation, equation 6.7 and its component z, which gives equation 6.10. In this way, we may eliminate v_r. Calling $\tilde{l} = (l_x, l_y, 0)$, we obtain

$$\tilde{l}\partial t + \left(v_r + \frac{3\nu}{2r}\right)\frac{\partial\tilde{l}}{\partial r} = \vec{\omega}_{LT} \wedge \tilde{l} + \frac{1}{\Sigma r^3\omega}\frac{\partial}{\partial r}\left(\nu_2\Sigma r^3\omega\frac{\partial\tilde{l}}{\partial r}\right) \tag{6.106}$$

Here I have already used $\omega = \omega_K \propto r^{-3/2}$. Since we are considering a stationary situation, and equation 6.39 holds, we shall only consider internal regions of the disk, where $R(r) \approx 1$. We finally find

$$\vec{\omega}_{LT} \wedge \tilde{l} + \frac{1}{\Sigma r^3\omega}\frac{\partial}{\partial r}\left(\nu_2\Sigma r^3\omega\frac{\partial\tilde{l}}{\partial r}\right) = 0 \tag{6.107}$$

The best way to solve this equation is to define a new quantity $w = l_x + \imath l_y$, which allows us to rewrite the first term as $\vec{\omega}_{LT} \wedge \tilde{l} \rightarrow \imath\omega_{LT}w$. Namely, we have

$$\imath\omega_{LT}w + \frac{1}{\Sigma r^3\omega}\frac{\partial}{\partial r}\left(\nu_2\Sigma r^3\omega\frac{\partial w}{\partial r}\right) = 0 \tag{6.108}$$

This is a linear (because we have linearized it!) ordinary differential equation; the boundary conditions at $r \rightarrow \infty$ are easily identified. Indeed, we can safely take $l_x = 1, l_y = 0$. The second one, $l_y = 0$, simply corresponds to orienting the axes in the plane xy so that at infinity, the perturbation to the normal is oriented only along l_x, and this, of course, may always be done. On the other hand, the boundary condition $l_x = 1$ makes sense because the equation is linear, and therefore we can always multiply an arbitrary solution by an appropriate coefficient, which gives $l_x = 1$ at infinity. As a consequence, for $r \to \infty$, $w = 1$.

In order to determine the remaining boundary condition, we need to specialize the solution. For the sake of simplicity, let us first fix $\nu \propto \nu_2$. Let us then consider the solution for the C zone of the disk, for which the solution is given by equation 6.65. We find

$$\nu_2 \Sigma r^3 w \propto r^{3/2} \tag{6.109}$$

$$\Sigma r^3 w \propto r^{3/4} \tag{6.110}$$

If we define new variables x, z, obtained by rescaling w and r, respectively, in a suitable way, we find ($\dot{x} = dx/dz$, and so on)

$$\ddot{x} + \frac{3}{2z}\dot{x} + \imath \frac{x}{z^{15/4}} = 0 \tag{6.111}$$

We can easily see that the point $z = 0$ (corresponding to $r = 0$) is a singular irregular point of the system (see Bender and Orszag 1978). If we look for solutions of the form $x = e^S$, the standard technique in these cases, we find

$$S = \pm \frac{e^{-\imath \pi/4}}{z^{7/8}} + \ldots \tag{6.112}$$

where the next terms in the development are *less* singular than $z^{-7/8}$ and may therefore be neglected. Thence we see that the solution with a positive sign has a real part >0 and is thus unacceptable for $z \to 0$. The only acceptable solution has a negative sign; thus we see that the only acceptable solution tends to zero exponentially for $z \to 0$. In order to obtain the complete solution, we must now vary $\dot{w}$ at $z \to \infty$, until we obtain *only* the converging solution for $z \to 0$.

A numerical solution of this equation is shown in figure 6.5; it represents the only finite solution of this equation

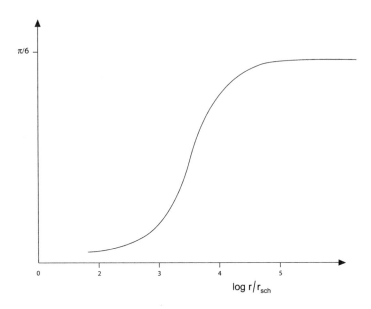

Figure 6.5. Numerical solution of equation 6.111: the horizontal axis is the logarithm of the coordinate r in Schwarzschild units, while on the vertical axis we plot θ, given by equation 6.113.

for w. The quantity in the figure is θ, given by

$$\cos \theta = \frac{1}{\sqrt{1 + l_x^2 + l_y^2}} \qquad (6.113)$$

The unit of distance on the horizontal axis is arbitrary, the solution of the unperturbed disk is that of the C zone, and we have used $\nu_2 = \nu$, the value of Shakura and Sunyaev (the exact value of α is irrelevant because it is absorbed in arbitrary constants).

It is easy to show that this process takes place (almost entirely) inside the C zone, so that the analysis we have presented is adequate. What happens is, briefly, that the precession is so fast that large velocity gradients arise, at large radii; in their turn, large velocity gradients lead to strong dissipation. In the end, viscous dissipation wins on LT forcing. Summing up, this phenomenon brings the disk to lie in the equatorial plane of the compact object, even though, at first, it lies on a different plane.

6.10 Problems

1. Consider N point masses in the gravitational field of a heavy object with a mass M. Show that for a given total angular momentum, the problem has a minimum of the total energy when the masses are all on the same circular orbit. (Hint: Use Lagrange multipliers.) However, this is a *local* minimum; can you imagine the *absolute* energy minimum? (Hint: Study fig. 6.1.)

2. Sometimes, in the literature, the local dissipation rate Q (eq. 6.21) is integrated between $r_1 = r_{\rm m}$ and ∞ and is thus found:

$$L = \frac{3}{2}\frac{GM\dot{M}}{r_1} \tag{6.114}$$

Can you explain where the coefficient $3/2$ comes from? In other words, what is the physical meaning of $\beta = 0$? (Hint: Consider what happens at smaller radii!)

3. In writing the temperature of an accretion disk as $T = T_{\rm c}(R(x)/x^3)^{1/4}$, show that the maximum value of T is $T_{\max} \ll T_{\rm c}$.

4. We have neglected the disk self-gravity in studying its structure. First derive a criterion to evaluate whether this is a good approximation; then derive the density stratification for an isothermal disk in the z direction. (Hint: You will need some drastic approximation to compute the disk force in the z direction, but remember that in any case, $H \ll r$.) Last, check that in the solution of 6.65, the disk is not self-gravitating for realistic values of the distance x.

5. Show that when the disk pressure is dominated by the radiation pressure,

$$\left(\frac{d\ln Q}{d\ln T_c}\right)_\Sigma = 8 \tag{6.115}$$

and

$$\nu\Sigma = \frac{64c^2 m_p^2}{27\alpha\sigma_T^2\omega\Sigma} \tag{6.116}$$

6. For the infinitely thin disk, we have derived an equation for mass conservation, equation 6.7, and one for the conservation of angular momentum, equation 6.10. It is also possible to derive an equation for energy conservation (Blandford and Begelman 1999). Show that

$$\frac{\partial}{\partial t}(2\pi r\Sigma\epsilon_b) + \frac{\partial}{\partial r}(2\pi r\Sigma v_r(\epsilon_b + w) + \omega\mathcal{G}) = \mathcal{G}\frac{\partial\omega}{\partial r} + 2\pi r\Sigma kT\frac{ds}{dt} \tag{6.117}$$

where $\epsilon_b = -\omega l/2$ is the binding energy of the circular orbit of specific angular momentum l, and h and s are the specific enthalpy and entropy, respectively.

7. We have seen that in equation 6.65, the disk is slightly concave and can therefore intercept a fraction of luminosity of the black hole. Estimate in this case the contribution of the central source to heating, and determine how the temperature varies with the radius.

8. It is also interesting to consider what happens when a *global* thermodynamic equilibrium holds, rather than the *local* equilibrium we have assumed so far. Neglecting gravity altogether, can you derive the rotation law for a gas with given angular momentum $J_x = J_y = 0, J_z \neq 0$? (Hint: What does viscosity do to the gas entropy?)

Chapter 7

Disk Accretion II

After introducing the standard theory of thin accretion disks, we consider here a few variations on the same theme, namely, interaction with the hot corona, properties of thick disks, nondissipative flows called advectron-dominated accretion flows (ADAFs), the fate of angular momentum at large radii, the accretion on magnetized objects, and boundary layers.

7.1 Other Disk Models

The solution for the A region, given in the preceding chapter, raises several perplexities. On one hand, the total optical depth is not $\tau \gg 1$, which implies that the energy transport is not correctly described by equation 3.35. Moreover, this solution is unstable, on thermal and on viscous time scales. Finally, the temperatures we have just determined for the accretion disk cannot explain in any way the X-ray emission observed both in objects of stellar dimensions, and in active galactic nuclei (AGNs); the observations show an abundant emission even in the hard X-ray region, beyond 100 keV. The standard model must therefore be modified, at least in the A zone.

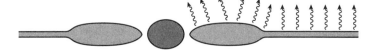

Figure 7.1. Schematic view of the model by Shapiro, Lightman, and Eardley (1976). From Thorne and Price 1975.

Shapiro, Lightman, and Eardley (1976) proposed the ear-
liest model to overcome these difficulties (fig. 7.1). They suggest that the innermost zones of the disk are neither thin nor
optically thick, because matter settles on a two-temperature
configuration, with protons at temperature $T_p \approx 10^{11}$ K, and
electrons at $T_e \approx 10^9$ K. The high temperatures in question
explain the thickness of the disks, whereas the low densities
justify the existence of two temperatures for the two components that cannot thermalize at a common temperature. The
main radiative mechanism of the model is the inverse Compton process (or Comptonization, if you like) of soft X-ray
photons coming from the outermost, cold thin zones, by relatively hot electrons ($T_e \approx 10^9$ K) of the A zone. The model is
built so that the Compton y parameter (eq. 3.160) is $y \approx 1$,
so that the resulting spectrum, given by equation 3.171, is
$I_\nu \propto x^3 \mathcal{N} \propto x^{-1} \exp(-x)$, with $x \equiv \hbar\omega/kT_e$. The resulting
power-law part of the spectrum, $\propto \nu^{-1}, x \lesssim 1$, is in good
agreement with the observations of some galactic compacts,
such as Cyg X-1 in the hard state. Besides, the gas parame-
ters ($kT \approx 0.2m_e c^2$ and $\tau_T \approx 1$, deriving from the requisite
$y \approx 1$) are justified by the following discussion about the
properties of pair gases.

There is, however, another solution to the problem of the
hard X-ray generation, by Ostriker (1977). The idea is that
above the disk, there is a component called the *hot corona*,
mainly, or perhaps wholly, made of electron-positron pairs,
which generate the hardest part of the spectrum. The really conclusive argument in favor of the existence of the
hot corona is that at least in one important case, we observe it directly; obviously, the Sun, which, though having

a surface temperature $T_s \approx 5500$ K, has a hot corona with $T \approx 2 \times 10^6$ K $\gg T_s$.

The observation of *loops* of hot matter emerging from the Sun's convective zone clearly shows that the Sun-corona interaction is of an essentially magnetohydrodynamic type, because matter is confined to the loops by something, which probably is the magnetic field. This makes the processes difficult to describe quantitatively even in the case of the Sun, and *a fortiori* in the more distant and therefore observation-poor case of accretion disks on compact objects.

The interaction of matter from the loops with the disk luminosity is very similar to the one in the model by Shapiro, Lightman, and Eardley. That is why, rather than focusing on the details of each model (both being quite uncertain), we shall present a general discussion, which will be qualitative rather than strictly quantitative, of the properties of pair plasmas. The relevance of this physics, in the framework of the above-discussed models, is that in cases where the gas has temperatures of the order of $kT/m_e c^2 \approx \mathcal{O}(1)$, pair creation is possible and likely.

7.1.1 The Origin of Particles

In the corona, we can expect to find at least two types of particles: thermal and nonthermal. As we have seen, the non-thermal particles always give origin to spectra that are simple power laws in frequency ν, whereas thermal distributions may give origin to the same spectra under specific conditions, as we have just seen. Since observations indicate the presence of spectra of the type $\propto \nu^{-\alpha}$ in the region of hard X rays, we may suppose that the hot corona can be made of both nonthermal and thermal particles.

The origin of nonthermal particles was discussed in chapter 4, where we showed that any shock can produce a population of nonthermal particles, with energy distribution $dN \propto E^{-\alpha}dE$, and $\alpha \approx 2 - 3$. We can easily imagine that there are shocks around compact sources; for example, near

AGNs, collisions between clouds responsible for the emission lines surely generate shock waves, since relative velocities, ≈ 5000 km s^{-1}, are largely supersonic for the clouds' temperatures $T \lesssim 10^5$ K. Near neutron stars or magnetic white dwarfs, there are shock waves close to the polar caps, where accretion flows are channeled by the magnetic field once the magnetic torques have destroyed the disk (see the last section of this chapter). Also, any disturbance in the disks must lead to the formation of a shock, again because the motion is highly supersonic; remember that for the disk to be thin, we must have $c_s \ll v_K$. In fact, it has been suggested more than once that accretion disks host perturbations, called *density waves*, which quickly assume the shape of a spiral and form a shock in an entirely natural way.

On the other hand, energetic thermal particles are formed by reconnection, exactly like near the Sun (Galeev, Rosner, and Vaiana 1979). Indeed we have seen in the preceding chapter that the magnetorotational instability generates magnetic fields with an energy density $B^2/8\pi \approx \rho c_s^2$. Flux freezing stretches these magnetic fields in a toroidal direction; however, please note that the polarity of the magnetic field differs depending on whether we consider the lower or the upper half-thickness of the disk. These are the ideal conditions for magnetic reconnection (see section 2.7.2), in which opposite polarities of magnetic field are pushed one against the other. Moreover, the magnetic bubbles in the disk can float (see section 2.7.1) and are pushed out of the disk. This makes reconnection possible; in fact (eq. 2.119), the reconnection speed is proportional to the Alfvén one, $v_A^2 = B^2/4\pi\rho$, which is small at $z = 0$ in the disk because ρ is relatively large, but increases outside the disk, where ρ decreases exponentially. Therefore, the magnetic field bubbles reconnect; in other words, they dissipate the magnetic energy into currents, which are dissipated by resistivity in the reconnection region. Therefore, the magnetic energy is transformed into heat by the component carrying the current, that is, by electrons only.

7.1.2 Dynamic Peculiarities of Pair Plasmas

When reconnection heats electrons up, they cannot escape from the disk because of the electrostatic attraction to protons; electrons and protons are strongly coupled. Since the energy content of the magnetic field, and thus also of the particles produced in the reconnection, is of the same order of magnitude as the disk's thermal energy, its equal distribution between electrons and protons cannot generate a hot corona (i.e., a structure with a scale height $\gg H$). Matter remains confined to the disk, $z \approx H$.

The situation changes if electron-positron pairs are formed, because, as we know, they are weakly tied to a normal gas. In fact, because of charge neutrality, the pairs do not have to drag protons. Since each electron (or positron) drags its positron (or electron), there will be no strong electric fields coupling protons to their electrons.

The main coupling between the normal gas and a pair plasma is represented by collisions. The Coulomb cross section for collisions, in the relativistic regime,[1] is $\sigma_C \approx 3\sigma_T/(8\pi\gamma^2)$ (see eq. 4.26), so that $\sigma_C \ll \sigma_T$. We shall show that $\tau_T \lesssim 1$ for the hot corona, so each pair has a negligible probability of thermalizing with the protons within the disk.

However, since the pair plasma is not coupled to the normal gas and since in the presence of pairs' formation the pairs' velocity is $\approx c$, it cannot be gravitationally confined and will therefore abandon the thin disk.[2] As a consequence, if we want to form a corona of hot gas from matter evaporated from the disk, this must *necessarily* be made up of relativistic pairs, with a low contribution of protons. This corona may also give origin to a wind leaving the compact object.

---

[1] If pairs are formed, they must necessarily be relativistic.

[2] This result remains true even in the limit of strong coupling, provided the pairs are so many that they dominate the total mass. In this case, indeed, the pair plasma behaves like a true relativistic gas, with sound speed $\approx c/\sqrt{3}$, and thus once again impossible to confine gravitationally.

Finally, we note that in the presence of pairs, the Eddington luminosity is much reduced. Since

$$L_E = \frac{4\pi c G M \mu_e m_p}{\sigma_T} \qquad (7.1)$$

we must introduce the new mass per electron, which is $\mu_{e\odot} = 0.615$ for solar abundances. In a plasma with an abundance of pairs z, where z is the number of positrons per baryon, we have instead

$$\mu_e = \mu_{e\odot} \frac{1 + 2z(m_e/m_p)}{1 + 2z} \qquad (7.2)$$

and, for $z m_e/m_p \gg 1$,

$$\mu_e \approx \mu_{e\odot} \frac{m_e}{m_p} \ll 1 \qquad (7.3)$$

Therefore, in the presence of pairs, we have a new Eddington luminosity:

$$L_{Ec} = \frac{4\pi c G M \mu_{e\odot} m_e}{\sigma_T} = \frac{\mu_{e\odot} m_e}{m_p} L_E \ll L_E \qquad (7.4)$$

This leads us to inquire about the thermodynamic and chemical conditions of the pair plasma. In other words, what sort of temperature and abundance of pairs (which, we remind, can be formed freely) will there be in a gas mainly composed of pairs, with a temperature $\gtrsim m_e c^2$? The result is that the gas will assume an equilibrium configuration with $kT \lesssim q m_e c^2$, $q \approx \mathcal{O}(1)$, and $\tau_T \approx 1$ for almost any heat input from the outside. However, this important result must be demonstrated in two different limits: by assuming that the heat input in the gas is not in the form of photons, or, conversely, that it is.

7.1.3 The Pair Plasma without Input of External Photons

It was noticed a long time ago that there is a maximum temperature for a pair plasma, independent of the heat input. This maximum temperature was originally calculated

(Bisnovatyi-Kogan, Zel'dovich, and Sunyaev 1971 [BKZS]) in the limit in which only processes between particles can create pairs. The processes in consideration are then

$$e^{\pm} + Ze \rightarrow e^{\pm} + Ze + e^{+} + e^{-} \tag{7.5}$$

$$e^{\pm} + e^{\pm} \rightarrow e^{\pm} + e^{\pm} + e^{+} + e^{-} \tag{7.6}$$

$$e^{+} + e^{-} \rightarrow 2e^{+} + 2e^{-} \tag{7.7}$$

In this case, we find that the maximum temperature T_l satisfies

$$\theta_l \ln \theta_l = \frac{\pi}{\sqrt{32\alpha}} \tag{7.8}$$

where θ is the temperature in units of $m_e c^2$, $\theta \equiv kT/m_e c^2$, and $\alpha = e^2/(\hbar c) \approx 1/137$ is the usual fine structure constant. Therefore, we find that

$$kT_l \approx 24 m_e c^2 \tag{7.9}$$

It is easy to understand the existence of this limit. As long as we consider only processes among particles, the pairs' creation and destruction rates depend on the square of the density, but we see from equation 3.213 that the cross section for annihilation, in the hyperrelativistic regime, decreases with the particles' energy, whereas, on the other hand, the creation cross-section for the above-described processes (otherwise not used in this book),

$$\sigma = \frac{3\alpha^2}{8\pi} \sigma_T \, (\ln \gamma_e)^3 \tag{7.10}$$

(here γ_e is the Lorentz factor of electrons) flattens in the same limit. Therefore, beyond the temperature T_l, the rate of pair creation always exceeds the rate of pairs' annihilation, and any further heating of the gas is used for the production of more pairs. As a consequence, the temperature of the plasma becomes independent of the heating rate.

This argument neglects the importance of all the processes in which photons are involved, such as, for example

$$\gamma + \gamma \rightarrow e^{+} + e^{-} \tag{7.11}$$

Therefore, it applies to gas clouds from which photons disappear infinitely fast, *that is,* in the limit of vanishing optical depth. Let us now study what happens when there are photons (Svensson 1982), in which case we must add all the processes of pair creation and annihilation in which photons appear, such as, for pair creation,

$$\gamma+\gamma \to e^{+}+e^{-}, \quad \gamma+e \to e+e^{+}+e^{-}, \quad e+e \to e+e+e^{+}e^{-}$$
$$(7.12)$$

for photon creation,

$$e+e \to e+e+\gamma, \quad e+\gamma \to e+\gamma+\gamma, \quad e^{+}+e^{-} \to \gamma+\gamma$$
$$(7.13)$$

which also includes the reaction of pair annihilation, and, finally, the Comptonization reaction

$$e+\gamma \to e+\gamma \qquad (7.14)$$

which, however, only applies to photons produced within the cloud. In order to specify the cloud properties, we call

$$\tau_{N} = RN\sigma_{T} \qquad (7.15)$$

where now R is the cloud's radius, N the density of protons, so that τ_{N} only indicates the contribution to opacity due to the electrons associated to protons, that is, it excludes the pairs' contribution. The total opacity τ_{T} includes the pairs' effect:

$$\tau_{T} = \tau_{N}(1+2n_{+}/N) \equiv \tau_{N}(1+2z) \qquad (7.16)$$

where n_{+} is the density of positrons.

The presence of a nonvanishing optical depth modifies the creation rate of particles, because photons will remain trapped in the cloud for a time given by

$$t_{esc} = \frac{R(1+\tau_{T})}{c} \qquad (7.17)$$

therefore all the rates will increase, simply because there is one more channel. The presence of processes involving photons allows us to explain a paradoxical aspect of the BKZS limit. Indeed, we know that when the density of the pairs n_- tends to infinity, the situation must tend to the limit of thermodynamic equilibrium, in which case the energy lost by the cloud is given by $L = 4\pi R^2 \sigma_B T^4$. In this limit, the temperature, in conditions of equilibrium, increases with the fourth root of the heating rate (which is $= L$, in thermal equilibrium!), whereas, according to BZKS, the temperature is independent from the heating rate, in apparent contradiction. However, when $n_- \to \infty$, the cloud's total optical depth $\to \infty$ also, so that the photons' density increases; this is why, in this limit, the processes of pair creation are dominated by those involving photons, and the assumptions on which the BZKS result is based are no longer correct.

The reasonable question we can now ask is whether, when we take into account the photons' processes, there is still a maximum temperature, or whether we have a continuous transition to the limit of thermal equilibrium (for $n_- \to \infty$) with T that keeps increasing. The answer (Svensson 1982) is that the upper limit still exists, in some approximate sense, and that of course the thermodynamic limit is recovered, but only for astrophysically implausible heating rates. Before showing how this happens, let us discuss a technical point.

How to Calculate the Equilibrium

The equilibrium is calculated by equating the sum of the pair production rates (for all the processes in question) to the sum of all destruction processes. All processes involving particles are proportional to the square of density; but processes involving photons are also proportional to the square of density, because, as we shall show shortly, the energy density in photons is proportional to N, namely, the density of protons. As a consequence, when we impose the equality of pairs' creation and destruction rates, all terms are proportional to

N^2. When this factor is simplified, what is left is a relation among the various dimensionless quantities of the problem, $z = n_+/N, \tau_{\mathrm{N}}, \theta$. Therefore, in the plane $z - \theta$, the locus of the states of equilibrium, at a fixed τ_{N}, is a simple curve, which does not otherwise depend on N or R separately; in other words, the problem is, in a sense, self-similar.

We have to explain why the energy density in photons ϵ_γ is proportional to N. The photons produced escape from the cloud in a finite time, as discussed above; therefore, the energy density is given by the energy production rate in the form of photons, $\dot\epsilon_\gamma$, multiplied by t_{esc}. Moreover, since all the photon production processes involve two bodies, $\dot\epsilon_\gamma \propto N^2$. Thus we have

$$\epsilon_\gamma = \dot\epsilon_\gamma t_{\mathrm{esc}} \propto N^2 t_{\mathrm{esc}} \propto N\tau_{\mathrm{N}}(1 + \tau_{\mathrm{N}}) \qquad (7.18)$$

which is proportional to the density of protons N, as anticipated.

Results

In equation 3.160, we defined the Compton parameter y as

$$y \equiv 4\frac{kT}{m_e c^2}\tau_{\mathrm{T}}^2 \qquad (7.19)$$

The importance of y is that according to whether $y \ll 1$ or $y \gg 1$, the photons' spectrum is either equal or different, respectively, to the emission spectrum. In the case $\tau_{\mathrm{T}} \ll 1$, which corresponds to $y \ll 1$ because the pair plasmas almost automatically have $kT \approx m_e c^2$, it is therefore easy to calculate the spectrum, and we shall now concentrate on this case.

The best way to see why there is still a maximum temperature is to consider the total energy losses per unit volume Λ, which, since all radiative processes are two-body processes, scales like $\Lambda \propto N^2$, where N is, as usual, the protons' density in the cloud. The quantity Λ/N^2 is shown in figure 7.2. Notice that the total optical depth is not given by τ_{N}, which has the small values indicated in the figure; it must be supplemented with the pair density, missing from τ_{N}. In any case, for the

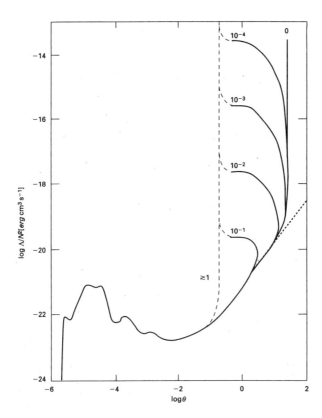

Figure 7.2. Power emitted per unit time and volume Λ, divided by the square of the protons' density N^2. We are not interested in the power at low temperatures, $\log\theta \lesssim -2$. The solid curves represent solutions labeled by their respective values of τ_N; the vertical line marked with 0 is the BKZS limit. The broken curve represents the zone with $\tau_T \gtrsim 1$, where these computations lose accuracy. Adapted from Svensson 1982.

solid curves of the figure, the total optical depth satisfies $\tau_N <$ 1. The broken-line curve represents the estimate of the region in which $\tau_T \gtrsim 1$, where a different treatment of the problem is necessary (see below). We can see that even in the case in which photon processes are included, there is a maximum temperature, which is given by the limit of equation 7.8 for

$\tau_N \ll 1$, but is given by $\theta \approx 0.3$ for $\tau_T \approx 1$. If we add pair creation processes, the maximum temperature we obtain can diminish.

However, we still have to understand how to reach the thermodynamic limit. In order to do this, consider figure 7.3. Here we can see that for each optical depth τ_N of electrons associated to protons, and temperature, there are *two* possible configurations of equilibrium: one with a low pair density, and the other with a high pair density. There is a transition from the one to the other when we reach the maximum temperature compatible with the given value of τ_N. The existence of these two distinct configurations of equilibrium can also be noticed in figure 7.2, where we also see that the power emitted, not just the pair abundance, distinguishes the two configurations.

If we keep N fixed (and therefore also τ_N) and increase the heating rate from a very cold configuration, we see that at first, the configuration's temperature increases, but it decreases after having reached the maximum temperature compatible with the value of τ_N. As a consequence, the configurations with a larger value of z have a *negative* thermal capability, namely, they get colder if we heat them more, because this excess heating is not used to increase the gas temperature but to produce more pairs, which, in their turn, will produce more cooling, since $\tau_T \lesssim 1$. Obviously, these configurations with a negative thermal capability are thermally unstable (see the discussion in section 6.8.2). They tend to form a two-phase fluid with a dense, cold phase (corresponding to larger values of z) immersed in a more tenuous, hotter fluid.

On the other hand, we must still determine what happens in the limit in which $y \gg 1$, or, as argued earlier, when $\tau_T \gtrsim 1$.[3] For $y \gg 1$, we know from the solution of the

---

[3] At this point, we should probably remind you that $\tau_T > 1$ does not necessarily imply that the cloud is optically thick, because, for many photons in these conditions, we must apply the Klein-Nishina cross section, which is $< \sigma_T$.

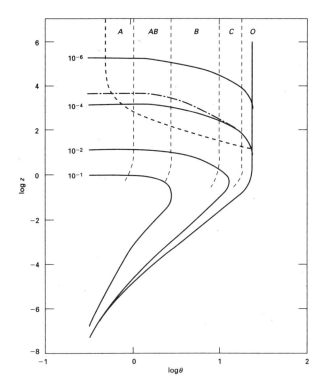

Figure 7.3. Relative abundance of pairs $z = n_-/N$ as a function of temperature. The solid curves are characterized by their values of τ_N. The A, AB, B, and C zones and the broken lines separating them indicate the regimes in which pair production is dominated by, respectively, annihilation of photons produced by annihilations, annihilation of photons produced by annihilation with photons produced by bremsstrahlung, by annihilation of photons produced by bremsstrahlung, and by photons' collisions with particles. Close to the line indicated by 0 (BKZS limit), the particle-particle collisions prevail. Adapted from Svensson 1982.

Kompaneets equation (eq. 3.166) that the occupation number of the energy states, for photons, is given by equation 3.168,

$$\mathcal{N} = \frac{1}{e^{(\mu + \hbar\omega)/kT} - 1} \tag{7.20}$$

which gives us a useful limit. For these distributions, we can (Svensson 1984) determine an analytical relation among L, R, and T, which is illustrated by the broken line in figure 7.4. Notice that even at a fixed τ_N and $\ll 1$, it is possible to reach this limit. Indeed, we should consider those solutions in figure 7.3 with $z \gg 1$, which necessarily give $\tau_T \gg 1$; therefore, a portion of this limit is the limit of the unstable part of the solutions with $\tau_N \ll 1$. At very large optical depths, $\tau_N \to \infty$, we know that the total luminosity must tend to

$$L = 4\pi\sigma_B R^2 T^4 \tag{7.21}$$

where the Stefan-Boltzmann constant has been indicated by σ_B, in order to distinguish it from the Thomson cross section. Therefore, we obviously expect that as τ_N increases, we have to pass from the above-described behavior for solutions with $\tau_N \ll 1$ (which show a maximum temperature), to solutions that instead have no maximum but closely approximate the blackbody solutions, $L \propto T^4$. This is exactly what happens (see fig. 7.4).

Now we must make an important comment: the units of the ordinates on the left-hand side of figure 7.4 are unusual and hide a relevant fact. The definition

$$l \equiv \frac{L\sigma_T}{Rm_e c^3} \tag{7.22}$$

conceals the fact that for most solutions of the figure, $L \gg L_E$. Using equation 7.4, we find

$$\frac{L}{L_{Ec}} = \frac{l}{4\pi} \frac{R}{R_{Sch}} \tag{7.23}$$

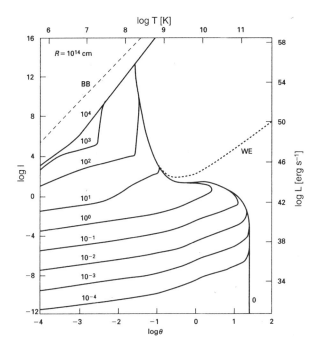

Figure 7.4. Luminosity-temperature diagram for plasmas in chemical equilibrium, marked by different values of $\tau_N = NR\sigma_T$, indicated near each line. The broken-line curve indicated with BB marks the blackbody limit, whereas the one called WE indicates a plasma in which photons satisfy equation 7.20. Svensson 1984.

If we assume that the hot corona occupies a volume with a radius $R \approx R_{Sch}$, we see that we must have $l \lesssim 10$; but if, in accordance with our idea that the hot corona must cover at least the whole A zone of the accretion disk, we require $R/R_{Sch} \approx 10^3$ (see the transition radius from the A to the B zone, eq. 6.70), and we see that $l \lesssim 10^{-2}$. We realize from figure 7.4 that $l \lesssim 0.01$ corresponds to the solutions with $\tau_N \lesssim 1$, namely, those studied in figures 7.2 and 7.3. These solutions have $\tau_T \lesssim 1$ and $T < T_1$, the BKZS limit, but even much less if τ_N is close to the maximum allowed by luminosity, $\tau_N \lesssim 1$.

7.1.4   The Pair Plasma with Input of External Photons

The above analysis can fail in two cases. *First*, various processes that try to establish the (chemical, or thermal Maxwell-Boltzmann) equilibrium may not be in equilibrium. This is, for instance, the case for the Maxwell-Boltzmann distribution, which, for $\theta \gtrsim 3.5$, cannot be reached either through Møller scattering $(e^{\pm} + e^{\pm})$ or through Bhabha scattering $(e^{\pm}+e^{\mp})$ (Gould 1982). When the equilibrium is not reached, the properties of the model can be inferred only from the initial conditions, specific to the model in question. In fact, the corona may be neither in chemical nor in thermal equilibrium, because its outflow (due to the fact that the typical thermal velocities of a pair plasma, $v_{\rm th} \approx c \gg v_{\rm esc}$, the escape speed from the hole's potential) is faster than the time necessary to reach equilibrium.

Second, the analysis does not take into account the fact that the corona is in contact with a cold wall, which emits a lot (i.e., the disk surface). In this case, the corona will become cold thanks to the inverse compton effect on the disk's soft photons, an effect that was left out of the above discussion.

Let us now consider in detail what happens in this case by introducing the argument (Pietrini and Krolik 1995) that illustrates that the gas in question contains a thermostat that strongly affects its temperature, its optical depth, and the spectrum of emitted photons (coming mainly from the process of Comptonization).

The main reason for this work comes from the remark that the X and γ observations in AGNs are always well fitted by a Comptonization spectrum (as in the above-mentioned model by Shapiro, Lightman, and Eardley [1976]), with free parameters (obviously fixed by the fit to observations) always included in very restricted intervals: for the temperature $\theta = 0.06-0.6$, and for the total optical depth $\tau_{\rm T} = 0.1-0.5$. The starting point of the argument is the just-discussed fact that a pair plasma already has a thermostat, consisting of the presence of pairs, which limit the range of accessible temperature and optical depth. The introduction of a source of external photons further reduces the available parameter space.

The Compactness Parameter

When we take into account the luminosity supplied to the cloud by an external medium, a new dimensionless quantity appears, l. If L is the total luminosity of the cloud, l is thus defined[4] as

$$l \equiv \frac{L\sigma_T}{4\pi R m_e c^3} \qquad (7.24)$$

and is often called the *compactness parameter*.

The origin of this definition is as follows. Let us consider a source with radius R, with L_γ as its total luminosity in photons with energy $\gtrsim m_e c^2$, which have a cross section $\approx \sigma_T$ for the creation of pairs with other photons with energy $\gtrsim m_e c^2$. The density of photons is therefore $n_\gamma \approx L/(4\pi R^2 c m_e c^2)$, and the optical depth for the creation of pairs $\tau = n_\gamma R\sigma_T$. This optical depth can also be rewritten as

$$\tau = \frac{L\sigma_T}{4\pi R m_e c^3} \qquad (7.25)$$

In other words, l and τ are the same thing: a measure of the opacity of the source to the production of pairs by photons.

However, the dedimensionalization in equation 7.24 is so handy that it is applied to any luminosity, not just to that of the photons above the pair production threshold; in fact, it will also be used for the luminosity in soft photons. However, in the literature, you can also find it applied to the luminosity to which a cloud is subject, in the form of nonthermal particles, Alfvén waves, magnetic reconnection, and so on. Keeping this in mind, we shall use it in the following section.

---

[4] According to the author, the factor 4π may or may not be present.

The Cloud's Thermal Structure

Our initial assumption is that the cloud is subject to illumi-
nation by (the remaining part of) the disk, as well as to some
form of heating. We shall express the luminosity of heating
(namely, the quantity of energy absorbed by the gas per unit
time) through a parameter of compactness l_h. In the same
way, we assume that a fraction of this luminosity of heating
is in the form of soft photons, with energy $\ll kT, m_e c^2$; once
again, l_s is the parameter of compactness corresponding to
this luminosity.

The initial energy of the photons (i.e., before being sub-
mitted to the Comptonization process) is not important, the
results are weakly dependent on it. The new idea consists in
assuming that the ratio l_s/l_h is a constant. The physical motivation for this fact (Haardt and Maraschi 1991) is that the gas
luminosity, which, in thermal equilibrium, equals the heating
rate l_h, is, at least in part, absorbed by the disk since this covers a fair fraction of the sky as seen from the corona; this energy is then thermalized and reemitted as lowfrequency photons toward the corona. If the gas is arranged like a corona,
then (roughly!) half of the whole emission of the corona is in-
tercepted by the disk and reemitted in the form of lowenergy
photons: $l_s/l_h = 1/2$. On the other hand, if the geometry is
that of figure 7.1, the ratio l_s/l_h will be smaller but still a
constant.

Let us start, however, *without* considering l_s/l_h as constant to compare the solutions of Pietrini and Krolik with
those of Svensson. Let us therefore fix l_s and impose that the
pairs' creation and destruction rates are equal; this fixes one
of the three dimensionless variables characterizing the chemical and thermal state of the plasma, τ_N, θ, z as a function
of the other two, as well as of l_s. At this point, let us also
calculate the energy loss rate (or total luminosity, if you prefer), and, if we assume that there is a thermal equilibrium, we
should assume that this equals the heating rate. For a specific
value $\tau_N = 1$, we find the results of figure 7.5. As in the preceding paragraph, there are two configurations of equilibrium

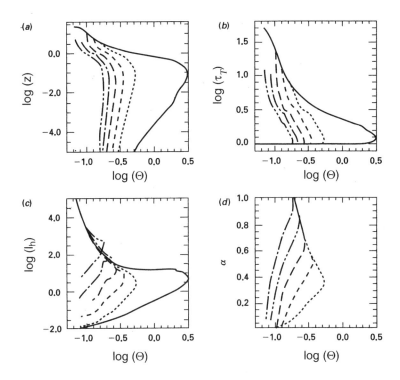

Figure 7.5. Pair abundance z, total optical depth τ_T, heating rate l_h, and spectral index as a function of temperature θ for a cloud with $\tau_N = 1$. The various curves indicate different luminosities l_s in soft photons: the solid curve has no input ($l_s = 0$), whereas the others have $l_s = 3 \times 10^{-3}$ (dots), then increase by a factor of 10 (short dash, long dash, dots and short dash, dots and long dash). From Pietrini and Krolik 1995.

for each temperature, one with a low pair content, the other with a large pair content. Naturally, the addition of external photons always lowers the temperature with respect to the case without external photons, since the latter are subject to Comptonization and effectively remove energy. On the other hand, when the abundance of pairs z becomes huge, the emission l_h converges to the curve without input of photons, since, obviously, the total luminosity produced within the cloud, in this case, exceeds that supplied from the outside, l_s, by a large factor.

Perhaps the most important point is that the maximum temperature of these models is now appreciably reduced, so that they occupy a very restricted range of θ; it is more or less a factor 3, compared to a really huge variation (six orders of magnitude) of l_h (see the third box of fig. 7.5). The reason is that the excess of external photons increases the cooling, in the presence of many target pairs; and allows adjustment of the number of pairs; this increases z in the low z solution, and decreases it in the high z one. This is therefore a thermostatic action.

We can see exactly the same thing in a slightly different way. Let us specify l_s and l_h and impose that the equations of chemical and thermal equilibria are verified. Since these are two equations, we can specify two of the three dimensionless quantities that characterize the cloud as a function of one of the others and of l_s and l_h. In order to do this, let us fix the ratio $l_s/l_h = 1$ and let τ_N vary; we can see the corresponding quantities in figure 7.6. The remarkable thing in this figure is that all quantities vary very little, whereas the heating rate varies by 6 orders of magnitude. The variation range is very small, at a fixed τ_T and when τ_T varies by about 1 order of magnitude.

Therefore, the main point of these computations is represented by this conclusion: the temperature of a pair plasma, in the presence of an input of external photons, varies very little around values of $\theta \approx 0.3$, almost independently from the value of the heating rate, provided the conditions (Maxwell-Boltzmann, thermal, and chemical) of the equilibrium and the approximate equality $l_s \approx l_h$ are satisfied.

7.2   Thick Accretion Disks

The physical motivation for considering thick accretion disks consists in the fact that they leave an approximately conic

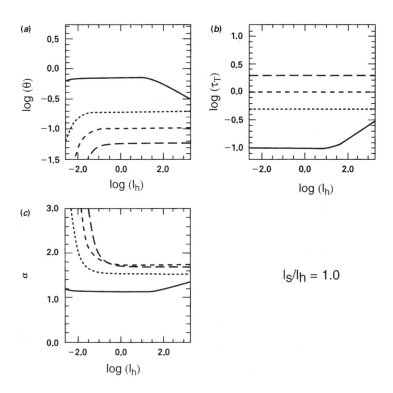

Figure 7.6. Temperature θ, total optical depth τ_T, and spectral index as a function of the cloud's heating rate; the various curves (solid, dots, short dash, long dash) correspond to $\tau_N = 0.1, 0.5, 1, 2$, respectively. From Pietrini and Krolik 1995.

region around the rotation axis, called a *funnel*, void of matter, with a large radiation density; in this environment scientists initially thought jets could be easily accelerated. However, this hope was never fulfilled, and, to make things worse, at least some important classes of thick disks are dynamically unstable. Despite these difficulties, thick disks are obviously omnipresent in astrophysics—just think of the coalescence of two neutron stars in a binary after the formation of the central black hole, or in the *collapsar* models for gamma ray bursts (Proga et al., 2003). It seems therefore important to study these models in any case, even though they do not apply to AGNs.

Unlike thin disks, there is no *standard* model for thick disks, and attention has so far focused on analytically tractable models. This is why we shall present here only a series of useful and general results and a simple model. In particular, there is so far no wholly self-consistent model showing an accretion speed.

7.2.1   Some General Properties

For the above reason, let us start by considering a fluid model in pure rotation, where, unlike in thin disks, the contribution of the pressure gradient cannot be neglected. The equations of motion are

$$\frac{1}{\rho}\nabla p = -\nabla\phi + \omega^2\vec{R} \qquad (7.26)$$

The last term on the right-hand side is the centrifugal force, and $R = (x^2 + y^2)^{1/2}$ is the radial coordinate in cylindrical coordinates. The first useful result is that in general, the surfaces with $p = $ constant (*isobars*) differ from those with $\rho = $ constant (*isopycnae*). In fact, if we take the curl ($\nabla\wedge$) of the equation, since the curl of a gradient vanishes, we find

$$\nabla\left(\frac{1}{\rho}\right)\wedge\nabla p = \nabla\wedge(\omega^2\vec{R}) = 2\frac{\partial\omega}{\partial z}\vec{v} \qquad (7.27)$$

Now, if the isopycnae and isobars coincided everywhere, the left-hand side would vanish. But this is possible only if

$$\frac{\partial\omega}{\partial z} = 0 \qquad (7.28)$$

A rotation law with this property is called *cylindrical*. When this happens, on each surface with $p = $ constant we also have $\rho = $ constant, and there is therefore a unique relationship between the two:

$$p = p(\rho) \qquad (7.29)$$

Models of this kind are called *barotropic* and are distinct from the *polytropic* models: each polytropic model is obviously barotropic, but the inverse is not true. The reason for this is that a barotropic relation is not due to some general thermodynamic property linking p to ρ, but to a *dynamic* relation, equation 7.28. Therefore, an arbitrary equation of state may have both barotropic models (in which eq. 7.28 holds), and nonbarotropic models. Needless to say, in the literature one finds almost strictly polytropic models, which are necessarily barotropic, because of their relative mathematical simplicity.

Let us therefore assume a cylindrical rotation law; in this case, the centrifugal force can be expressed by means of a potential,

$$\phi_{\rm rot} = \int \omega^2 R dR \qquad (7.30)$$

and the equation of motion becomes

$$\frac{1}{\rho}\nabla p = -\nabla(\phi - \phi_{\rm rot}) \equiv -\nabla \phi_{\rm eff} \qquad (7.31)$$

The equation for the isobars is obviously given by

$$\phi_{\rm eff} = \phi - \phi_{\rm rot} = {\rm constant} \qquad (7.32)$$

Also in the case of thick disks, the primary source of gravity is given by the external object, in which case $\phi = -GM/r$. Then the above equation immediately gives isobars as a function of the parameter $\phi_{\rm eff}$:

$$\phi_{\rm eff} = -\frac{GM}{(R^2 + z^2)^{1/2}} - \phi_{\rm rot}(R) \qquad (7.33)$$

This is an equation giving z as a function of R, for a given $\phi_{\rm eff}$. In this way, we obtain the equation for all isobars. A nice thing, furthermore, is that we need to know the form of *only one isobar* in order to find the law of rotation. In fact, if we know that $z_{\rm R} = z(R)$ is an isobar, and put this result in the equation above, we immediately find ϕ_{rot}. The most important isobar is the one where $\rho = 0$ (remember that isobars and isopycnae coincide in this case), which is the outer surface of the disk. Therefore, if we know just the form of the outer surface of the disk, I can find, first of all, the law of rotation and then, through equation 7.31, pressure and density everywhere.

We present the following quantitative example because it often happens that numerical simulations show thick disks with a very simple outer surface:

$$z_{\text{R}} = \pm qR \qquad (7.34)$$

In this case, the model of cylindrical rotation we are discussing is used to compare the simulations. Since the value of ϕ_{eff} on a given surface is arbitrary, we may take $\phi_{\text{eff}} = 0$ on the outermost surface. We must therefore have

$$\phi_{\text{eff}} = 0 = -\frac{GM}{(R^2 + z^2)^{1/2}} - \phi_{\text{rot}} \qquad (7.35)$$

when equation 7.34 holds. Thus,

$$\phi_{\text{rot}} = -\frac{GM}{R}\frac{1}{(1 + q^2)^{1/2}} \qquad (7.36)$$

From this, we immediately find the law of rotation,

$$\omega^2 = \frac{1}{R}\frac{d\phi_{\text{rot}}}{dR} = \frac{GM}{R^3}\frac{1}{(1 + q^2)^{1/2}} \qquad (7.37)$$

which is a Keplerian rotation law, for a mass smaller than the real one by a factor $1/\sqrt{1 + q^2}$. Naturally, the pressure gradient compensates the gravitational attraction of the remaining fraction of the mass, $1 - 1/\sqrt{1 + q^2}$. The form of all isobars may also be found parametrically, with ϕ_{eff} as the parameter. We easily find

$$r = \frac{GM}{-\phi_{\text{eff}}}\left(1 - \frac{1}{\sin\theta\sqrt{1 + q^2}}\right) \qquad (7.38)$$

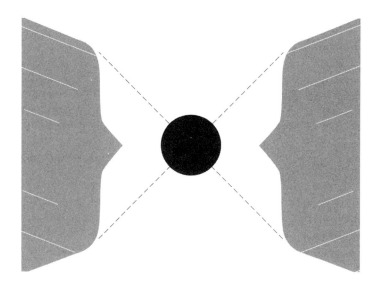

Figure 7.7. Geometry of a thick disk, with its cusp.

We have followed a purely Newtonian treatment; it is possible to give an entirely general-relativistic treatment (Abramowicz, Jaroszynski, and Sikora 1978), and also one based on the Paczynski and Wiita potential (1980). When models are built in either of these two ways, an interesting property appears, which does not hold for Newtonian gravity: all thick accretion disks have a cusp, and the cusp is occupied by a circular orbit, placed between the marginally stable orbit and the marginally bound one (see fig. 7.7). When matter reaches this orbit, it may then fall freely on the black hole.

7.2.2 The Inapplicability of the Eddington Limit

One of the reasons for proposing thick accretion disks is that although they are very close to a Bondi-like spherical accretion flux, they can nevertheless emit a luminosity larger than the Eddington limit. We illustrate this result here, following the original discussion by Abramowicz, Calvani, and Nobili (1980).

We have seen in section 3.1.1, equation 3.35, that the radiation flux is given by

$$\vec{F}_{\rm R} = -\frac{ac}{3\alpha_{\rm R}}\nabla T^4 = -\frac{c}{\alpha_{\rm R}}\nabla p_{\rm R} \qquad (7.39)$$

where, as usual, $\alpha_{\rm R}$ is the Rosseland average of the absorption coefficient, and we have used the equation of state for the radiation, $p_{\rm R} = aT^4/3$. Obviously, in order to maximize the flux from the disk, we may take a pressure gradient exclusively due to radiation. Normally in these conditions, the absorption coefficient $\alpha_{\rm R}$ is due mostly to scattering, $\alpha_{\rm R} = \sigma_{\rm T}\rho/m_{\rm p}$, so equation 7.26 becomes

$$\vec{F}_{\rm max} = \frac{cm_{\rm p}}{\sigma_{\rm T}}(\nabla\phi - \omega^2\vec{R}) \qquad (7.40)$$

In order to know the total luminosity, we need to integrate this quantity on the whole surface of the thick disk:

$$L_{\rm max} = \frac{cm_{\rm p}}{\sigma_{\rm T}}\left(\int \nabla\phi \cdot d\vec{A} - \int \omega^2\vec{R} \cdot d\vec{A}\right)$$

$$= \frac{cm_p}{\sigma_{\rm T}}\left(\int(\nabla^2\phi - \nabla \cdot (\omega^2\vec{R}))dV\right) \qquad (7.41)$$

where we have used Gauss's theorem. We remind the reader that $\nabla^2\phi = 4\pi G\rho$, so the first integral simply becomes

$$\frac{cm_{\rm p}}{\sigma_{\rm T}}\int \nabla^2\phi dV = \frac{4\pi GM_{\rm d}cm_{\rm p}}{\sigma_{\rm T}} \qquad (7.42)$$

which is the Eddington luminosity *for the disk mass only.* Indeed, the disk surface does not contain the black hole, so the volume integral extends to the disk mass only; however, since this is still small compared with that of the black hole, we shall neglect it from now on.

In order to rewrite the second term, we use

$$\nabla \cdot (\omega^2\vec{R}) = \omega^2\nabla \cdot \vec{R} + \vec{R} \cdot \nabla\omega^2 = 2\omega^2 + R\frac{\partial\omega^2}{\partial R}$$

$$= \frac{1}{2}\left(\left(\frac{1}{R}\frac{\partial}{\partial R}(R^2\omega)\right)^2 - \left(R\frac{\partial\omega}{\partial R}\right)^2\right) \qquad (7.43)$$

We use partial derivatives for R, instead of total derivatives, because we have *not* assumed so far that the fluid is barotropic (eq. 7.28). Therefore, when we replace this expression in equation 7.41, we find a result which applies to any model in pure rotation, not just to barotropic ones:

$$L_{\max} = \frac{cm_p}{2\sigma_T} \left(\int \left(R\frac{\partial\omega}{\partial R} \right)^2 dV - \int \left(\frac{1}{R}\frac{\partial}{\partial R}(R^2\omega) \right)^2 dV \right) \quad (7.44)$$

The first term is the energy in differential rotation, or shear, whereas the second one is a term that vanishes when the specific angular moment, $l = \omega R^2$, is constant in radius. Soon we shall see that this case, $\partial l/\partial R = 0$, is particularly important; there is only the first term. This term is dominant in any case and is also responsible for the fact that the thick disk may exceed the Eddington luminosity.

An obvious example is the disk defined by equation 7.34. For the total luminosity emitted between two radii, R_1 and R_2, we can easily find from the above equation

$$L = \frac{4\pi GMcm_p}{\sigma_T} \frac{q}{\sqrt{1+q^2}} \ln\frac{R_2}{R_1} \quad (7.45)$$

Here M is the central mass of the black hole, not of the disk. Thus, we should simply take a sufficiently large dynamic range, $R_2 \gg R_1$, in order to obtain $L \gg L_E$. The fact that *one* model violates the Eddington limit means that it is not a general constraint any more. Indeed, many models proposed in the literature have this feature. In general, this very large luminosity can be a problem for continuous sources, because it can be maintained only in the presence of very large accretion rates. On the other hand, for short-term sources, such as the coalescence of neutron stars, and collapsars, this simple exercise indicates that there is no conceptual problem in obtaining very high luminosities.

7.2.3   Polytropic Models

The polytropic models play an important role within the barotropic category. Particular attention has been paid to models with rotation law

$$\omega = \omega_{\mathrm{K}} \left(\frac{R_{\mathrm{K}}}{R} \right)^q \qquad (7.46)$$

where R_{K} is a fiducial radius where the angular velocity equals the Keplerian one. For $q = 3/2$, the disk is in centrifugal equilibrium; the model with this q must then reduce to the thin disk studied above. We also know that according to Rayleigh's criterion, the models with $d(\omega R^2)/dR < 0$ (corresponding to $q > 2$) are unstable to axially symmetric perturbations. The models with $q < 3/2$ must also be discarded, because for $R > R_{\mathrm{K}}$ their rotation speed is higher than the Keplerian one, so that beyond the force of gravity, there must be other forces that provide a centripetal acceleration. Since this seems unlikely, it follows that configurations with the following property,

$$3/2 \le q \le 2 \qquad (7.47)$$

have mainly been investigated so far.

In the case of a polytropic fluid, we can immediately apply Bernoulli's theorem, equation 1.26,

$$\nabla \left(w + \phi + \phi_{\mathrm{rot}} \right) = 0 \qquad (7.48)$$

where the specific enthalpy is $w = \gamma p/((\gamma - 1)\rho)$ for a polytropic equation of state, $p \propto \rho^\gamma$. Now we shall specialize (Papaloizou and Pringle 1984) to the case $q = 2$, which corresponds to constant specific angular momentum, $l = \omega R^2 = l_{\mathrm{o}} = $ constant; sometimes, in the literature this is even called the zero vorticity case, because, if $q = 2$, we can easily check that vorticity vanishes, $\nabla \wedge \vec{v} = 0$. In any case, from equation 7.30 we find

$$\phi_{\mathrm{rot}} = \frac{l_{\mathrm{o}}^2}{2R^2} \qquad (7.49)$$

We should also define a distance $R_\circ \equiv l_\circ^2/GM$, which is simply the distance at which a particle with specific angular momentum $l_\circ$ would rotate in the potential of a pointlike object with mass M. From the equation of motion, equation 7.26, we see that this point is where the pressure maximum occurs. Using the above equation, as well as the definition of $R_\circ$ in Bernoulli's theorem, we find

$$w = \frac{GM}{R_\circ}\left(\left(\frac{R_\circ^2}{R^2+z^2}\right)^{1/2} - \frac{1}{2}\left(\frac{R_\circ}{R}\right)^2 - C\right) \qquad (7.50)$$

with C an integration constant. The surface of the thick disk is the place where $p = 0$, whose equation is

$$\left(\frac{R_\circ^2}{R^2+z^2}\right)^{1/2} - \frac{1}{2}\left(\frac{R_\circ}{R}\right)^2 = C \qquad (7.51)$$

The points where the surface of the disk cuts the plane $z = 0$ are given by

$$R_\pm = \frac{R_\circ}{1 \mp \sqrt{1-2C}} \qquad (7.52)$$

This equation shows that we always have $2C \le 1$ for a physical solution. A measure of the distortion is given by the quantity $(R_+ + R_-)/(2R_\circ) = 1/(2C)$. When $2C \to 1$, the torus degenerates to an infinitely thin ring, while, when $2C \to 0$, the two extremes tend to $R_+ \to R_\circ/2$ and $R_- \to \infty$.

One of the features of this and other similar models is a thin funnel that is completely empty, directed along the z axis; we have just seen that the innermost point of the outer surface, R_+, varies in the interval $R_\circ/2 \le R_+ \le R_\circ$ and is always $R_+ > 0$. The funnel, of course, is full of radiation emitted by the disk surface; the energy density has an obvious peak at $r = z = 0$ and decreases as $|z|$ increases. Thence the hope to use this funnel, as well as its radiation pressure, like a sort of slide along which jets are accelerated. We may also find the opening θ of the funnel (see fig. 7.7). Indeed, it is only a matter of finding the point of tangency between the disk surface, equation 7.51, and the straight lines of the type $z = \tan \theta R$. We find

$$\sin^2 \theta = 2C \qquad (7.53)$$

which gives us another geometrical interpretation of the quantity C. In fact, we see that for $2C \to 1$, the funnel is completely open ($\theta = \pi/2$), whereas it tends to close more and more as $2C \to 0$.

7.2.4 Properties of Thick Disks

The stability of the model we have just introduced was studied by Papaloizou and Pringle in 1984. It satisfies the Rayleigh criterion for the stability against small axially symmetric perturbations, but Papaloizou and Pringle studied small perturbations of type $\delta X \propto e^{im\phi - i\omega_p t}$, where ω_p, the *eigen-frequency of the mode*, can be complex. They found, with a certain surprise, that these modes are unstable, and the time scale for the growth of the instability is dynamic, $t_{din} \approx 1/\omega$. Therefore, the instability is violent and fast. Later studies have shown that for $q > \sqrt{3}$ there are always unstable modes, and their time scale of growth is always dynamic. In general, the nonlinear regime of this instability leads to the fragmentation of the disks into many small *counterrotating* planets, namely, rotating in the direction opposite to the disk.

Attempts to stabilize thick disks have used the disk self-gravity and matter transport (Blaes 1987), but they have not been very successful. The physical mechanism of instability is easy to explain. A mode depending on ϕ and t of the type $\delta X \propto e^{i(m\phi - \omega_p t)}$ can also be seen as a wave propagating in radius. Let us call r_{cr}, or the *corotation radius*, the point where the Keplerian frequency equals ω_p/m, namely, $\omega_K(r_{cr}) = \omega_p/m$. The unusual thing is that these waves, for $r > r_{cr}$, have a positive energy density, but, for $r < r_{cr}$, they have a negative energy! When a wave arrives at the corotation radius from the region $r > r_{cr}$, it is partly transmitted in the region $r < r_{cr}$, and partly reflected in the same region $r > r_{cr}$. However, in order to conserve the total energy, the amplitude of the reflected wave will be larger than that of the incident one, since the transmitted wave has a negative energy! It is now easy to imagine various scenarios in which this phenomenon increases indefinitely; for example, if the transmitted wave is reflected at the inner edge of the disk, when it goes back to the corotation radius it generates a wave in the region $r > r_{cr}$, and, in order to assure conservation of energy it will increase in amplitude. This mechanism is the same which Toomre proposed for the generation of spiral waves in the disks made of stars (Binney and Tremaine 1987). We can now understand why the presence of a strong accretion speed may perhaps stabilize thick discs, as suggested by Blaes: if matter falls quickly on the black hole, in the region of corotation the waves may be subject to advection by the fluid and no longer manage to go back to the region $r > r_{cr}$.

The other hope, which had initially caused the investigation of thick disks, was the possibility of accelerating matter to large Lorentz factors along the central funnel, using the radiation pressure gradient. From numerical simulations, however, it emerged that the final Lorentz factor, even in the absence of protons, is never larger than $2-3$, and, when protons are introduced, the outflow speed is reduced to $\lesssim c/3$. It is easy to understand, at least qualitatively, how this happens: there is a Compton drag effect, due to the interaction with photons produced by the disk, which limits acceleration.

In order to see this, we calculate, following an argument by Krolik (1999), the loss (or gain) of energy by an electron moving with a Lorentz factor γ in a radiation field with a specific intensity $I_\nu(\hat{n}_i)$, where $\hat{n}_i$ identifies the photon's direction of motion *before* colliding with the electron. The photon's change of energy ϵ during a collision is given by (having used eqs. 3.121 and 3.124)

$$\triangle\epsilon = \epsilon_i \left(1 - \gamma^2(1 - \beta\cos\theta)(1 + \beta\cos\theta_1)\right) \qquad (7.54)$$

where θ is the angle between the direction of motion of the electron and of the incident photon in the laboratory reference frame, while θ_1 is the angle between the direction of motion of the electron and of the outgoing photon, measured, however, in the electron's frame of rest.

We now recall that the Thomson scattering has a back-front symmetry, so that the average value of $\langle \cos \theta_1 \rangle = 0$. Moreover, the relative speed between photons and electron is given by $c(1 - \beta \cos \theta)$, so that the electron's energy gain is given by

$$m_e c^2 \frac{d\gamma}{dt} = \int d\nu \int d\Omega I_\nu(\hat{n}_i) \sigma_T (1 - \beta \cos \theta)(1 - \gamma^2(1 - \beta \cos \theta))$$

$$= \int d\nu \int d\Omega I_\nu(\hat{n}_i) \sigma_T \beta \gamma^2 (1 - \beta \cos \theta)(\cos \theta - \beta) \quad (7.55)$$

The last term in parentheses is crucial, since it tells us that the contribution to the integral is positive (i.e., the electron gains energy) only if the photon's direction of motion is well aligned with the electron's. In fact, since $1 - \beta \approx 1/(2\gamma^2) \ll 1$, in order to have an acceleration of the electron, we need to have $\theta \lesssim 1/\gamma$, an almost perfect alignment.

For the radiation field, we now make an extreme assumption: let us assume that it is made up of two components, an isotropic one and another one perfectly focused forward:

$$I(\hat{n}_i) = \frac{F}{2\pi} \delta(\hat{n}_i - \hat{z}) + J \quad (7.56)$$

If we now put this formula into the preceding one, we find

$$m_e c^2 \frac{d\gamma}{dt} = \sigma_T \left(\frac{\beta(1 - \beta)}{1 + \beta} F - \frac{16\pi}{3} \beta^2 \gamma^2 J \right) \quad (7.57)$$

which, in the limit $\beta \to 1$, gives

$$m_e c^2 \frac{d\gamma}{dt} = \sigma_T \left(\frac{F}{4\gamma^2} - \frac{16\pi}{3} \gamma^2 J \right) \quad (7.58)$$

whence we find that there is a limiting Lorentz factor, given by

$$\gamma_{\text{lim}} = \left(\frac{3F}{64\pi J} \right)^{1/4} \quad (7.59)$$

Whence we see that in order to reach a modest Lorentz factor $\gamma \approx 10$, we must have $F/J \gtrsim 10^6$, which seems, at the moment, unlikely. For example, near a black hole, we expect $F \approx J$. That is why it seems reasonable to conclude that radiation *decelerates* the jet, instead of accelerating it.

Naturally enough, the disks we have discussed so far are not true accretion disks, but only figures of equilibrium: in the absence of viscosity, or, in general, of angular momentum transport processes, nothing really falls on the black hole. When dissipative effects are included in the discussion, we find that there is a paradox: in the presence of cylindrical rotation, it is *locally* impossible to satisfy the energy equilibrium. This situation is well known in the physics of stellar interiors, where it is called the *van Zeipel paradox*. Its solution is that the equation of local thermal equilibrium we have used assumes a fluid in pure rotation, but that in general, motions superposed to pure rotation are necessary, also along closed trajectories. In stellar interiors, these motions lead to the so-called *meridional circulation*. It is obvious that this complicates the analysis of the configurations, so much so that of the interesting physical problems like the merging of a binary pulsar or the collapsar model, there exist only numerical simulations.

7.3   Nondissipative Accretion Flows

When we computed the local temperature of thin accretion disks, we remarked that it was necessary to assume that the heat produced by the viscosity was released in loco. This approximation fails for low-density, low-accretion-rate flows $\dot{M}$. As we saw in chapter 3, this is due to the fact that the main radiative mechanisms in the disk (bremsstrahlung and the excitation of forbidden or semi forbidden lines) are collisional, so that the cooling time scales as $1/\rho$. For low-density flows, this cooling time may exceed the accretion time, $t_a \equiv r/v_R$, in which case the heat due to viscosity is absorbed *in place*, but is removed by advection (i.e., by the accretion process) before being radiated away. In this case, the gas cannot radiate away the heat absorbed. Even in the other limit, though, the limit of large accretion rates, the heat produced locally is not radiated away because, as seen in chapter 5 (section 5.5) for spherical accretion, the scattering time for photons exceeds their advection time.

Since protons cannot radiate, they must get hotter until they reach the so-called virial temperature $T_{\rm vir}$:

$$T_{\rm vir} \equiv \frac{GMm_{\rm p}}{kr} = 10^{13} \; {\rm K} \frac{r_{\rm s}}{r} \qquad (7.60)$$

This derives only from energy conservation. The fate of electrons is much less clear: a gas of electrons, at this temperature, would be hyperrelativistic and would therefore have no difficulty in radiating huge quantities of energy. As a consequence, electrons are likely to have a temperature different from that of protons, with $T_{\rm p} \gg T_{\rm e}$. The exact value of $T_{\rm e}$ determines the total emissivity of the flow and, therefore, its observability, but it is extremely difficult to predict because the processes of energy exchange between electrons and protons are (probably) not due to Coulomb scattering, which, as we saw in chapter 2, is very slow (Coulomb Scattering) but is due instead to collective processes, such as the interactions between particles and waves described by Begelman and Chiueh (1988). Unfortunately, these processes are very uncertain and have been a topic of lively discussion in the literature, without a general consensus emerging.

It is, however, worthwhile to study an idealized model in which protons and electrons have different temperatures. The main argument supporting this approximation is that the magnetohydrodynamic turbulence generated in the disk heats protons preferentially, rather than electrons, because of their heavier mass. In this way, we can assume $T_{\rm e} = 0$. At the same time, protons do not radiate, so we can assume that all the heat produced locally is absorbed by protons, which then carry it toward the compact object through advection.

The original idea for a model of this type was proposed by Paczysnki and Wiita (1980), and this idea was taken up with new impetus by Narayan and Yi (1994, 1995), who coined the acronym ADAFs (advection-dominated accretion flows). We shall present here the self-similar solution for ADAFs (Narayan and Yi 1994). It is an approximate solution, based on three hypotheses. The first one is that we can treat plasma like a one-temperature fluid. This is possible if electrons and protons have the same temperature (in which case, electrons will emit copiously), or when electrons have a negligible temperature compared with that of protons, $T_e \ll T_p$. In this case, the only fluid we use for the description is that of protons, whereas the electrons' fluid is dynamically negligible (apart from providing charge neutrality, obviously). The second one is that the flow may be assumed to be geometrically thin, even though it is not. In fact, Narayan and Yi use the equations integrated on the disk thickness, as if they were dealing with a thin disk, when the ions' temperature is almost virial, so that the disk cannot be thin. In a later work (Narayan and Yi 1995) this hypothesis is abandoned in order to find (numerically) the exact solution; this solution differs very little from the approximate one, which is therefore presented here because of its simplicity and accuracy. The third hypothesis will be discussed later on.

The equations for mass and radial momentum conservation are

$$\frac{d}{dR}(\Sigma R v_R) = 0 \qquad (7.61)$$

$$v_R \frac{dv_R}{dR} - \omega^2 R = -\omega_K^2 R - \frac{1}{\rho}\frac{d}{dR}(\rho c_s^2) \qquad (7.62)$$

where c_s^2 is the isothermal sound speed: $c_s^2 \equiv p/\rho$. Here ω is defined as $\omega \equiv v_\phi/R$, and this cannot be taken equal to the Keplerian rotation speed $\omega_K \equiv (GM/R^3)^{1/2}$, because we are assuming that the gas may be hot, in which case (see also the case of thick disks) its rotation speed must be sub-Keplerian. The equation of angular momentum transport is equation

6.10, which, thanks to equations 6.11 and 6.37 and to the definition of kinematic viscosity with the α prescription,

$$\nu = \alpha c_{\mathrm{s}} H = \alpha \frac{c_{\mathrm{s}}^2}{\omega_K} \tag{7.63}$$

becomes

$$v_{\mathrm{R}} \frac{d}{dR}(\omega R^2) = \frac{1}{\rho R H} \frac{d}{dR} \left(\frac{\alpha c_{\mathrm{s}}^2 R^3 H}{\omega_K} \frac{d\omega}{dR} \right) \tag{7.64}$$

Finally, we need an equation for energy conservation, which we take in the form

$$\frac{\partial s}{\partial t} + \vec{v} \cdot \nabla s = v_{\mathrm{R}} \frac{ds}{dR} = q^+ - q^- \tag{7.65}$$

Here we have used the assumption of stationarity ($\partial/\partial t = 0$) and of axial symmetry, so that only the derivative ds/dr does not vanish. The coefficients q^+ and q^- take into account viscosity heating and as radiative losses. For a polytropic gas with an index γ,

$$\delta s = \frac{\delta p}{p} - \gamma \frac{\delta \rho}{\rho} \tag{7.66}$$

Thence we deduce that heating, per unit time and mass, is given by

$$T v_{\mathrm{R}} \frac{ds}{dR} = \frac{3 + 3\epsilon}{2} v_{\mathrm{R}} \frac{dc_{\mathrm{s}}^2}{dR} - T v_{\mathrm{R}} \frac{d\rho}{dR} \tag{7.67}$$

where

$$\epsilon \equiv \frac{5/3 - \gamma}{\gamma - 1} \tag{7.68}$$

Now, we already know that the local heating rate Q^+ (i.e., the average energy gain per unit surface and time, which differs from q^+, the average entropy gain per unit mass and time), in the presence of viscosity is given by equation 6.18, which can be used together with equation 6.37 to obtain

$$Q^+ = \frac{2\alpha c_{\mathrm{s}}^2 \rho R^2 H}{\omega_k} \left(\frac{d\omega}{dR} \right)^2 \tag{7.69}$$

Following the discussion above, we know that we cannot assume that the local cooling rate, Q^-, equals $Q^- = Q^+$. We therefore define a new parameter f as

$$\frac{3+3\epsilon}{2}\Sigma v_R \frac{dc_s^2}{dR} - \Sigma T v_R \frac{d\rho}{dR} = Q^+ - Q^- \equiv fQ^+ \qquad (7.70)$$

If we now assume that the quantity $\epsilon' \equiv \epsilon/f$ is independent of radius, we can find an extremely simple solution of the system of equations 7.61, 7.64, and 7.70; this solution depends only on ϵ', not on f or ϵ separately. Obviously, ϵ' need not be constant with radius; it is only a handy assumption that allows us to find a particularly simple solution for our problem. This is the last of the above-mentioned three assumptions.

The system of equations 7.61, 7.64, and 7.70 has a self-similar solution, which was discovered by Spruit et al. (1987), with the following scalings:

$$\rho \propto R^{-3/2}, \quad v_R \propto R^{-1/2}, \quad \omega \propto R^{-3/2}, \quad c_s^2 \propto R^{-1} \qquad (7.71)$$

To give the exact solution, it is convenient to define a quantity

$$g \equiv \left(1 + \frac{18\alpha^2}{(5+2\epsilon')^2}\right)^{1/2} - 1 \qquad (7.72)$$

The complete solution is therefore

$$v_R = -\frac{5+2\epsilon'}{3\alpha} g\omega_K R \qquad (7.73)$$

$$\omega = \left(\frac{2\epsilon'(5+2\epsilon')g}{9\alpha^2}\right)^{1/2}\omega_K \qquad (7.74)$$

$$c_s^2 = \frac{2(5+2\epsilon')g}{9\alpha^2}\omega_K^2 R^2 \qquad (7.75)$$

The easiest way to check these solutions is by replacement in the original equations.

Originally, this solution was discussed by Spruit et al. only in the limit $f \ll 1$, corresponding to the case of efficient radiative cooling. Not surprisingly, in this case we obtain once again the standard solution, corresponding to the case of thin accretion disks, studied in the preceding chapter. On the other hand, the most interesting limit is the opposite one, when $f \to 1$. We can easily see that $\omega < \omega_K$; the rotation is sub-Keplerian. The radial accretion speed, v_R, is larger than in the case of the thin disk, where $v_R = \alpha c_s^2/(\omega_K R)$, by a factor $\omega_K R/c_s$; matter falls quickly on the compact object. However, the flow is never transonic, since the ratio v_R/c_s does not depend on the radius; the solution is either subsonic or supersonic, everywhere. As a consequence, the limiting solution for $f \to 1$ cannot tend to any of the Bondi solutions, because they have a sonic point, unless $\gamma = 5/3$, which corresponds to $\epsilon = \epsilon' = 0$, when the newly found solution is exactly Bondi's for $\gamma = 5/3$.

These solutions are also convectively unstable; the generated entropy gradient is so large, that a strong convection is triggered. However we can easily take into account this extra heating source for the outermost layers, without changing the discussion so far (Narayan and Yi 1994).

The new aspect of these solutions is their (potential) capacity to generate a jet from the disk accretion. Here is the argument. Once we know the solution, we can determine the constant that appears in Bernoulli's theorem:

$$\frac{v_R^2 + \omega^2 R^2}{2} - \omega_K^2 R^2 + \frac{\gamma}{\gamma - 1}c_s^2 \equiv \mathrm{Be} \qquad (7.76)$$

Here the term $-\omega_K^2 R^2$ is equal to the gravitational potential ϕ. If we use equation 7.73, we find

$$\mathrm{Be} = \omega_K^2 R^2 \frac{3\epsilon - \epsilon'}{5 + 2\epsilon'} \qquad (7.77)$$

Thence we see that provided $f \geq 1/3$, Be > 0. The Bernoulli constant is conserved in isentropic motions, namely, in the absence of viscosity. The fact that Be > 0 means that if only we could reverse the flow, this would have enough energy to go to infinity. This is not in contradiction with the fact that the material is bound and is accreting on the compact object. In fact, we know that the viscous stress transfers energy from small to large radii, so this energy excess at each radius is exclusively due to viscous phenomena. This is just what happens in the disks in which the energy radiated at each radius is 3 times larger than the one released locally, for $\beta = 0$, equation 6.22. Unfortunately, we cannot discuss what happens at the edge of the disk (namely, when $R \to 0$ and $R \to \infty$), but we are pretty sure that if we were capable, we would obtain a system that strictly respects energy conservation.

One may suppose that this energy excess available in loco may be used in some way to start a wind or bipolar jets (i.e., directed along the two directions of the rotation axis). The exact mechanism through which this may happen is not understood yet (however, see the simulation by Proga and Begelman [2003], and the discussion in chapter 8), but without any doubt the situation is, at least potentially, interesting.

The ADAF models have been much criticized in the literature, *first* because they assume that electrons are not heated by viscous effects, unlike protons. A few authors (Bisnovatyi-Kogan and Lovelace 1997; Blackman 1999) have indeed emphasized how a range of effects, such as ohmic heating in quasi-equipartition magnetic fields and noncompressional dissipation modes, succeed, perhaps, in heating electrons, thus making accretion flows more similar to the thin disks of the preceding chapter, than to ADAFs. When the magnetic field is far from equipartition ($\beta \equiv 8\pi p_g / B^2 \gtrsim 10^3$), Gruzinov (1998) shows that it is indeed possible to heat protons only, but for larger magnetic fields electrons are heated by viscosity, so that the radiative effectiveness of the accretion flow again reaches $\eta \approx 0.1$, the typical value for thin accretion disks; therefore, ADAFs seem to be possible only for small values of the magnetic field.

On the other hand, the most favorable hypothesis for ADAFs (Quataert 1998) is that the magnetohydrodynamic turbulence generated in the viscous process is of the type postulated by Goldreich and Sridhar (1995), in which case the cyclotron resonance becomes irrelevant, and the dissipation of the Alfvén turbulence takes place to the exclusive advantage of protons, thus reversing all preceding conclusions, even in the regime of parameters already analyzed.

The presence of a wind has not been included in a self-consistent way, so densities may be very high initially, satisfying the conditions for ADAFs, but the reduction of the density once the wind starts might lead to a violation of these conditions. These considerations have been developed by Blandford and Begelman (1999), who explicitly include the effects of mass, angular momentum and energy losses, even though in a parametric form. In fact, at the moment, we are totally unable to specify exactly the properties of the coupling between winds (or jets) and accretion flows. The solutions thus derived (Blandford and Begelman 1999) are called advection-dominated input-output solutions.

7.4   Further Developments of the Theory

Because of the ubiquity of the angular momentum, accretion from a disk is the main accreting mode for almost all astrophysical sources, including planets, young stars, and white dwarfs, not just neutron stars and black holes. Needless to say, this problem has received a great deal of attention in the literature and many important questions cannot be adequately treated in this book. Therefore, we shall only briefly discuss a few relevant questions, referring the reader to the original sources for a more detailed presentation: general-relativistic corrections to the standard theory, and the eventual fate of angular momentum.

7.4.1 General-Relativistic Corrections

The above theory is strictly Newtonian, but there is also a general-relativistic version by Page and Thorne (1974) and Abramowicz, Lanza, and Percival (1997).

All the effects of general relativity are encapsulated in three correction factors, that replace the Newtonian reduction factor $R(x)$ and slightly redefine the acceleration perpendicular to the disk, and the amount of energy dissipated locally. The three corrections may be written explicitly for a rotating black hole with arbitrary specific angular momentum a. If we compare nonrotating black holes with the Newtonian treatment, the difference between the two cases is typically of the order of 10%; if, on the other hand, we compare the Newtonian treatment with a maximally rotating black hole, the corrections are ≈ 2 when $x = r/r_\mathrm{s} \lesssim 3$. In view of the fundamental uncertainty about α, these corrections unfortunately have more of a theoretical importance than an observational relevance.

7.4.2 The Fate of Angular Momentum at Large Radii

We have seen from equation 6.14 that at radii large compared with r_m, the advective part of the angular momentum transport is balanced by the torque $\mathcal{G}$, which is entrusted, so to speak, with the transport of the angular momentum toward the outermost parts of the disk. It seems reasonable to assume that at large radii there would be a sink of angular momentum.

Active Galactic Nuclei: Magnetic Winds

In the case of an AGN, where the disk forms at the center of a galaxy, the problem of how this angular momentum is absorbed remains totally open. The best hypothesis has been made by Blandford and Payne (1982, however, see also Blandford [1976] and Lovelace [1976]; a more extensive discussion will be presented in chapter 8), according to whom, if the disk has a large-scale magnetic field with a strong poloidal component, the matter still attached to each magnetic field line (which, therefore, is forced to co-rotate with it and with the disk while freely sliding down along the line) can reach a speed larger than the escape velocity; thence, a magnetic wind is originated, with ensuing removal of the disk's angular momentum. Thus the sink of angular momentum is not an object that absorbs it, but the self-generated wind.

Roughly speaking, this effect can be explained as follows. Let us consider a disk threaded by a large-scale poloidal magnetic field, taking into account the disk's vertical structure. Specifically, let us assume that close to the disk plane, $|z| \ll H$, the energy density in the gas is much larger than that of the magnetic field, whereas the opposite occurs when $|z| \approx H$. This means that the center of the disk drags the poloidal line, because of flux freezing while approaching the compact object and will thus have the geometry of figure 7.8. In the outer parts, however, the motion of the gas is strongly constrained by the presence of the powerful magnetic field; it can only slide along the magnetic field line like pearls along a bead. Since the line corotates with the central parts of the disk, a fluid element bound to move along the line is subject to an effective potential:

$$\phi_{\text{eff}} = -\frac{1}{2}\omega_{\circ}^2 R^2 - \frac{GM}{r} \qquad (7.78)$$

Here R is the distance from the rotation axis (the distance in cylindrical coordinates), while r is the distance (in spherical coordinates) from the origin, where the compact object sits. The effective potential ϕ_{eff} is displayed in Fig. 7.9. In order to have an equilibrium for $r = R_{\circ}$, we need to have $\omega_{\circ}^2 = GM/R_{\circ}^3$. Furthermore, r, in terms of Cartesian coordinates, is given by

$$r^2 = R^2 + (R - R_{\circ})^2 \tan^2\theta \qquad (7.79)$$

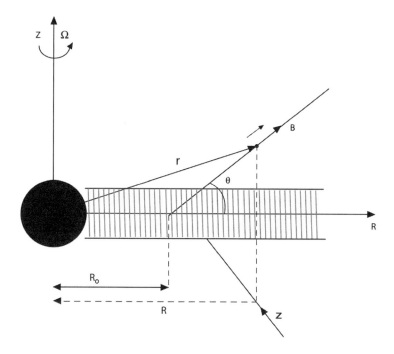

Figure 7.8. Geometry explaining the Blandford and Payne mechanism.

where θ is the angle between the plane of the disk and the direction of the magnetic field crossing the disk in r_0. Therefore,

$$\phi_{\text{eff}} = -\omega_o^2 \left(\frac{R^2}{2} + \frac{R_o^3}{\sqrt{R^2 + (R - R_o)^2 \tan^2 \theta}} \right) \qquad (7.80)$$

We can easily check that ϕ_{eff} has an extremum ($\partial \phi_{\text{eff}}/\partial R = 0$) for $r = R_o$, but, just as easily, we find that $\partial^2 \phi_{\text{eff}}/\partial R^2 < 0$ when $\theta < \pi/3$. Therefore, the extremum we have just found is a *maximum* for sufficiently inclined magnetic field lines. Thus the particles, which can freely slide along the magnetic field lines, move away from R_o, all the while keeping $\omega = \omega_o$ constant, since they remain attached to the line dragged by the disk. In doing this, the specific angular momentum of the

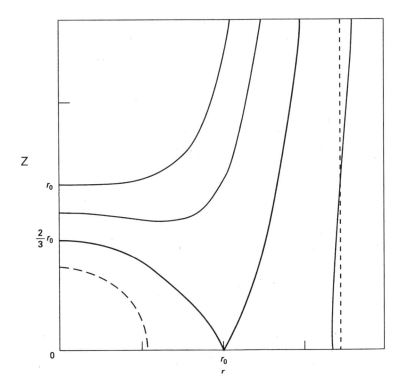

Figure 7.9. Level curves for the effective potential of equation 7.78. The bold line separates the dynamically stable zones (above) from the unstable ones (below). From Blandford and Payne 1982.

particles, $l \approx \omega_o R^2$, increases. This effect is called *centrifugal acceleration*.

We should note that this effect is possible only if there are open magnetic field lines; if the lines were closed, the particles leaving the disk would return to their starting point, obviously with the same angular momentum with which they left. The way in which open lines are generated will be discussed in detail in the next chapter. However, in order to get an idea of the mechanisms at stake, just consider a gas surrounding a compact object in which the magnetic field is in conditions of flux freezing. Sufficiently far from the central object, the gas density exceeds the magnetic energy; if the gas escapes to infinity, it advects the field, stretching the lines, which become asymptotically open. The point where corotation yields to advection is obviously where the local speed equals the Alfvén speed.

The rate $\dot{J}_{\rm w}$ for the angular momentum loss through the magnetohydrodynamic wind is given by

$$\dot{J}_{\rm w} = \dot{M}_{\rm w}\omega_{\rm o}r_{\rm A}^2\cos^2\theta \qquad (7.81)$$

where $\dot{M}_{\rm w}$ is the mass lost in the wind, and $r_{\rm A}$ the radius at which the outflow velocity equals the Alfvén speed. Unfortunately, detailed theoretical estimates are difficult because of our ignorance of the dependence of the magnetic field on radius, and of the mass loss rate. In stellar-mass compact objects, such as cataclysmic variables, it seems that for each 100 g of accreted matter, there is 1 g expelled in the wind, but Blandford and Begelman (1999) and Proga and Begelman (2003) calculate that most matter, which is at first accreting, may be expelled before being swallowed by the black hole.

In fact, this mechanism removes angular momentum from the whole disk and has been originally proposed as an alternative theory to viscosity to explain the torque $\mathcal{G}$; its limit is that it obviously works only in the presence of a sufficiently ordered magnetic field, and we are not sure at all that this is a realistic condition. As discussed above, disks have a mechanism for the creation of a magnetic field (the magnetorotational instability) on modest scales, of the order of the disk thickness, but it is not obvious that this field may acquire coherence on longer scales. Mechanisms of this type are discussed in the literature (Tout and Pringle [1996] discuss an *inverse fall*) but are not universally accepted. However, as discussed in section 5.4, the creation of a magnetohydrodynamic wind seems to be also a prerogative of rather (but not completely!) disordered field configurations (see section 6.7) (see Stone and Norman 1994; Miller and Stone 1997; Proga and Begelman 2003). At this point, it is perhaps worth recalling that the only explicit disk/wind models are the idealized

self-similar ones by Blandford and Payne (1982), Wardle and
Königl (1993), and Ferreira (1997).

Stellar Compact Objects

In objects of stellar dimensions, the problem of the removal
of the angular momentum carried by $\mathcal{G}$ at large radii is better understood, thanks to the presence of companion stars.
This mechanism is also subject to observational checks but is
complex enough to force us to provide a brief summary only.

The fundamental idea is very simple: all accretion disks
in objects of stellar dimensions are in binary systems, with
two tightly bound stars (one of which is the compact one
around which the accretion disk forms). The outer parts of
the disk are about halfway between the two stars, and interact gravitationally not just with the compact star, but
also with the secondary. Since we are considering internal,
conservative central forces, in the inertial (i.e., noncorotat-
ing) reference frame the total angular momentum and energy
must be conserved, but still, internal gravitational forces can
cause a redistribution of angular momentum from the disk to
the secondary and therefore remove the angular momentum
from the disk after $\mathcal{G}$ has carried it to the outermost disk regions.

Even in the outermost zones, the gravitational forces
dominate on all other forces, so that the problem is exquisitely ballistic. The total gravitational potential is not simply $1/r$, since it is the sum of two components $1/r$, centered
on different points. Any potential that is *not* of the form $1/r$
or r^2 does not support closed orbits, apart from a few lucky
cases.[5] However, these closed orbits, and those with initial
conditions close to them, remain nearly closed for a long period of time and thus play a fundamental role in this process.
In fact, if an orbit is almost periodic, it always crosses the

---

[5] The correct term is *for a set of zero measure of the initial conditions.*

same regions with almost identical velocities, and if the gravitational effect of the secondary is the subtraction of angular momentum from the orbit, it will still be subject to this subtraction at each passage; the effects accumulate. This makes a net transfer of angular momentum possible on relatively short time scales. In fact, the relative time scale is the orbital one, not the (much longer) viscous time scale. Notice that this mechanism is absolutely identical to the one explaining dynamic friction for a satellite galaxy falling into a larger galaxy (Weinberg 1986); here too, the interaction with closed orbits (sometimes also called *resonant*) determines the *net* momentum loss.

This phenomenon can be directly observed in a subclass of dwarf novae, called SU UMa systems, among which Z Cha is the most studied one. These novae normally have luminosity variations called *humps*, modulated on the orbital period, but some of them, called *superhumps*, occasionally appear, with periods longer than the orbital one by a few percent. Numerical simulations (Whitehurst and King 1991) have showed that the disk assumes, because of the above interaction, an elliptical shape in its outermost regions, with a precession period slightly longer than the orbital period. The dissipation excess, due to gas compression where the disk is smaller, leads to a modulation on the observed time scale.

7.5 Accretion Disks on Magnetized Objects

Accretion on magnetized objects has great importance in astrophysics. This category includes X-ray pulsars, neutron stars with quasi-periodic oscillations, the rapid burster, magnetic white dwarfs, and the cataclysmic variables, T Tauri stars.

Black holes are not among these sources; it is worthwhile to explain why. Obviously, black holes cannot carry currents unless they are both rotating and electrically charged. However, we have seen (chapter 2, problem 1) that the total charge that can be realized on a black hole in astrophysical conditions is but an infinitesimal fraction ($\approx 10^{-9}$) of the maximum charge that can be theoretically allowed. As a consequence, black holes do not give rise to significant fields. But matter falling onto the black hole has its own magnetic field, and we expect that this leads to a magnetic configuration generated by currents *outside* the black hole. The typical magnetic field inside the disk is generated by magnetorotational instability (section 6.7), which, as we have seen, stops growing when the energy density in the magnetic field $B^2/8\pi$ becomes comparable to the thermal energy density in the gas, ρc_s^2. On the other hand, for thin discs, $\rho c_s^2 \ll \rho v_\phi^2$; therefore, the typical magnetic field surrounding a black hole cannot influence the matter's motion in the accretion disk.

For pulsars and the other above-mentioned objects, the situation is completely different. These objects can carry currents even though they are electrically neutral (unlike black holes!) and, therefore, have such intense magnetic fields (up to $\approx 10^{15}$ G in the most extreme case) that the whole accretion flow changes and is controlled by the magnetic field.

7.5.1 The Alfvén Radius

Many neutron stars have extremely strong magnetic fields, in the range $10^7 \, \mathrm{G} \lesssim B_d \lesssim 10^{15} \, \mathrm{G}$, which can deeply alter the structure of the accretion flow at small radii. The interaction between matter and magnetic field is extremely complex, so we present here only the major and, hopefully, best-established results.

The magnetic field of neutron stars certainly has an important dipolar component, with a dependence on distance from the neutron star, r:

$$B \approx \frac{\mu}{r^3} \tag{7.82}$$

where the *dipole moment*, μ is given by

$$\mu = B_\mathrm{d} R_\mathrm{NS}^3 \qquad (7.83)$$

with $R_\mathrm{NS} \approx 10$ km for the radius of the neutron star. The Maxwell tensor contains a pressure $p_\mathrm{B} = B^2/8\pi$ due to this magnetic field, with steep spatial dependence, $\propto r^{-6}$. Therefore, at large distances from the star, the magnetic pressure will be wholly negligible, but it becomes important at small radii. Thus, at large distances, the accretion proceeds exactly as described in chapter 5 for the spherical case, and in chapter 6 for the case of thin disks.

Let us compute the distance from the star where the magnetic pressure becomes dominant, for a spherical accretion flow. In that case, we know that

$$\dot{M} = 4\pi r^2 \rho v_\mathrm{r} \qquad (7.84)$$

is a constant, and, considering regions inside the sonic radius, we know that the speed is essentially that of free fall:

$$v_\mathrm{r} = \left(\frac{2GM}{r}\right)^{1/2} \qquad (7.85)$$

In order to know where the magnetic pressure becomes dominant, we compare it with the density of kinetic energy (the prevailing energy component, beyond the sonic point) of the accretion flow. We have thus

$$\frac{B^2(r_\mathrm{A})}{8\pi} = \frac{\rho v_\mathrm{r}^2(r_\mathrm{A})}{2} \qquad (7.86)$$

where we have called r_A the *Alfvén radius*, where the equality holds. Using the preceding equations, we find

$$r_\mathrm{A} = 2.3 \times 10^8 \text{ cm } \dot{m}^{-2/7} m^{6/7} \eta^{-1} \mu_{30}^{4/7} \qquad (7.87)$$

where the only new term is

$$\mu_{30} \equiv \frac{\mu}{10^{30} \text{ G cm}^3} \qquad (7.88)$$

We note that a normal young pulsar, with magnetic field 10^{13} G, has $\mu_{30} \approx 10$. If we want to use the luminosity in units of the Eddington luminosity $l \equiv L/L_E$, in the above estimate, we should remember that $\dot{m} = l$. This estimate of r_A, the radius within which the magnetic field becomes dynamically important, is obviously rather rough, but, given the dependence of the pressure on r^{-6}, we expect it to be accurate. For $r \lesssim r_A$, we expect that the accretion flux is no longer radial, but follows the magnetic field lines and, as a consequence, is channeled toward the polar caps. The details of this process are not well understood yet.

For disk accretion the situation is more complicated because there is an exchange of angular momentum with the neutron star, due to the presence of the star's magnetic field. In order to see this, we note that at large distances, the magnetic field of the neutron star will essentially be a dipole field, with radial dependence

$$B \propto r^{-3} \qquad (7.89)$$

Therefore its pressure scales like r^{-6} and is very weak. So we can expect that matter in the disk succeeds in dragging the lines of the stellar magnetic field in its rotatory motion, thus generating a toroidal field that would not exist if B were purely dipolar. However, since the magnetic lines give rise to the magnetic tension, the force produced in bending them generates a torque that acts on the disk material. Since the star-disk system is closed, the torque on the disk must appear as a torque of the opposite sign on the star, slowing it down or accelerating it. This torque (often called *external torque*) is one of two torques to which the disk is subject, the second one obviously being the viscous (or *internal torque*), as discussed in the preceding chapter. Since we have seen that the evolution of an accretion disk is driven by its capacity to lose angular momentum, we may wonder whether the new torque will ever be more powerful than the viscous one. The radius where the two torques have the same intensity is called, once again, the *Alfvén radius*.

In order to calculate the total torque acting on the region of the disk (supposed to be infinitely thin) inside a radius r (see fig. 6.2), let us consider a cylinder with radius r and height h and compute the flux of the z component of the angular momentum through the two bases, $rM_{z\phi}$, and across the lateral surface, $rV_{r\phi}$. The second flux, $rV_{r\phi}$, was discussed in the preceding chapter, section 6.7; it is the usual viscous torque. Instead, $rM_{z\phi}$ represents the magnetic part of the flux, which *does not* flow toward the outermost parts of the disk, but to the neutron star. We are now ready to move on to the explicit computation.

We shall assume that the neutron star has a dipole magnetic field, with axis parallel to the disk rotation axis, a case called *the aligned rotator*; we neglect the much more difficult and still unresolved case of the misaligned rotator. The poloidal field at the disk ($z = 0$) is

$$B_z = -\eta \frac{\mu}{r^3} \qquad (7.90)$$

The minus sign is present because the field lines cross the disk from north to south; the rough factor η takes into account the fact that the disk may carry currents that at least partially, screen the field; however, Ghosh and Lamb (1978) show that $\eta \approx$ constant in the radial interval of interest to us.

We shall now determine the condition that determines the Alfvén radius for disk geometry. We know that a small ring of radius r and width dr is subject to a viscous torque given by

$$t_{\text{vis}} dr = dr \frac{\partial \mathcal{G}}{\partial r} \qquad (7.91)$$

Here we can use equations 6.37 and 6.40 to obtain, for the torque on a ring of unit thickness,

$$t_{\text{vis}} = \frac{\partial \mathcal{G}}{\partial r} = \dot{M} \frac{d}{dr}(r^2 \omega_{\text{K}}) \qquad (7.92)$$

where we have already specialized to the Keplerian case, and we have used $R(r) \approx 1$ because the Alfvén radius is always very large compared with the minimum radius the disk would have in the absence of a magnetic field.

Moreover, the torque on a ring of surface $2\pi r$ due to the Maxwell stress, which flows through the two bases of the cylinder, is

$$t_{\text{mag}} = 2 \times 2\pi r (r M_{\phi z}) = -r^2 B_z B_\phi \qquad (7.93)$$

where the extra factor 2 comes from the fact that the disk has two sides, the upper one and the lower one.

It is now obvious that at large radii, $t_{\text{mag}} \ll t_{\text{vis}}$, and the magnetic torque is only a small perturbation. However, when $t_{\text{mag}} \approx t_{\text{vis}}$, it can no longer be neglected; imposing equality, we find the definition of the Alfvén radius for a disk,

$$B_z(r_A) B_\phi(r_A) r_A^3 = -\frac{1}{2}\dot{M}\sqrt{GMr_A} \qquad (7.94)$$

We shall soon develop this further.

The real problem here is the value of B_ϕ. Following Wang (1995), we can write the *empirical* equation

$$\frac{\partial B_\phi}{\partial t} = -\frac{B_\phi}{\tau_\phi} + \gamma(\Omega - \omega_K)B_z \qquad (7.95)$$

The second term on the right-hand side is the field generation term; it is linear in B_z, because we expect, of course, that the stronger the initial poloidal field, the stronger will be the toroidal one too. Moreover, there is the coefficient $\Omega - \omega_K$, which represents the difference of angular velocity between the star (Ω) and the disk (ω_K); this term appears because, if the disk and the star corotated, no toroidal field would be generated. On the other hand, this field grows with the delay (or the advance) with which the disk follows (or precedes) the star. Finally, γ is a term that takes into account the details of the radial structure; it is ≈ 1 (Ghosh and Lamb 1979). The first term on the right-hand side, on the other hand, is a term that limits the growth of the magnetic field; if it were absent, the field might become unrealistically large. There are several

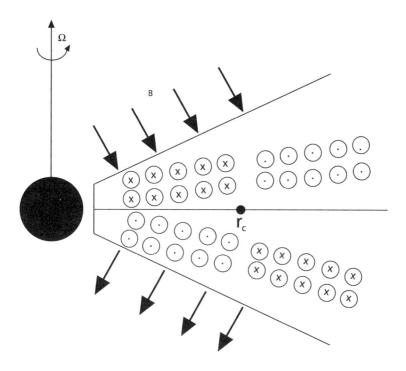

Figure 7.10. Geometry of the magnetic field inside the accretion disk. Notice the field reversal across the plane $z = 0$, and beyond the corotation radius.

mechanisms that oppose the field growth, as we will explain shortly.

Naturally, we are interested in the stationary solution, so

$$\frac{B_\phi}{\tau_\phi} = \gamma(\Omega - \omega_K)B_z \qquad (7.96)$$

From this we see that since the torque is proportional to $-B_\phi B_z$, the disk gains or loses angular momentum at the expense of the star, according to whether $\omega_K > \Omega$ or vice versa. This is illustrated in figure 7.10.

We must now specify which process determines τ_ϕ in equation 7.96. Ghosh and Lamb assumed that there was reconnection within the disk; indeed, the field is tightly wrapped, with northern and southern hemispheres having opposite polarities. These are the ideal conditions for reconnection (section 2.7.2). Another possibility is flux expulsion by magnetic buoyancy (section 2.7.1), which leads to the same equations as reconnection. Alternatively, the field might reconnect outside the disk. However, the most realistic hypothesis is that the main effect is anomalous diffusivity due to turbulence (Yi 1995); in the presence of magnetohydrodynamic turbulence, there is a larger magnetic diffusivity than that due to atomic processes (eq. 2.15). In analogy with the turbulent kinematic viscosity (Shakura and Sunyaev 1973), the anomalous diffusivity is

$$\frac{c^2}{4\pi\sigma} = \chi\lambda v_t \qquad (7.97)$$

where v_t is a typical turbulence speed, λ the maximum turbulent scale, and χ a numerical factor ≈ 1 (Parker 1979). For λ, in a thin disk, we can safely take $\lambda \approx H$, and for v_t the Alfvén speed is sometimes taken; but we have seen in section 6.7 that the magnetic field does not really grow up to the equipartition value, so that the sound speed c_s is (perhaps marginally) larger than the Alfvén speed. We shall therefore take $c^2/(4\pi\sigma) = \alpha c_s H$. The time scale on which the magnetic field is destroyed is therefore $\tau_\phi = H^2/(\alpha c_s H) = H/(\alpha c_s) = 1/(\alpha\omega_K)$, thus,

$$\frac{B_\phi}{B_z} = \frac{\gamma}{\alpha}\frac{\Omega - \omega_K}{\omega_K} \qquad (7.98)$$

For reasons to be described in the next paragraph, it is necessary to have $\Omega < \omega_K$; now, assuming $\Omega \ll \omega_K$ and using equations 7.90 and 7.94, we find that

$$r_A = 1.7 \times 10^8 \text{ cm } \gamma^{2/7}\alpha^{-2/7}\dot{m}^{-2/7}m^{6/7}\eta^{-1}\mu_{30}^{4/7} \qquad (7.99)$$

which is almost the same as equation 7.87, the difference being a factor of the order 1. In particular, the dependence from uncertain factors, α and γ, is modest. Obviously, this is a rather rough estimate; however, thanks to the steep dependence ($\propto r^{-6}$) of the Maxwell stress tensor, this is once again a sufficiently good estimate. In practice, in the estimate of the Alfvén radius, we always use equation 7.87, also for disk accretion, even though its derivation in the case of axial symmetry is considerably different from the spherical case.

7.5.2 Interaction between the Disk and the Magnetosphere

It is extremely important to note here that the interaction between the disk and the star's magnetic field (i.e., its *magnetosphere*) generates a torque on the star, and this torque is absolutely fundamental for an understanding of X-ray pulsar evolution. The theory of this phenomenon comes from Ghosh and Lamb (1979), but we shall follow the treatment of Wang (1995).

Of course, the star gains angular momentum at the rate

$$\dot{J}_{+} = \dot{M} l(r_A) = -2 B_\phi(r_A) B_z(r_A) r_A^3 \qquad (7.100)$$

because the matter falling on it, when leaving the disk, is no longer subject to viscous torques and thus conserves its specific angular momentum at the Alfvén radius; we have already used the expression for the Alfvén radius, equation 7.94. However, the star, to which the magnetic field lines are attached, gains angular momentum at the rate

$$\dot{J}_{-} = -\int_{r_A}^{\infty} r^2 B_z(r) B_\phi(r) dr \qquad (7.101)$$

where $\dot{J}_{-}$ is the angular momentum lost by the disk per unit time. Notice that at this point, the sign of $\dot{J}_{-}$ is unclear. In fact, looking at equation 7.98, we see that near r_A, we must have $\Omega - \omega_K < 0$, otherwise we cannot satisfy equation 7.94. However, when $r \to \infty$, $\omega_K \to 0$, and thus, $\Omega - \omega_K > 0$. Therefore, the sign of $\dot{J}_{-}$ can be determined only by computing the integral. Before calculating it, we note that this rate is an implicit function of the star's angular velocity Ω, so we may wonder whether there is a value Ω_{crit} such that the star does not change angular momentum, namely,

$$\dot{J}_+ + \dot{J}_- = 0 \tag{7.102}$$

Ω is often expressed in dimensionless units:

$$f \equiv \frac{\Omega}{\omega_{\text{K}}(r_{\text{A}})} = \left(\frac{r_{\text{A}}}{r_{\text{c}}}\right)^{3/2} \tag{7.103}$$

The parameter f is often called the *fastness parameter*, and r_{c} is the *corotation radius*, that is, the radius where the Keplerian angular velocity equals the star's:

$$r_{\text{c}}^3 \equiv \frac{GM}{\Omega^2} \tag{7.104}$$

Obviously, we must have $f < 1$; for $f > 1$ the magnetic field is so intense that it starts dominating the motion where the star's angular velocity is larger than Keplerian; thus, matter trying to corotate with the field and with the star feels an unbalanced centrifugal force. Indeed, since $\Omega > \omega_{\text{K}}$, the centrifugal force $\Omega^2 r_{\text{A}}$ is larger than the gravitational force, $\omega_{\text{K}}^2 r_{\text{A}} = GM/r_{\text{A}}^2$. At that point, $\dot{J}_+$ vanishes because nothing is accreted anymore, and the sum of the two terms, $\dot{J}_+$ and $\dot{J}_-$ is <0. We talk in this case of a *propeller effect* (Illarionov and Sunyaev 1975). But what is the value of equilibrium of f, thanks to which the star neither accelerates nor slows down?

First of all, let us rewrite the expression for B_ϕ with this definition of the parameter f. We have

$$B_\phi = B_{\phi 0} \frac{f^2}{1-f} \left(1 - \left(\frac{r}{r_{\text{c}}}\right)^{3/2}\right) \left(\frac{r_{\text{c}}}{r}\right)^3 \tag{7.105}$$

Now we introduce this equation into equation 7.101, together with equation 7.90, to obtain

$$\dot{J}_- = \dot{J}_+ \frac{1 - 2f}{6(1 - f)} \tag{7.106}$$

Thence we finally see that the total torque on the star is

$$j = j_+ \frac{7 - 8f}{6(1 - f)} \qquad (7.107)$$

which leads us to estimate when the star is accelerated as it reduces speed; the neutral point is reached for

$$f = f_{\text{crit}} = \frac{7}{8} = 0.875 \qquad (7.108)$$

and the star gains angular momentum for $f < f_{\text{crit}}$ and loses it for $f > f_{\text{crit}}$. This value of f_{crit} is specific to this model for the dissipation of B_ϕ, but all the other above-mentioned models give, in any case, $f_{\text{crit}} \approx 0.9$ (Wang 1995), which is therefore a good, robust estimate.

7.5.3   Accretion Columns

Once the radius r_A is reached, the fluid is channeled by the magnetic field toward the polar caps, see Fig. 7.11. It enters the magnetosphere with a typical plasma process, the so-called *exchange instability*, but we shall not concern ourselves with this short phase. At the distance r_A we see that the sound speed is $\approx 10^6$ cm s^{-1}, from equation 6.69; since matter is free to slide along magnetic field lines, we expect a fall time $\approx r_A/c_s \lesssim 10^3$ s.

One of the features of this accretion mode is that matter does not evenly cover the whole surface of the neutron star but is channeled only to the polar caps. We can obtain an estimate of the fraction of the star surface covered as follows. For a dipolar field, a magnetic field line is identified by the simple equation

$$\frac{\sin^2 \theta}{r} = \text{constant} = C \qquad (7.109)$$

The magnetic field line to which matter is attached when it leaves the disk is the line passing through $\theta = \pi/2$, $r = r_A$, which has thus as a constant $C = 1/r_A$. This same line meets the star surface at a latitude

$$\sin^2 \theta = r_{\text{ns}} C = \frac{r_{\text{ns}}}{r_A} \qquad (7.110)$$

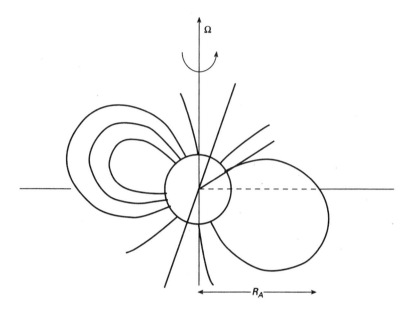

Figure 7.11. Accretion geometry for a magnetized neutron star.

The fraction f of the involved area of the star is therefore

$$f_\Omega = 2\frac{\pi r_{ns}^2 \sin^2 \theta}{4\pi r_{ns}^2} = \frac{r_{ns}}{2r_A} \approx 10^{-3}-10^{-2} \qquad (7.111)$$

where the factor 2 is present because the star accretes on both the north and the south poles. A further point of interest is that when we follow this argument to the end, the accretion flow is not really distributed over the whole polar cap, $\sin^2 \theta \leq r_{ns}/r_A$, but is confined to a thin circle with radius θ.

   This relation has been derived for the case of the *aligned rotator*, namely, the case in which the rotation axis and the dipole vector coincide. As a general rule, this is not a realistic situation (suffice it to think of the simple case of Earth!), but the preceding estimate does not change in order of magnitude. The actual difference in the case of the *oblique rotator* is that since the zones of energy release are close to the two poles, which rotate with an angular frequency Ω, the flux emitted by the accreting gas is modulated on the time scale $T = 2\pi/\Omega$, which explains the mechanism of emission for X-ray pulsars.

After leaving the accretion disk at an almost sonic velocity, the matter falls freely in the gravitational field of the neutron star, thus reaching a free-fall velocity

$$v_{\text{ff}}^2 \approx \frac{2GM}{r} \tag{7.112}$$

which, on the star surface, is

$$v_{\text{ff}}^2(r = r_{\text{ns}}) \approx c^2 \frac{r_{\text{S}}}{r_{\text{ns}}} \approx 0.3c^2 \tag{7.113}$$

The density is given by

$$\dot{M} = 4\pi f_\Omega \rho v_{\text{ff}} r^2 \tag{7.114}$$

where we have taken into account the fact that the flux is not evenly distributed over the whole star surface but covers a fraction f_Ω of the available solid angle.

There are various circumstances that make the study of accretion columns extremely difficult. The first one is the presence of the strong magnetic field of the neutron star, which has a (maybe surprising) consequence: we cannot tell whether a shock wave is formed when the accretion flux reaches the star surface. The reason is very simple: we have seen that most shock waves in astrophysics are noncollisional, since the protons' mean free paths, at low astrophysical densities, are very long (chapter 1). In these conditions, the redistribution of momentum vectors is made possible by transient electromagnetic fields, which deflect randomly the perfectly ordered motions with which fluid particles enter the shock. However, in the presence of magnetic fields that close to neutron stars almost always exceed 10^{12} G, how can we be sure that transient fields amount to more than a small perturbation? In fact, if transient fields remained $\ll 10^{12}$ G, they could not accomplish their task, and there would be no shock. We cannot find refuge in theoretical predictions, because, from first principles, we have no answer to this question and *cannot* predict whether shock waves form or not. There is weak observational evidence that seems to indicate, at low accretion rates, *no* shock wave is formed, while there is a shock wave at high accretion rates.

Apart from this essential difference, such intense magnetic fields modify the rates for almost all fundamental physical processes and alter the particles' mean free paths.

Another source of concern is the possible relevance of the radiation pressure. If the accretion rate $\dot{M}$ is distributed on a fraction f_Ω of the total solid angle, there is an Eddington luminosity corresponding to $f_\Omega L_E$, above which the radiation pressure becomes comparable to the force of gravity and must be included in the discussion. The limit $\dot{M} \lesssim \dot{M}_E$ (eq. 5.39) here becomes

$$\frac{\dot{M}}{f_\Omega} \lesssim \dot{M}_E = 1.4 \times 10^{18} \text{ g s}^{-1} \frac{M}{M_\odot} \frac{0.1}{\eta} \tag{7.115}$$

Incidentally, this explains the observational relevance of the Eddington luminosity for neutron stars emitting in the X-ray band. In fact, these sources are disk sources in the outermost regions, for which there is no maximum luminosity, but later become column sources, hence the relevance of L_E.

Let us then consider two models for the accretion column. The first one, *subcritical*, will have $\dot{M}/f_\Omega \ll \dot{M}_E$, in which case, we expect to be able to neglect the dynamic importance of radiation pressure, and we shall assume that there is no shock. The second one, *critical*, will have $\dot{M} f_\Omega \approx \dot{M}_E$ and will contain a shock, but will also include the radiation transport, because this is what makes the shock possible.

Low Accretion Rates

Let us therefore consider a one-dimensional system, stratified in z, and leave out the lateral variation of all quantities. At first sight, this may seem wrong, but this is an excellent approximation because of the presence of the strong magnetic field, which prevents motion across the field lines. For the typical temperatures of the order $10^8 - 10^9$ K that we shall soon establish, corresponding to kinetic energies $\lesssim 100$ keV, even electrons are subrelativistic. In this case, we know from quantum mechanics that in the very strong magnetic fields of pulsars, motions perpendicular to the magnetic field are quantized with energy levels called *Landau* levels, separated by

$$\triangle \epsilon = \hbar \omega = \hbar \frac{eB}{mc} \approx 100 \text{ keV} \frac{B}{10^{13} \text{ G}} \qquad (7.116)$$

This indicates that electrons cannot even accede to the first excited Landau level, and, since they are forced to remain at the ground level, they cannot move from one line to another. The same argument does not apply to protons because of their larger mass; however, they are obviously tied to the electrons by the Coulomb attraction. This effect prevents the particles' motion perpendicularly to the field. Furthermore, the field cannot even be compressed, since its transverse pressure, $B^2/8\pi$, exceeds that of the gas by several orders of magnitude, as we can check post facto. Therefore, it is reasonable to assume that the accretion column does not expand sideways.

We shall also assume that the vertical extension of the column, $\triangle z$, is small in comparison with the star radius, $\triangle z \ll r_{\rm ns}$, so that the acceleration of gravity does not vary with z. The equations for the accretion column are easily identified. There must be hydrostatic and thermal equilibria (we shall check *later* that the accretion speed is low); the latter is a bit complicated, therefore we shall start with that one. The accretion column receives heat because of the accretion flow impact; the ensuing kinetic energy input is not thermalized by a shock, but by collisions with other particles. This heating inside the column must be balanced by radiative heat transfer (not convection: why?).

Protons arrive on the star with a velocity $v_{\rm ff} \approx 0.5c$, equal to that of the electrons, but, since $m_{\rm p} \gg m_{\rm e}$, they

carry almost all the kinetic energy. How much matter is necessary to halt them, if the prevailing form of collisions is
Coulomb scattering? In analogy with the kinetic coefficients
discussed in chapter 2, we can define the characteristic *slowdown* time t_s of a particle with a mass m_p and energy
$E \gg kT$, which moves in a medium of particles with density n_f and a mass m_f. Spitzer (1962) gives

$$t_s = \frac{m_f}{m_p + m_f} \frac{m_p^2 v^3}{4\pi e^4 n_f \ln \Lambda} \tag{7.117}$$

where $\ln \Lambda$ is the usual *Coulomb logarithm*, which is $\ln \Lambda \approx$
$6-30$. The arriving particle can collide against either elec-
trons or protons, and from the preceding equation we see
that electrons are much more effective than protons in stopping the high-energy proton. The *stopping power* is defined
as the amount of matter crossed during the time t_s. It is
obviously given by $\Sigma_s = n_f m_p t_s v$:

$$\Sigma_s(E) = \frac{m_e m_p^2 v^4}{4\pi e^4 \ln \Lambda} = \frac{m_e E^2}{\pi e^4 \ln \Lambda} \approx 52 \text{ g cm}^{-2} \tag{7.118}$$

where the numerical value has been computed for $v = v_{ff} =$
$c/2$. The presence of the strong magnetic field changes this
estimate by only $\approx 20\%$. Therefore, by assuming, for the sake
of simplicity, that protons deposit their energy evenly in the
whole region with

$$\Sigma = \int_z^\infty \rho(z')dz' \le \Sigma_s(E) \tag{7.119}$$

and that this has an area $A = \pi r_{ns}^2 \sin^2 \theta = \pi r_{ns}^3/r_A$, we find
that protons heat up the region $\Sigma < \Sigma_s(E)$ at the rate

$$\Gamma = \frac{2GM\dot{M}}{r_{ns}} \frac{\rho}{A} \frac{1}{\Sigma_s(E)} \tag{7.120}$$

whereas there is no energy release outside this interval in Σ.
This heating rate must be balanced by a cooling rate, which is
mainly due to bremsstrahlung and inverse Compton, but we

shall consider only the limit, which is analytically easier to treat, in which radiation thermalizes at temperature T_R. In this case, the radiation flux is given, as usual, by

$$F_R = -\frac{c}{\alpha_R} \frac{dp_R}{dz} \tag{7.121}$$

where, as usual, the mean Rosseland coefficient reduces to pure scattering, $\alpha_R = \sigma_T \rho / m_p$. This radiation flux balances heating:

$$F_R = -\frac{c m_p}{\sigma_T} \frac{dp_R}{d\Sigma} = \frac{2GM\dot{M}}{r_{ns} A} \frac{\Sigma - \Sigma_s}{\Sigma_s} \tag{7.122}$$

which, as we can easily see, is similar to the equation for the transport and creation of heat in the stars. Here too, we require that the *local* rate, at which energy is removed, be identical to the rate at which it is released.

At the same time, the hydrostatic equilibrium must hold, in which case, if we assume the pressure to be due to the gas (i.e., neglecting radiation pressure), we find

$$\frac{dp}{dz} = -\frac{GM}{r_{ns}^2} \rho - F_c \tag{7.123}$$

The term F_c represents the force per unit volume exerted by the matter falling on the column. We have assumed that this matter releases its kinetic energy evenly over the whole region with $\Sigma < \Sigma_s$; the momentum arriving per unit time (and thus, the total force) is $\dot{P} = \dot{M} v_{ff}$, and this is spread evenly over the mass $\Sigma_s A$. As a consequence, the force per unit volume is

$$F_c = \frac{\dot{M} v_{ff} \rho}{\Sigma_s A} = \frac{4\pi f_\Omega \rho_a v_{ff}^2 r_{ns}^2 \rho}{\Sigma_s A} = \frac{\rho_a v_{ff}^2}{\Sigma_s} \rho \tag{7.124}$$

where we have indicated with ρ_a the density of accreting matter, to distinguish it from that of the column. This relation can now be introduced in the above equation, which can then be integrated to obtain

$$p = \frac{\rho k T}{\mu m_p} = \frac{GM}{r_{ns}^2} \Sigma + \frac{\rho_a v_{ff}^2 \Sigma}{\Sigma_s} \tag{7.125}$$

This is a differential equation, since both sides depend on ρ. The system consisting of the previous equation and of equation 7.122 can be solved numerically. The numerical computation gives $T \approx 10^{8.5-9}$ K, and $\triangle z \approx 10^{2-3}$ cm $\ll r_{\text{ns}}$.

The total mass is given by $\Sigma_c A \approx 10^{13}$ g. Since the solution has an accretion rate of $\dot{M} = 10^{15}$ g s^{-1}, the time it takes matter to cross the column is $t_c = M/\dot{M} \approx 10^{-2}$ s. This corresponds to a velocity $v_a = \triangle z/t_a \approx 10^{4-5}$ cm s^{-1}, which is much lower than the thermal velocity, $\gtrsim 10^8$ cm s^{-1}. The accretion velocity is therefore fully subsonic, and our approximation of hydrostatic equilibrium is excellent.

High Accretion Rates

In the opposite limit of high accretion rates, we expect that radiation pressure, due to the luminosity released by the accretion of the gas itself, becomes an important dynamic factor. Indeed, since there are many electrons, the optical depth for Thomson scattering increases, $\tau_T \gg 1$, and the accretion luminosity transfers its momentum to the gas. However, most of the luminosity emitted by the column material escapes from the column perpendicularly to the column's z axis that is, in the x, y direction, the reason being simply that there is less material in these two directions than the z direction. Since the radiative flux for optically thick material is proportional to the gradient of the temperature, $\partial T/\partial x, \partial T/\partial y$, we must take into account in this case the structure of the accretion column in the directions perpendicular to the z axis. This makes the problem a bit more difficult than that for low accretion rates, where we assumed very simply that any photon produced could easily escape without further interaction with columnar material.

We have the usual equation for mass conservation,

$$\dot{M} = -A\rho v_z \qquad (7.126)$$

It is important to notice that the product ρv_z is constant in z but, individually, ρ and v_z depend also on the transversal coordinates, x and y, as well as on z. We also have the equation for momentum conservation along the z direction,

$$\rho v_z \frac{\partial v_z}{\partial z} = -\frac{\partial p_R}{\partial z} - \frac{GM}{r^2}\rho \qquad (7.127)$$

in which we have already introduced our main assumption that $p_R \gg p_g$, because the luminosity is close to the Eddington value. Let us take, as usual, $p_R = aT^4/3$. If we assume that the accretion column is once again small relative to the star radius, $\triangle z \ll r_{ns}$, this equation can be immediately integrated in z,

$$\rho v_z^2 = -p_R - \frac{GM}{r_{ns}^2}\Sigma + \text{constant} \qquad (7.128)$$

where we have used the mass conservation equation 7.126. Where the column begins, $v_z^2 = 2GM/r_{ns}$, so that the first term is much larger than the third one because $\triangle z \ll r_{ns}$. If we neglect this term and choose the constant of integration to make the radiation pressure $p_R = 0$ where $v_z^2 = 2GM/r_{ns} \equiv v_{ff}^2$, we obtain

$$p_R = -\rho v_z (v_z + v_{ff}) \qquad (7.129)$$

Since the radiation pressure depends on the fluid temperature, in order to determine it we must make an assumption about the nature of the radiation transport. We can assume, once again, as we have often done before, that the optical depth is large; radiative energy transport ensues, but we *cannot* assume equation 3.35, which has been derived under the assumption that the fluid is at rest. The correct form we need is instead

$$\vec{F}_R = -\frac{c}{\alpha_R}\frac{\partial p_R}{\partial z} + 3\vec{v}p_R \qquad (7.130)$$

Of course, this is a first-order approximation in the small parameter v/c, in which we have kept only the first two terms of the series, those of order $(v/c)^0$ and $(v/c)^1$. It is easy to show that the equation is dimensionally correct, and we must only prove that the numerical coefficient is correct. This task is left for problem 1.

This equation describes how the radiation is transported, not how it is generated. We have a fluid that is slowed down by the radiation flux, so the disappearance of the kinetic energy generates the flux of photons. Let us make a simple hypothesis, namely, that the generation is *prompt*; in other words, we neglect the processes mediating the transformation of kinetic energy in radiation. Instead, we assume that these are fast processes, and that therefore, as much kinetic energy disappears here and now, as energy in the form of photons appears here and now. Mathematically, this is expressed in the following way. The total flux of energy, in photons plus kinetic energy, across a closed surface, is zero. The radiation flux is $\vec{F}_{\rm R}$, whereas the kinetic energy flux is $\rho\vec{v}v^2/2$, so that

$$\nabla \cdot \left(\vec{F}_{\rm R} + \rho\vec{v}\frac{v^2}{2} \right) = 0 \tag{7.131}$$

The energy flux, in general, is $\rho\vec{v}(v^2/2 + w)$, with w as specific enthalpy. The hypothesis of prompt release means that the released energy *does not* heat the gas, and therefore w can be neglected. Using equation 7.126, we find

$$\nabla \cdot \vec{F}_{\rm R} = \rho v_z \frac{\partial}{\partial z}\left(\frac{v_z^2}{2} \right) \tag{7.132}$$

Now we define the quantity $\varpi \equiv v_z^2/v_{\rm ff}^2$, thanks to which we find, from equation 7.129,

$$p_{\rm R} = -\rho v_z v_{\rm ff}(1 - \varpi^{1/2}) = \rho v_{\rm ff}^2(\varpi^{1/2} - \varpi) \tag{7.133}$$

which we can now introduce into equation 7.130 to obtain

$$\vec{F} = \frac{m_{\rm p}v_{\rm ff}^2 c}{2\sigma_{\rm T}}\nabla\varpi + 3\rho v_{\rm ff}\,\vec{v}(\varpi^{1/2} - \varpi) \tag{7.134}$$

and this can be inserted into equation 7.132 to obtain, eventually,

$$\nabla^2\varpi = -\frac{\rho v_z \sigma_{\rm T}}{m_{\rm p}c}(6\varpi^{1/2} - 5\varpi) \tag{7.135}$$

which is our fundamental equation (Davidson 1973). The Laplacian is conveniently expressed in cylindrical coordinates:

$$\nabla^2 = \frac{\partial^2}{\partial r^2} + \frac{1}{r}\frac{\partial}{\partial r} + \frac{\partial^2}{\partial z^2} \tag{7.136}$$

In the Laplacian, we can obviously neglect the dependence on the angle ϕ, because the problem is axially symmetric. The boundary conditions require a discussion. For $z \to \infty$, of course we expect $\varpi \equiv v_z^2/v_{\text{ff}}^2 \to 1$. For $z \to 0$, in the same way, we expect $\varpi \to 0$. We have, as the boundary of the accretion cylinder, $r_c = r_{\text{ns}}\sin\theta$, with θ given by equation 7.110, and as boundary condition, we can take $T(r = r_c) = 0$; this implies $p_R = 0$ at the lateral surface, and thus $\varpi = 1$ (eq. 7.133). This is the same approximation we make when, in a stellar model, we take the photospheric temperature as $=0$; it is wrong, but it is a mistake that is relevant only for $r \approx r_c$.

The solution, computed for the parameters $\dot{M} = 10^{17}$ gs^{-1}, $r_c = 10^5$ cm and $v_{\text{ff}} = c/2$, gives $\epsilon_\gamma = 3p_R = 1.4 \times 10^{17}$ erg cm^{-3}, and therefore a photospheric temperature

$$T_{\text{f}} = \left(\frac{c\epsilon_\gamma}{4\sigma}\right)^{1/4} = 6.6 \times 10^7 \text{ K} \tag{7.137}$$

where σ is the usual Stefan-Boltzmann constant. Furthermore, since $\rho = 2 \times 10^{-4}$ g cm^{-3}, the optical *lateral* depth is $\tau_T \approx 7$, and this justifies our hypothesis of thermalized radiation.

However, the most interesting part of the solution is the very fast transition in velocity that occurs at a distance of about 10^5 cm from the star surface; this steep transition of velocity increases density by about a factor of 3, comparable with the strong hydrodynamic shocks studied in chapter 1. These are the typical numbers of shock waves, apart from the fact that there are no collisions here or transient electromagnetic fields to cause the slowdown, just radiation pressure. That is why we say that this transition contains a *radiative shock*; we actually refer to this, when we say that the solution at high accretion rates has a shock.

This result is obviously approximated, because we have treated in a very rough manner the radiation transport and generation, and we have completely neglected the influence of the strong magnetic field on all physical processes. More realistic simulations have now become possible (Kraus, et al. 2003, and references therein); they include, besides these effects, also the exact form of the accretion column, the heating of the neutron star's surface, and the relativistic effects in the Schwarzschild metric. As a consequence, it has now become possible to study such details as the time profile of the light emitted during rotation, and its dependence on the photons' energy, as well as on viewing angle.

7.6 Boundary Layers

It is evident, from what we have said so far, that accretion disks around black holes terminate at the radius of the marginally stable orbit. For neutron stars, they may terminate at the Alfvén radius, if the stars are sufficiently magnetized; or at the radius of the marginally stable orbit, if the neutron stars are smaller than this radius; or, lastly, at the surface of the neutron star, if the latter has a radius larger than the marginally stable orbit, which is certainly the case for some admissible equations of state (see Shapiro and Teukolsky 1984). Also, accretion disks around white dwarfs terminate at the star surface, the exception being the rare cases of magnetized dwarfs with a large Alfvén radius.

When the disk extends to the star's surface, a new phenomenon occurs: the formation of a region not in Keplerian rotation. We call this region a *boundary layer*. Indeed, the star will have its own rotation rate $\Omega_\star$, which differs from the disk angular speed at the star's edge (because the star need not be in centrifugal equilibrium):

$$\Omega_\star < \omega_{\rm K}(R_\star) \qquad (7.138)$$

where, obviously, $\omega_{\rm K}(R_\star)$ is the angular speed for Keplerian rotation at the star radius $R_\star$. There must thus be a transition zone between the disk and the star in which the disk is no longer in centrifugal equilibrium. If we consider once again equation 6.55,

$$\frac{1}{2}\frac{\partial v_{\mathrm{r}}^2}{\partial r} - \frac{v_\phi^2}{r} = -\frac{1}{\rho}\frac{\partial p}{\partial r} - \frac{GM}{r^2} \tag{7.139}$$

Since, in the zone of transition, $v_\phi < v_{\mathrm{K}} = \sqrt{GM/r}$, the gravitational term must be balanced either by $1/2\, \partial v_{\mathrm{r}}^2/\partial r$ or by the pressure gradient, $\partial c_{\mathrm{s}}^2/\partial r$. However, we know that in the unperturbed disk, $v_{\mathrm{r}} \ll c_{\mathrm{s}}$, so that we expect the pressure gradient to dominate. Let us now assume that the transition zone has thickness δr; the above equation then simplifies to

$$\frac{c_{\mathrm{s}}^2}{\delta r} \approx \frac{GM}{R_\star^2} \tag{7.140}$$

which of course determines δr. In order to evaluate δr more precisely, we notice that the disk thickness H is

$$H = \frac{c_{\mathrm{s}}}{v_{\mathrm{K}}}R_\star \tag{7.141}$$

so that we can rewrite

$$\delta r = H\frac{H}{R_\star} \ll H \tag{7.142}$$

(see fig. 7.12). Therefore, the layer where the disk slowdown (from the Keplerian velocity to the star's) occurs is very thin, much more than H, thence the name *boundary layer* for this region.

Deceleration takes place thanks to the presence of a significant pressure gradient, which implies an increase of the local temperature with respect to adjacent regions where the support against external gravity is assured by the centrifugal force. The temperature excess leads to a radiation excess. In the case in which the disk is optically thick, it is easy to calculate the temperature in the boundary layer (BL). The surface from which the excess of radiation is emitted is

$$\Sigma = 2 \times 2\pi R_\star H \tag{7.143}$$

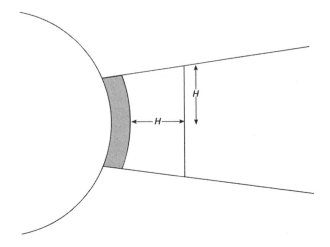

Figure 7.12. Geometry of the boundary layer.

On the other hand, we have already seen in the preceding chapter that the energy released per unit time is

$$L = \frac{GM\dot{M}}{R_\star} \qquad (7.144)$$

Notice that this differs from equation 6.22 by a factor of 2 ($\beta = 1$ necessarily, here). The reason is that the energy radiated is *only* $GM\dot{M}/(2R)$ when the matter is on an elliptical orbit, whereas here the matter dissipates all its kinetic energy. For the temperature, we find, therefore,

$$4\pi R_\star H \sigma T_{\mathrm{BL}}^4 = \frac{GM\dot{M}}{R_\star} \qquad (7.145)$$

In other words,

$$T_{\mathrm{BL}} = \left(\frac{2R_\star}{H}\right)^{1/4} T_\star \qquad (7.146)$$

where $T_\star$ (eq. 6.24) is the temperature the disk would have in the *absence* of the boundary layer. Thus we see that T_{BL} is larger than $T_\star$, but not by much because of the weak dependence of $T_{\rm BL}$ on the ratio $R_\star/H$. In white dwarfs we typically find $T_{\rm BL} \approx 10^5$ K.

However, an excess of rather hard X emission (around 10 keV) has been observed in some white dwarfs, which is certainly not compatible with the modest temperatures derived above. In order to solve this problem, two solutions have been proposed. The first one is that the gas forms a strong stationary shock. This idea is certainly made plausible by the fact that the motion in the direction ϕ is highly supersonic, $v_\phi \gg c_{\rm s}$; however, in motions of this type, it seems much more likely that the accreting matter suffers a series of weak oblique shocks, which transform only a small fraction of the bulk kinetic energy into internal energy. The other possibility is that the disk is condemned to the same fate described earlier for the A zone of the disk. Indeed, in this case too, we can show that the disk is thermally unstable, and we may thus suppose that here, too, it tends to a two-phase equilibrium, in which the photons emitted from the cold phase are Comptonized by the hot phase at the observed energies, ≈ 10 keV.

7.7 Problems

1. Derive the second term on the right-hand side of equation 7.130. (Hint: The angular distribution for Thomson scattering by an electron at rest is given by equation 3.114. What if the electron is moving instead?).

2. In discussing magnetohydrodynamic winds, we have not written the effective potential in the usual form, $\phi_{\rm eff} = l^2/2r^2 - GM/R$, where l is the specific angular momentum, but instead in the form of equation 7.78. Why?

3. In section 7.1.3, we said that a fluid with a negative thermal capacity is thermally unstable. Can you prove this?

Chapter 8

Electrodynamics of Compact Objects

This chapter is devoted to the electrodynamics of pulsars and black holes, with or without an accretion disk. Its goal is to present a unified treatment of these topics, which, at the moment, does not exist in the literature. Electrodynamic phenomena are credited by some to have the potential to explain the structure of almost all astrophysical sources, from pulsars to active galactic nuclei (AGNs), including gamma ray bursts and soft gamma ray repeaters. Even if we do not want to accept such an extreme opinion, we have to note that electrodynamics plays an absolutely essential role for pulsars. As far as AGNs are concerned, and, in general, the sources that contain both an accretion disk and a moderately collimated linear jet of matter, it provides the only coherent picture for understanding disk-jet coupling. Moreover, Electrodynamics provides a further way to produce relativistic particles and thus explain γ and charged-particles emission, which has been observed, more or less directly, from compact objects. I shall start with a description of the electrodynamics of pulsars and black holes; I shall then discuss a few models for disk-jet coupling.

We can discuss the electrodynamics of compact objects in a qualitative manner by developing an analogy with batteries.

This is done in section 8.3.4, which is intended as a further aid to understand the detailed computations that precede them.

8.1 The Gold-Pacini Mechanism

Before starting the discussion of the electrodynamics of collapsed objects, we should briefly discuss the Gold-Pacini mechanism (Pacini 1967) for energy loss by pulsars. The fundamental idea is that pulsars may possess a magnetic moment $\vec{\mathcal{M}}$ inclined by an angle α with respect to the pulsars' rotation axis. This is not surprising; without going too far, Earth has a magnetic axis that is slightly different from the rotation axis,[1] and it seems therefore wholly plausible that a pulsar, having probably conserved the original magnetic moment of the parent star, still possesses an inclined moment.

In these conditions, an observer external to the pulsar sees a magnetic field that varies with time and, thus necessarily, there is emitted electromagnetic radiation. The rate of energy loss is (Landau and Lifshitz 1987a)

$$\dot{E} = -\frac{2}{3c^3}(\ddot{\mathcal{M}})^2 \tag{8.1}$$

The magnetic field generated by the dipole at a distance R along a direction $\vec{n}$ is

$$\vec{B} = \frac{3\vec{n}(\vec{n} \cdot \vec{\mathcal{M}}) - \vec{\mathcal{M}}}{R^3} \tag{8.2}$$

and the one exactly at the magnetic north pole is given by

$$\frac{1}{2}R_{\text{ns}}^3 B_{\text{p}} = \mathcal{M} \tag{8.3}$$

where R_{ns} is the radius of the neutron star. We can easily see that

$$(\ddot{\mathcal{M}})^2 = \Omega^4 \sin^2 \alpha \mathcal{M}^2 \tag{8.4}$$

---

[1] Strictly speaking, Earth's magnetic moment does not even cross the center of Earth but is slightly shifted.

and therefore

$$\dot{E} = -\frac{1}{6}\sin^2\alpha c B_{\rm p}^2 R_{\rm ns}^2 \left(\frac{\Omega R_{\rm ns}}{c}\right)^4 \tag{8.5}$$

where Ω is the star's angular velocity. Notice that the term $cB_{\rm p}^2 R_{\rm ns}^2$ is determined by purely dimensional arguments.

This is the well-known Gold-Pacini model, showing that even isolated neutron stars could be easily detected. The total amount of energy released is in fact $\approx 10^{35}$ erg s^{-1} for the Crab pulsar ($T = 0.03$ s, $B_{\rm p} = 4 \times 10^{12}$ G), almost two orders of magnitude larger than the solar luminosity.

The source of the lost energy is, of course, the pulsar's rotational energy, $E = I\Omega^2/2$, where I is the star's moment of inertia. As a consequence, the change in the pulsar's rotation period must satisfy

$$\dot{E} = I\Omega\dot{\Omega} = -\frac{1}{6}\sin^2\alpha c B_{\rm p}^2 R_{\rm ns}^2 \left(\frac{\Omega R_{\rm ns}}{c}\right)^4 \tag{8.6}$$

Therefore, whenever we can directly observe the variation of the period in pulsars, namely $\dot{\Omega}$, we can immediately derive the intensity of the magnetic field.

Moreover, the electromagnetic radiation must remove angular momentum; indeed, since $I\dot{\Omega} < 0$, the star loses angular momentum. In this model, the star is in a vacuum, so that the only possible sink for the angular momentum is the emitted electromagnetic radiation.

Notice that the dependence on Ω is typical of the magnetic dipole field. It is easy to demonstrate that magnetic fields of order l ($l = 1$ corresponds to the magnetic dipole, $l = 2$ to the quadrupole, and so on) have energy losses scaling as $(\Omega R_{\rm ns}/c)^{2l+2}$. Since the quantity $\Omega R_{\rm ns}/c \ll 1$, the energy losses will be dominated by the dipole term. As a consequence, when we observe $\dot{\Omega}$, the magnetic field we directly observe is not the one at the surface of the pulsar, but *only* the dipole component, since the other components lead to energy losses smaller by at least a factor of $(\Omega R_{\rm ns}/c)^2$. The field at the surface will have other components with $l > 1$, but these cannot be inferred from these observations. The only way we have to measure the magnetic field at the surface is to use cyclotron lines (see chapter 3).

However, the Gold-Pacini formula, equation 8.5, contains two subtleties. On one hand, since the pulsar rotates with an angular velocity Ω, we may expect its luminosity to vary with a frequency $\nu = \omega/(2\pi)$, but, strictly speaking, pulsars are observed in the radio region of the electromagnetic spectrum, with frequencies $\gg \omega/(2\pi)$. This means that around the star, there are electrons transforming the energy contained in the large-amplitude electromagnetic wave, in kinetic energy of the electrons that may then emit the radio waves observed at Earth. Therefore, the environment surrounding the pulsar, far from being empty, hosts important physical phenomena. Furthermore, the plasma surrounding pulsars does not play a purely passive role, transforming unobservable waves into observable ones for our convenience, but plays an absolutely key role in determining the electromagnetic properties of pulsars. What we are going to discover is that a pulsar *cannot* be surrounded by void, but must be placed in a high-density medium.

8.2 The Magnetospheres Surrounding Pulsars

Let us start our study of the electrodynamics of compact objects from pulsars' magnetospheres, even though there is no fully satisfactory model for this phenomenon. The model we are going to introduce, indeed, has a serious defect, as we shall demonstrate toward the end of this section. The reasons for introducing this model, despite its serious deficiency, are the following: First of all, it provides the basic ideas on which all models in the literature are built. Second, some features are certainly right. Third, it is the prototype on which the black hole model is built, a model that is immune to the criticism appropriate for pulsars. Last, it allows us to discuss, in a simple way, thanks to the Newtonian context, the approximations we usually make in the discussion of these phenomena.

The reason why we are talking about the electrodynamics of compact objects, rather than hydrodynamics or magnetohydrodynamics, is that the electromagnetic forces on matter outside a compact object are much larger than any other force. We now discuss this point quantitatively, focusing on the case of pulsars, neutron stars with a radius $R_p \approx 10$ km, a mass $M \approx 1.4 M_\odot$, periods of rotation around their axis $T \approx 0.03-10$ s, endowed with strong magnetic fields, $B_p \approx 10^{12} - 10^{14}$ G. We shall suppose, for sake of simplicity, that the magnetic field is purely dipolar:

$$\vec{B} = \frac{3\hat{n}(\vec{M} \cdot \hat{n}) - \vec{M}}{R^3}, \qquad B_p = \frac{2M}{R_p^3} \qquad (8.7)$$

Here B_p is the magnetic field at the pole. Let us also assume that the pulsar's rotation axis and magnetic axis are parallel; in the literature, this case is often called the *aligned rotator* case. Judging from equation 8.5, we might think that this case is devoid of interest, but we shall see that is not so.

Before starting, we note a circumstance that will considerably simplify our work: in such intense magnetic fields, particles slide freely along the magnetic field lines, but their motion perpendicular to the field itself is strongly impaired. For nonrelativistic particles, the Larmor radius is given by

$$r_L = \frac{mvc}{eB} = 3 \times 10^{-11} \text{ cm} \left(\frac{T}{10^6 \text{ K}}\right)^{1/2} \frac{10^{12} \text{ G}}{B} \qquad (8.8)$$

We have assumed that the particles in question are electrons, with a temperature comparable to the surface temperature of the neutron star. It follows that the particles, even for field values much lower than 10^{12} G, are strictly bound to their magnetic field line and will tend to follow it closely. For this reason, we often say that the magnetic field lines behave like *live wires*.

The approximation according to which, in the absence of electric fields, the charges remain close to their magnetic field line is valid as long as $r_{\rm L} \ll R$, the distance from the compact object. Since, at a large distance from currents, the magnetic field reduces to a dipolar field, so that $B \approx B_{\rm p}(R_{\rm p}/R)^3$, we can easily see that $r_L \ll R$ implies

$$R \ll R_{\rm m} \equiv \sqrt{eB_{\rm p}R_{\rm p}^3/mvc} \qquad (8.9)$$

Therefore, at sufficiently large distances, particles can migrate from one line to another. However, the quantitative estimate of R_m has assumed that particles conserve their initial energy and that the field is a pure dipole, although we shall soon see that both assumptions are wrong. For this reason, the above definition of $R_{\rm m}$ is an upper limit.

We are now ready to demonstrate that pulsars must be surrounded by an electrically charged medium. Indeed, let us naively suppose that they are placed in vacuum and calculate the electric force acting on the pulsar's surface charges. In order to do this, we note that each pulsar is an excellent conductor, so that electrons must assume a distribution that cancels the magnetic force acting on them. As a consequence,

$$\vec{E} + \frac{\vec{v}}{c} \wedge \vec{B} = 0, \text{where} \quad \vec{v} = \vec{\Omega} \wedge \vec{r} \qquad (8.10)$$

Here $\vec{v}$ is the star's velocity of rotation. This equation shows that there must be an electric field inside the pulsar; therefore, the electric field outside the pulsar cannot be strictly zero because of the presence of surface charges. The electrostatic potential ϕ outside the pulsar satisfies Laplace's equation,

$$\nabla^2 \phi = 0 \qquad (8.11)$$

which, as we know, has solutions (for each integer l) of the type

$$\phi = \phi_\circ \left(\frac{R_{\rm p}}{R}\right)^{l+1} P_{\rm l}(\cos\theta) \qquad (8.12)$$

with P_l a Legendre polynomial. In order to specify l and ϕ_o, we must use the boundary conditions: on the neutron star surface, equation 8.10 applies, in the presence of a dipolar magnetic field, equation 8.7. However, we are not interested in the radial component of the electric field, but only in the tangential component (in this case, only the component directed along $\hat{e}_\theta$). In fact, we know from elementary courses of electromagnetism that the component of the electric field perpendicular to the surface of a conductor may be discontinuous because of the presence of a surface distribution of charges, whereas the component of the electric field tangent to the conductor's surface is always continuous. Therefore, we have from equations 8.10 and 8.7

$$E_\theta = -\frac{B_p \Omega R_p}{c} \cos\theta \sin\theta \qquad (8.13)$$

If we recall that $E_\theta = -(1/R)\partial\phi/\partial\theta$, comparing the above equation with equation 8.12, we find that

$$\phi = -\frac{B_p \Omega R_p^5}{3cR^3} P_2(\cos\theta) \qquad (8.14)$$

In problem 1, we show that there are surface charges. Let us now calculate the quantity

$$\vec{E} \cdot \vec{B} = -\frac{\Omega R_p}{c} \left(\frac{R_p}{R}\right)^7 \cos^3\theta \qquad (8.15)$$

We want to calculate this quantity[2] because, as we said before, the magnetic field lines act just like live wires, along which the charges are free to move. Therefore, the presence of a component $E_\parallel$ along B tells us that the surface charges are subject to an electric force given by $eE_\parallel = e\Omega R_p B_p \cos^2\theta/c$, which cannot be canceled by the magnetic field. This electric force must be compared to the gravitational force, to which

---

[2] Apart from the obvious fact that this a relativistic invariant, which, therefore, can never vanish as a consequence of a Lorentz transformation.

electric charges are also subject. We find

$$\frac{eE_\parallel}{GMm/R_p^2} = 8 \times 10^{11} \frac{B_p}{10^{12} \text{ G}} \frac{1 \text{ s}}{T} \cos^2 \theta \qquad (8.16)$$

for an electron. As a consequence, the electrostatic forces on a charged particle, whether a proton or an electron, cannot be offset by gravitational forces. Therefore, if a pulsar is placed in the void, the surface electric fields are so intense that they can strip off surface charges and fill the vacuum.

Since electromagnetic forces are much larger than gravitational forces, we say that pulsars are surrounded by magnetospheres, rather than atmospheres (in which gravitational forces prevail).

In this section, we assume we have, outside pulsars, the charges necessary to achieve a stationary situation. We shall discuss in section 8.4 how these charges can be generated. Right now, however, we should ask ourselves how to obtain a stationary situation.

It is rather easy to realize that as long as there is a force acting on electric charges, these will be stripped from the pulsar's surface. Therefore, the stationary condition is one where the total force on electric charges vanishes:

$$\rho\vec{E} + \frac{\vec{j}}{c} \wedge \vec{B} = 0 \qquad (8.17)$$

Here, ρ and $\vec{j}$ are, respectively, the charge and the current densities *outside* pulsars, and they satisfy $\rho, \vec{j} \neq 0$ because we have just shown that if these vanish, there can be no stationary situation. The vanishing of the Lorentz force is (nearly!) equivalent to the vanishing of the total force, since, as we showed above, the gravitational forces, and therefore the inertial forces too, are negligible with respect to electromagnetic forces. Thus we use the approximation $m = 0$ for each particle in question. From the above equation, we immediately find

$$\vec{E} \cdot \vec{B} = 0 \qquad (8.18)$$

which tells us that there will be no electric field parallel to the magnetic field.

Physically, this is a very simple requirement, namely, that so many electric charges are created outside pulsars that the sum of magnetic and electric (repulsive) forces generated by these charges vanishes. The only complication is that external charges will contribute to generate both electric and magnetic fields, whose effect on the motion of the charges themselves must be included in a self-consistent way.

In general, we must therefore solve equation 8.17, together with Maxwell's equations, outside a conducting sphere in which we know (eq. 8.10) the components $E_\parallel$ and $B_\perp$ at the surface of the sphere.[3]

8.2.1 Quasi-Neutral or Charge-Separated Plasma?

Before going into mathematical details, we need to discuss an important approximation we need.

We are going to discover (problem 4) that the pulsar must be surrounded by a charge density that corotates with the pulsar itself, given by

$$\rho = -\frac{\Omega B_z}{2\pi(1 - \Omega^2 r^2/c^2)} \qquad (8.19)$$

We notice that the minimum density of particles satisfying this equation is given by

$$n_{\min} = \frac{\rho}{-e} = 7 \times 10^{10} \text{ cm}^{-3} \frac{B_z}{10^{12} \text{ G}} \frac{1 \text{ s}}{T} \frac{1}{1 - \Omega^2 r^2/c^2} \qquad (8.20)$$

where T is the pulsar's rotation period. From this we see that the density of particles near the pulsar is certainly not small, rather, it exceeds that typical of the interstellar medium by more than 10 orders of magnitude.

However, we note that this is a minimum density; in principle, we can add to this density an equal number of positive

---

[3] We recall that because of the possible presence of surface currents in the conductor, the component of the magnetic field parallel to the surface may be discontinuous, whereas the perpendicular component must necessarily be continuous.

and negative charges $N_+ = N_-$, which can be even much larger than n_{min}, without changing the density of the required electric charge. Thus we face the following problem: is the plasma outside the pulsar almost neutral (by which we mean $N_+ = N_- \gg n_{min}$), or is it in a regime of *complete charge separation*, $N_+ = N_- = 0$?

A crucial point depends on this answer. In the next section, we shall show that current density is given by

$$\vec{j} = \rho \Omega r \hat{e}_\phi + \kappa \vec{B} \tag{8.21}$$

where ρ is the charge density, and κ a local function that at least so far, the theory cannot identify. Let us consider first the case $\kappa = 0$. Now, if we are in a regime of complete charge separation, the quantity $\vec{j}/\rho = \vec{v}$ is the average velocity of electric charges, which turns out to be $\vec{v} = \Omega r \hat{e}_\phi$; this, for $r > c/\Omega$, exceeds the speed of light. Therefore, in a regime of charge separation, there cannot be corotating charges beyond the *light cylinder*, $r = c/\Omega$. However, this is false for a quasi-neutral plasma, as the following example shows. Let us consider a plasma made of equal densities of positive N_+ and negative $N_- = N_+$ charges. Of course, the positive charges will move in the opposite direction of the negative ones, $V_+ = -V_-$. Thence it follows that $\rho = eN_+ - eN_- = 0$, whereas $j = eN_+V_+ + (-e)(-V_+)N_- = 2eN_+V_+ \neq 0$. So

$$\vec{v} \equiv \frac{\vec{j}}{\rho} = +\infty \tag{8.22}$$

and the problem at the light cylinder disappears! In other words, when, in the same element of volume dV, there are sufficiently many charges of both signs, the velocity $\vec{v} \equiv \vec{j}/\rho$ has no physical meaning.

If now $\kappa \neq 0$, in a regime of charge separation, we must impose that

$$\left| \frac{\vec{j}}{\rho} \right| < c \tag{8.23}$$

which establishes a heavy constraint on the unknown function κ.

The usual answer is that the regime of charge separation seems physically more motivated for pulsars. When we consider the motion of electric charges, since we have discovered that electric forces exceed the gravitational and inertial ones (i.e., $m\vec{a}$) by several orders of magnitude, we have decided to suppose the latter equal to zero, because we think that in the correct solution, there will be a small electric field balancing gravitational and inertial terms. This is certainly correct for the electric charges of one sign, whereas it is certainly false for those of the opposite sign: the small electric field will double gravitational and inertial forces. Then why should the charges of this sign remain suspended above the pulsar, without falling on it? The only possible answer is, of course, pressure, but this requires thermal velocities of the order of

$$v_{\text{th}}^2 \approx \frac{GM}{R} \tag{8.24}$$

which, for the small radii of neutron stars ($R \approx 10$ km), corresponds to temperatures of the order of $T \approx 10^9$ K. These temperatures are very high, and it would be extremely difficult to keep electrons (or positrons) constantly heated at such temperatures, against the huge radiative losses in such strong magnetic fields. Moreover, the typical surface temperatures of neutron stars are of the order of 10^6 K $\ll 10^9$ K, so that any interaction with photons emitted by the surface would cool particles, because of the obvious inverse Compton effect. As a consequence, it seems impossible to keep particles of *both* signs of charge out of a pulsar; the plasma will be in a regime of separation of charge.

8.2.2 The Goldreich and Julian Magnetosphere

Goldreich and Julian (1969) were the first to try to derive the properties of magnetospheres. They proposed a model that is illustrated in figure 8.1.

They assumed complete charge separation, as discussed above, and derived the following main features. On one hand,

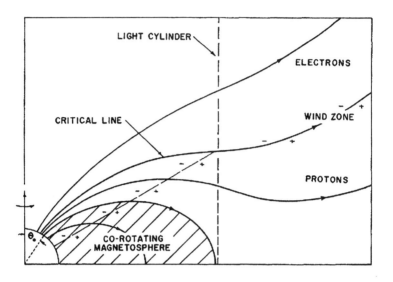

Figure 8.1. The Goldreich and Julian magnetosphere. From Goldreich and Julian 1969.

the light cylinder (i.e., the surface with $r = c/\Omega$ in cylindrical coordinates) is important because inside it, there may be charges that corotate strictly with the star, so that the magnetic field lines are equipotential, in agreement with equation 8.18, whereas, outside it, there cannot be such corotating charges. Therefore, the magnetic field lines are no longer equipotential. The magnetic field lines that close within the cylinder of light give origin to the so-called *corotating magnetosphere*; on the other hand, the lines crossing the cylinder of light will no longer close. This is certainly a plausible hypothesis, but, strictly speaking, it has not been proved. The lines acquire a toroidal component (i.e., along $\hat{e}_\phi$), which becomes dominant at large radii and gives origin to a decrease of the magnetic field of the type

$$B_\phi \propto \frac{1}{r} \quad \text{for } r \to \infty \qquad (8.25)$$

The last magnetic field line that manages to close within the cylinder of light is estimated to make an angle θ_1 with the rotation axis, at the pulsar surface, given by

$$\sin^2 \theta_1 \approx \frac{\Omega R_{ns}}{c} \qquad (8.26)$$

where R_{ns} is the pulsar's radius. This estimate is obtained assuming a pure dipolar field inside the light cylinder; its derivation is left as an exercise to the reader.

The particles placed on the open magnetic field lines, once they have crossed the light cylinder, are accelerated and will never return to the pulsar. To insure that the star does not lose charge, Goldreich and Julian assume that there is a *critical* magnetic field line (which leaves the pulsar surface with an angle θ_c) having electrostatic potential equal to that at infinity. In this way, the particles initially at small latitudes ($\theta < \theta_c$) feel a negative potential with respect to the critical line (see eq. 8.14), whereas those with $\theta > \theta_c$ feel a higher potential than the critical line. As a consequence, negative charges must flow toward infinity along the lines closer to the pole, whereas the positive charges must flow from the lines closer to the last closed line. Goldreich and Julian estimate that the total potential difference between the critical line and the most positive line is

$$\Delta\Phi \approx \frac{1}{2} \left(\frac{\Omega R_{ns}}{c} \right)^2 B_p R_{ns} \qquad (8.27)$$

an estimate obtained by halving the potential difference between the north pole and the last closed line, identified above.

While many authors are still inclined to think that this description of the magnetosphere is, at least roughly, correct, its shortcoming is clear. Indeed, from equation 8.19, we can see that the corotating charge changes sign where $B_z = 0$; assuming that the field outside the star is roughly dipolar (i.e., the corrections due to the corotating charge are still small), we see from equation 8.7 that we have $B_z = 0$ where $\cos^2 \theta = 1/3$, which is the broken line in figure 8.1. We thus see that most open lines cross regions of both positive and negative charge. Thus, no matter what the sign of the outflowing charges is, there is a breakdown of the initial assumption of charge separation somewhere in the solution; the solution is not self-consistent.

8.2.3 The Pulsar Equation

Of course, we can be more quantitative than this. Let us go back to our original problem. Let us assume we want to solve Maxwell's equations, subject to equation 8.17 and to the usual boundary conditions on the pulsar's surface, namely, the continuity of the normal (for $\vec{B}$) and tangential (for $\vec{E}$) components, knowing that the magnetic field inside the star is a pure dipole (eq. 8.7), and for $\vec{E}$ inside the star equation 8.10 holds.

For arbitrary symmetries, the above problem cannot be solved analytically, but, in the case of the aligned rotator, it has a solution, which can be easily calculated (Michel 1973; Scharlemann and Wagoner 1973), and which we describe here. It is convenient to use cylindrical coordinates, r, z, and ϕ, and note that because of the axial symmetry, $\partial X/\partial\phi = 0$ for each physical quantity X; besides, because of the stationarity, $\partial X/\partial t = 0$.

The Electric Field

Let us start by showing that $E_\phi = 0$. The reason is very simple. Because of axial symmetry, $\partial E_\phi/\partial\phi = 0$, so that $E_\phi = k$, a constant. However, let us now consider (fig. 8.2) a circle, $z = z_o, r = r_o, 0 \le \phi \le 2\pi$, and integrate the Maxwell equation

$$\nabla \wedge \vec{E} = -\frac{1}{c}\frac{\partial \vec{B}}{\partial t} \tag{8.28}$$

on the surface subtended by this circle. For the left-hand side, we immediately obtain

$$\int (\nabla \wedge \vec{E}) \cdot d\vec{A} = \oint \vec{E} \cdot d\vec{l} = r \int_0^{2\pi} E_\phi d\phi = 2\pi r k \tag{8.29}$$

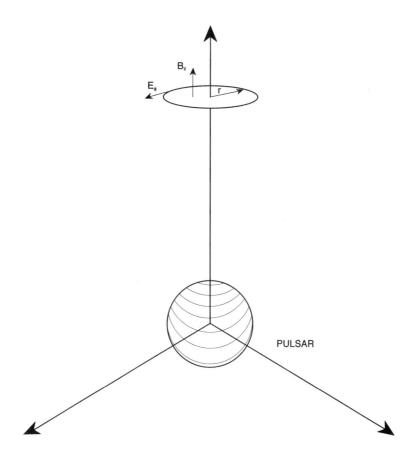

Figure 8.2. Geometry for the proof that $E_\phi = 0$.

Here we have used the theorem $\int \nabla \wedge \vec{X} \cdot d\vec{A} = \oint \vec{X} \cdot d\vec{l}$, where the second integral is a line integral along the surface contour. On the other hand, the integral of $\partial \vec{B}/\partial t$ on the same surface vanishes, since we are considering a stationary situation, for which $\partial X/\partial t$ for each physical quantity X. Therefore, we must have

$$2\pi r E_\phi = 0 \Rightarrow E_\phi = 0 \tag{8.30}$$

Now we know that $\vec{E}$ has no toroidal component (namely, $E_\phi = 0$) and is also perpendicular to $\vec{B}$. A vector perpendicular both to $\hat{e}_\phi$ and to $\vec{B}$ can always be written as

$$\vec{E} = -h\hat{e}_\phi \wedge \vec{B} \qquad (8.31)$$

where h is a scalar that for dimensional reasons must be dimensionless. This can be rewritten as

$$\vec{E} = -\frac{\omega r}{c}\hat{e}_\phi \wedge \vec{B} \qquad (8.32)$$

where ω is, for the moment, an arbitrary function of position. We shall soon prove that $\omega = \Omega$, the pulsar's angular rotation.

We can now determine the current density $\vec{j}$. In order to do this, we eliminate the electric field between equations 8.17 and 8.32. We find

$$\left(\vec{j} - \frac{\rho \Omega r}{c}\hat{e}_\phi\right) \wedge \vec{B} = 0 \qquad (8.33)$$

Thence we easily find that

$$\vec{j} = \rho \Omega r \hat{e}_\phi + \kappa \vec{B} \qquad (8.34)$$

Here κ is a function, at the moment unknown, of position (but not of ϕ, as usual), whose physical meaning will be illustrated shortly. Therefore, we see that the current density is the sum of two parts: one ($\rho \Omega r \hat{e}_\phi$) corotates with the pulsar, whereas the other is due to currents parallel to the magnetic field lines, often called *field-aligned currents* in the literature.

The Magnetic Field

Let us now concentrate on

$$\nabla \cdot \vec{B} = \frac{1}{r}\frac{\partial}{\partial r}(rB_r) + \frac{1}{r}\frac{\partial B_\phi}{\partial \phi} + \frac{\partial B_z}{\partial z} = 0 \qquad (8.35)$$

where the divergence has been explicitly written in cylindrical coordinates to show that since $\partial B_\phi/\partial \phi = 0$ (once again because of the assumption of axial symmetry), the above equation reduces to

$$\nabla \cdot \vec{B} = \nabla_p \cdot \vec{B}_p = 0 \qquad (8.36)$$

where $\vec{B}_p = B_r \hat{e}_r + B_z \hat{e}_z$ is the poloidal part of the magnetic field. Moreover, we now write

$$\vec{B}_p = \frac{1}{r} \nabla f \wedge \hat{e}_\phi \qquad (8.37)$$

It is easy to see that this $\vec{B}_p$ automatically (i.e., for each scalar f) satisfies equation 8.35. Indeed,

$$B_r = -\frac{1}{r} \frac{\partial f}{\partial z} \equiv -\frac{f_z}{r}; \quad B_z = \frac{1}{r} \frac{\partial f}{\partial r} \equiv \frac{f_r}{r} \qquad (8.38)$$

and, introducing these formulae into equation 8.35, we find

$$\nabla_p \cdot \vec{B} = \frac{1}{r} \left(-\frac{\partial^2 f}{\partial z \partial r} + \frac{\partial^2 f}{\partial r \partial z} \right) = 0 \qquad (8.39)$$

Therefore, if we write $\vec{B}_p$ in this form, we have reduced the number of our unknown quantities. Indeed, instead of having simultaneously B_r and B_z, connected by equation 8.36, we have one scalar only, f, which automatically satisfies the condition $\nabla \cdot \vec{B} = 0$. In this way, we have considerably simplified our problem. The quantity f is well known in the literature, it is often called *Euler's potential*, but it is just the component A_ϕ of the potential vector. Euler's potential has this physical interpretation: by definition, equation 8.37, we see that the gradient of f is always perpendicular to $\vec{B}_p$, and thus the poloidal part of each flux line of $\vec{B}$ is tangent to the lines $f = $ constant.

We can now identify the unknown function ω that appears in equation 8.32. Indeed, introducing equation 8.37, we find

$$\vec{E} = -\frac{\omega}{c} (f_z \, \hat{e}_z + f_r \, \hat{e}_r) = -\frac{\omega}{c} \nabla f \qquad (8.40)$$

Now we want to make sure that $\vec{E}$ satisfies the Faraday-Neumann-Lenz equation,

$$\nabla \wedge \vec{E} = 0 \qquad (8.41)$$

In order to do this, let us take the curl of the above equation

$$\nabla \wedge \vec{E} = 0 = -\frac{1}{c}(\nabla \omega \wedge \nabla f + \omega \nabla \wedge \nabla f) = -\frac{1}{c}\nabla \omega \wedge \nabla f \quad (8.42)$$

Thence we see that ∇f and $\nabla \omega$ are always parallel. Therefore, the surfaces $\omega = $ constant coincide with the surfaces $f = $ constant, and we must have $\omega = \omega(f)$. The angular velocity is constant on magnetic surfaces, namely, the surfaces with $f = $ constant. We continue by considering equation 8.10; it is valid at the pulsar's surface, so we must have $\omega = \Omega$ at the star's surface. On the other hand, ω is constant on all lines with $f = $ constant, thus $\omega = \Omega$ everywhere. As a consequence,

$$\vec{E} = -\frac{\Omega r}{c}\hat{e}_\phi \wedge \vec{B} \quad (8.43)$$

or even

$$\vec{E} = -\frac{\Omega}{c}\nabla f \quad (8.44)$$

whence we see that apart from a multiplicative constant Ω/c, f is also the electrostatic potential! Moreover, for the charge density, we find

$$\nabla \cdot \vec{E} = 4\pi\rho = -\frac{\Omega}{c}\nabla^2 f = -\frac{\Omega}{c}\left(f_{\rm rr} + f_{\rm zz} + \frac{f_{\rm r}}{r}\right) \quad (8.45)$$

Let us now try to satisfy the last Maxwell equation, namely, Ampère's law,

$$\nabla \wedge \vec{B} = \frac{4\pi}{c}\vec{j} \quad (8.46)$$

in which $\vec{j}$ is given by equation 8.34, $\vec{B}_{\rm p}$ by equation 8.37, and ρ by equation 8.45.

Let us consider the toroidal component of this equation. We easily find

$$(\nabla \wedge \vec{B})_\phi = -\frac{1}{r}\left(f_{\rm zz} + f_{\rm rr} - \frac{f_{\rm r}}{r}\right) \quad (8.47)$$

Using this and equations 8.45 and 8.34 in equation 8.46, we find

$$f_{zz} + f_{rr} - \frac{1}{r}\frac{r_c^2 + r^2}{r_c^2 - r^2}f_r = -\frac{4\pi}{c}\frac{\kappa r_c^2 r B_\phi}{r_c^2 - r^2} \tag{8.48}$$

where $r_c \equiv c/\Omega$ is the corotation radius; namely, the distance from the pulsar where a rotating body at the pulsar's angular velocity reaches the speed of light. This equation shows that we must derive a relation between B_ϕ and f in order to have a closed system of equations (i.e., containing as many unknowns as equations).

In order to do this, we use the z component of equation 8.46 in integral form. Let us integrate equation 8.46 on the surface of the circle in figure 8.2, with radius r and $z = z_o, 0 \leq \phi \leq 2\pi$. The left-hand side gives

$$\int \nabla \wedge \vec{B} \cdot d\vec{A} = \oint \vec{B} \cdot d\vec{l} = 2\pi r B_\phi \tag{8.49}$$

The right-hand side, however, gives

$$\frac{4\pi}{c}\int \vec{j} \cdot d\vec{A} \equiv \frac{4\pi}{c}I \tag{8.50}$$

where I is the whole current crossing the surface in question. Therefore we find

$$rB_\phi = \frac{2I}{c} \tag{8.51}$$

This equation is well known from elementary courses. Indeed, it is the relation between the toroidal component of the magnetic field and the current carried by a wire, in axial symmetry.

Before reintroducing this result into equation 8.48, let us make a small, useful manipulation. We note that the magnetic flux F through the surface of the same circle is

$$F = 2\pi \int_0^r r dr B_z \tag{8.52}$$

If we introduce equation 8.38 here, we find

$$F = 2\pi \int_0^r rdr\frac{1}{r}f_r = 2\pi f(r) \qquad (8.53)$$

from which we see that apart from a factor 2π, Euler's potential f can be identified with the magnetic flux through the surface of the circle described above. This is the easiest physical interpretation of f.

At this point, we need to learn something more about κ. Let us impose charge conservation, which requires, in stationary conditions, $\nabla \cdot \vec{j} = 0$. From equation 8.34, we easily find that

$$\nabla \cdot \vec{j} = \vec{B} \cdot \nabla \kappa = 0 \qquad (8.54)$$

In order to derive this result from equation 8.34, we have used $\nabla \cdot (\rho r \hat{e}_\phi) = 0$, which comes from the fact that no physical quantity can depend on ϕ because of the hypothesis of axial symmetry; moreover,

$$\nabla \cdot (\kappa \vec{B}) = \kappa \nabla \cdot \vec{B} + \vec{B} \cdot \nabla \kappa = \vec{B} \cdot \nabla \kappa \qquad (8.55)$$

thanks to Maxwell's equation, $\nabla \cdot \vec{B} = 0$. Also, since $\partial \kappa / \partial \phi = 0$ because of axial symmetry, we see from the above equation that

$$\vec{B}_{\mathrm{p}} \cdot \nabla \kappa = 0 \qquad (8.56)$$

Therefore, $\nabla \kappa$ is a purely poloidal vector (because $\partial \kappa / \partial \phi = 0$) and perpendicular to the poloidal part of $\vec{B}$; from equation 8.37, we see that ∇f also has these features, so that the two vectors, $\nabla \kappa$ and ∇f, are always parallel. The surfaces with $\kappa = $ constant and $f = $ constant are the same, so that

$$\kappa = \kappa(f) \qquad (8.57)$$

The best, and most usual way in the literature, to state this result is the following: the current flux through two surfaces, whose perimeter is determined by the same magnetic field lines, is identical.

This relation between κ and f now allows us to proceed. We easily find

$$\delta F = 2\pi \nabla f \cdot d\vec{A} \qquad (8.58)$$

and

$$\delta I = 2\pi \kappa(f) \nabla f \cdot d\vec{A} \qquad (8.59)$$

Combining the two, we find

$$\kappa = \frac{dI}{dF} = \frac{1}{2\pi} \frac{dI}{df} \qquad (8.60)$$

Finally, we can put this equation, as well as equation 8.51, into equation 8.48 to obtain

$$f_{zz} + f_{rr} - \frac{1}{r} \frac{r_c^2 + r^2}{r_c^2 - r^2} f_r = -\frac{4r_c^2}{c^2 \left(r_c^2 - r^2\right)} I \frac{dI}{df} \qquad (8.61)$$

This is the equation we were looking for, the so-called *pulsar equation*, which determines Euler's potential f and, from this, the poloidal magnetic field (eq. 8.37), the toroidal magnetic field (eq. 8.51), the electric charge density (eq. 8.45), the current density (equation 8.34), and the electric field (eq. 8.43).

How to Determine $I(f)$

One might think naively that the function $I(f)$ must be determined from physical considerations, but this is not so. It is univocally determined by a mathematical condition, namely, that f and its radial derivative f_r are continuous across the light cylinder, $r = r_c$. The argument showing that these conditions of continuity are not automatically satisfied has an exquisitely mathematical nature; the readers who have no interest in the mathematical details may immediately move on to the next section, simply remembering that the solution of equation 8.61 is made of f and the only form of $I(f)$ that makes f and f_r continuous across the light cylinder.

Let us start by assuming a specific form for $I(f)$, among the infinite possible ones. We shall show that for an arbitrary $I(f)$, f will not be continuous to the cylinder of light. Now we note that equation 8.61 is an elliptical equation; for elliptical equations, we know that we can give boundary conditions of the Dirichlet type (i.e., the value of the function f on the surface) or of Neumann type (the value of the normal derivative on the surface, $\partial f/\partial n$), but not both on the same surface. However, we can choose whether to give, on each surface bounding the region of integration, either the conditions of Dirichlet or Neumann, changing our choice according to the surface.

Finally, we notice that equation 8.61 has a *critical surface*, namely, a surface on which the coefficient of the highest-order derivative vanishes. We need simply rewrite it in the form

$$\left(r_c^2 - r^2\right)(f_{zz} + f_{rr}) - \frac{r_c^2 + r^2}{r} f_r = -\frac{4r_c^2}{c^2} I \frac{dI}{df} \quad (8.62)$$

in order to see that the critical surface is the light cylinder, $r = r_c$. Note that there is a close similarity with the theory of the critical (or sonic, if you prefer) point discussed in chapter 5. In order for equation 8.62 to have a regular solution at the light cylinder, it is necessary that for $r = r_c$,

$$f_r = \frac{2r_c}{c^2} I \frac{dI}{df} \quad (8.63)$$

This means that since we have assumed to know $I(f)$, we also know the value of f_r on the light cylinder, and we notice that the knowledge of f_r may be used as a Neumann condition on the light cylinder, since, obviously, $\partial f/\partial n = f_r$. This suggests that we solve equation 8.62 separately, first within the light cylinder, then outside it.

The zone within the light cylinder is bounded (see fig. 8.3) by the curves $r = 0$, $z = 0$ (but only for $r \leq r_c$), $r = r_c$,

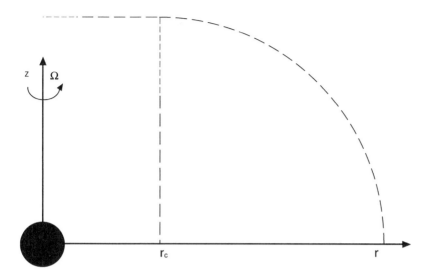

Figure 8.3. Domain of the solution of equation 8.61.

$z = +\infty$, and $R = \sqrt{r^2 + z^2} = R_{\text{ns}}$, where R_{ns} is the neutron star radius. On the surface $r = 0$, we can use the Dirichlet condition $f = 0$. This is obvious because $f(r, z)$ is the magnetic field flux through a circle at height z, with radius r; if $r = 0$, the flux through the circle must obviously vanish. In the same way, the flux f must vanish on the surface at infinity, $z \rightarrow \infty$; at large distances, for finite radii (since $r \leq r_{\text{c}}$), the magnetic field decreases to zero, the circle surface remains finite, and $f = 0$ (Dirichlet condition) for $z \rightarrow \infty$. On the star surface, f must of course tend to the value appropriate for the dipole, well known from elementary courses as $f = B_{\text{p}}r^2/R^3$, which is to be evaluated for $R = R_{\text{ns}}$. This is another Dirichlet condition. In order to decide which condition is to be applied on the straight line $z = 0$, $r \leq r_{\text{c}}$, let us follow the model of Goldreich and Julian, according to whom the magnetic field lines close to form a corotating magnetosphere. Since the lines must be closed, because of the symmetry of the problem on the straight line $z = 0$, $r \leq r_{\text{c}}$, we must have $B_r = 0$. Looking at equation 8.38, we realize, therefore, that $f_z = 0$ there. Notice that $f_z = 0$ is a Neumann condition on this line. Finally, we have already noted that equation 8.63 is a Neumann condition on the light cylinder. As a consequence, we have all the boundary conditions to integrate equation 8.62 within the light cylinder. Once we have done this, we obtain the value of f immediately inside the light cylinder, which we call f_-.

We also have all the necessary boundary conditions to integrate equation 8.62 outside the light cylinder. On the outer side of the light cylinder, equation 8.63 still holds. On the axis $z = 0$, $r > r_c$, the boundary condition has changed; indeed, if we follow the model of Goldreich and Julian, the field lines passing through the light cylinder close only at infinity. In this case, we easily realize that the axis $z = 0$, for $r \geq r_c$, must be a flux line, $f = $ constant, which is a Dirichlet boundary condition. The value of the constant is obtained by noting that the point $z = 0$, $r = r_c$ belongs both to the external and to the internal solution, for which f has been determined; its value at this point can therefore be imported from the internal solution. At infinity, we can assume that the external solution is bounded by a quarter circle with a radius $R \to \infty$, on which it is possible to determine the value of f as follows. Since we have assumed that the magnetic field lines do not close, they must close at infinity; in other words, they must become radial: when $R \to \infty$, we expect that f tends to a simple limit,

$$f \to f_0(z/r) \quad \text{for } R \to \infty \tag{8.64}$$

For $R \to \infty$, equation 8.61 reduces (introducing into it equation 8.64, and using the notation $x \equiv z/r$, $\dot{f_0} \equiv df_0(x)/dx$) to

$$(1 + x^2)\ddot{f_0} + x\dot{f_0} = \frac{4I}{\Omega^2}\frac{dI}{df} \tag{8.65}$$

which has the following first integral

$$I(f_0) = \frac{\Omega}{2}\sqrt{1 + x^2}\,\dot{f_0} = \frac{\Omega}{2\sin\theta}\dot{f_0} \tag{8.66}$$

This equation tells us that the asymptotic form to which f tends, $f_0(\theta)$, is determined by the form we assume for $I(f)$. In fact, having assumed a specific form for $I(f)$, this is a simple first-order differential equation for $f_0(\theta)$. Therefore we have the boundary conditions also for the region outside the light cylinder. We can therefore integrate equation 8.62 and find the value of f immediately outside the light cylinder, f_+.

At this point, we have reached our impass: there is no guarantee that for the form we have assumed for $I(f)$, f is continuous at the light cylinder, and thus that $f_- = f_+$. In fact, since we have to choose as boundary conditions on the light cylinder between the Dirichlet or Neumann type, we can impose either continuity of f_r or of f, but *not* the two simultaneously.

The discontinuity of f is a serious problem. Since the magnetic and the electric fields contain the derivatives of f, we see that they contain Dirac's deltas; their derivatives contain derivatives of the Dirac delta, placed at the light cylinder. This implies the existence of surface currents and charges at the light cylinder, which obviously cannot remain there. In fact, the light cylinder is a geometric locus, not a physical place where we can put some charges. As a consequence, for an arbitrary choice of $I(f)$, we cannot obtain a physically acceptable solution.

Our alternative is therefore to vary $I(f)$ with continuity until we find the only possible choice assuring the continuity of the Euler's potential at the light cylinder, $f_- = f_+$. It has been recently demonstrated by Contopoulos, Kazanas, and Fendt (1999) that this can be done; this procedure automatically determines the only form of $I(f)$ compatible with a stationary solution.

8.2.4 The Solution

Contopoulos, Kazanas, and Fendt (1999) have solved equation 8.61 numerically, identifying the only possible choice of

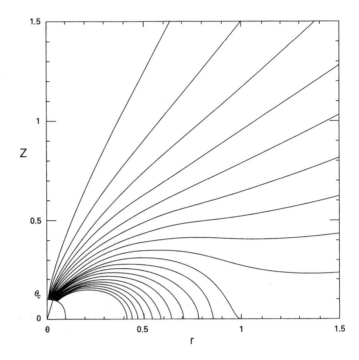

Figure 8.4. The curves $f = $ constant for the solution by Gruzinov (2005) of equation 8.61 in a regime of quasi-neutral plasma.

$I(f)$ such that f is continuous at the light cylinder. Their solution, however, is slightly incomplete, because the solution is weakly singular near the *separatrix* (i.e., the curve separating the open lines from the curved ones), and their numerical solution lacks the resolution to identify this singular behavior. Gruzinov (2005) has completed their work and showed that it is possible to obtain a completely satisfying solution, which is represented in figure 8.4.

8.2.5 The Transport of Angular Momentum

The above discussion on boundary conditions allows us to demonstrate an extremely important fact, namely, the pulsar loses energy and angular momentum. We shall first concentrate on the angular momentum.

This may happen in two ways. On one hand, the electric charges flowing toward infinity, besides gaining energy at the expense of the star, may acquire angular momentum along the path. On the other hand, we know that the electromagnetic field too, through the Maxwell stress tensor, can carry angular momentum to infinity. The flux of angular momentum is described by the total stress tensor

$$T_{ik} = R_{ik} + M_{ik} \qquad (8.67)$$

which is given by the sum of the Reynolds tensor (eq. 1.35), which contains the particles' contribution, and of the Maxwell tensor (eq. 6.72), which contains the fields' contribution. Here it is convenient to use spherical coordinates, R, θ, ϕ, rather than cylindrical coordinates, r, ϕ, z. We want to calculate the flux in the radial direction of the component of angular momentum along the z direction, which is obviously given by $rT_{R\phi}$. The angular momentum lost at infinity is given by the integral of $rT_{R\phi}$ on the surface of the sphere with a radius R

$$\frac{dL_z}{dt} = - \int R^2 d\phi \, d\cos\theta \, rT_{R\phi} \qquad (8.68)$$

letting then $R \to \infty$.

The contribution due to particles is negligible. Indeed, the total torque on the particles is given by gravitational forces, which, as we know, are central and cannot cause any torque, and electromagnetic forces, which, however, satisfy equation 8.17 and cannot therefore transfer angular momentum to particles. Thus, particles *conserve* the angular momentum with which they left the pulsar surface. There, they have the same specific angular momentum of the pulsar's material. However, since we obviously expect that the total amount of mass lost by the pulsar δM is much smaller than the mass of the pulsar itself, $\delta M \ll M$, also the total angular moment that has been lost, δL_z, will satisfy $\delta L_z \ll L_z$.

The contribution to the loss due to the electric field is zero, since $E_\phi = 0$ (eq. 8.30). However, the contribution to $rT_{R\phi}$ due to the magnetic field does not vanish. It is easiest to compute it at large distances from the pulsar, where

$$f \rightarrow f_0(z/r) \tag{8.69}$$

Introducing this expression into the equation 8.38, we find

$$B_r = -\frac{1}{r^2}\dot{f}_0, \quad B_z = -\frac{z}{r^3}\dot{f}_0 \tag{8.70}$$

where $\dot{f}_0(x) \equiv df_0(x)/dx$ is the derivative with respect to the argument z/r. From these equations we find

$$B_R = -\frac{R}{r^3}\dot{f}_0 \tag{8.71}$$

Now we can use this equation, as well as equation 8.51, in equation 6.72 to obtain

$$M_{R\phi} = \frac{1}{2\pi c}\frac{\dot{f}_0 I(f_0)}{r^3 \sin\theta} \tag{8.72}$$

The angular momentum along the z direction, which flows outside a sphere with a radius R per unit of time, $-dL_z/dt$, is thus obtained

$$-\frac{dL_z}{dt} = \int R^2 d\phi d\cos\theta \, r M_{R\phi} = \frac{\Omega}{2c}\int d\theta \sin\theta \left(\frac{df_0}{d\theta}\right)^2 \tag{8.73}$$

where we have used equation 8.66. It is important to notice that the result of the integral does not depend on R and is always >0; the star can only lose this amount of angular momentum per unit time.

Therefore, the star's angular momentum, L_z, decreases because there is a flux of angular momentum toward infinity, given by the above equation. This flux is carried neither by the plasma nor by the electric field, but only by the magnetic field. In problem 3, you are asked to prove that there is also a loss of kinetic energy, which is given by $\Omega dL_z/dt$.

We should emphasize the importance of this result. In the pulsars' classic model, Pacini (1967) assumed that the pulsars were placed in a vacuum and were endowed with a magnetic moment inclined by an angle α to the rotation axis. Any observer outside the pulsar will then see a magnetic field variable in time, and therefore also an electric field variable with time. This generates a flux of electromagnetic waves, which is proportional to $\sin^2 \alpha$ and which, therefore, in the case of the aligned rotator ($\alpha = 0$), must vanish. However, we have just shown that when we drop the assumption that the rotator be in a vacuum, this result is wrong: even the aligned rotator can radiate.

In the case of the Contopoulos Kazanas-Fendt-Gruzinov (CKFG) solution (Contopoulos, Kazanas, and Fendt 1999; Gruzinov 2005), it is possible to obtain the pulsar's total loss of energy, integrating numerically equation 8.73. We find

$$\frac{dE}{dt} = \Omega \frac{dL_z}{dt} = -(1 \pm 0.1)\frac{M^2\Omega^4}{c^3} \qquad (8.74)$$

where M is star's magnetic moment, and you are asked to verify the relationship $dE/dt = \Omega dL_z/dt$ in problem 3.

It is perhaps worthwhile to explain why, using equation 8.73, we appear to obtain $dE/dt \propto \Omega^2$ instead of $dE/dt \propto \Omega^4$. The reason is that the boundary conditions at infinity $f_0(\theta)$ are not given independently but are strictly connected to the current distribution $I(f)$ (see eq. 8.66); but neither is $I(f)$ independently known, as discussed above. The alert reader will remember that $I(f)$ is the *only* choice of current distribution that for the given stellar magnetic moment and rotation speed, allows f to be continuous across the light cylinder. So $I(f)$ and f_0 must be determined from M and Ω, and, on purely dimensional arguments, we must find $f \propto M\Omega/c$. The constant of proportionality between f and $M\Omega/c$, however, requires knowledge of the solution everywhere, which is exactly the reason why we had to wait for the CKFG solution to find the previous equation.

8.2.6 Discussion

The treatment based on the preceding equation is not completely satisfactory. In fact, looking at equation 8.276, we see that when $B_z = 0$, the velocity of electric charges inside the light cylinder must necessarily be $>c$! Indeed,

$$v \geq v_\phi = \frac{j_\phi}{\rho} = \frac{c^2}{\Omega r} \qquad (8.75)$$

and this, for all points with $B_z = 0$ inside the light cylinder, exceeds c. That these points exist in the CKFG solution can be seen directly from figure 8.4.

As we have seen above, this is not a problem if we accept a solution for quasi-neutral plasma, but it does become a problem if, as in the case of pulsars, we are looking for a solution in a regime of charge separation. This means that the CKFG solution is an acceptable solution for the problem in a regime of quasi-neutral plasma only and is *not* suitable for pulsars.

There is another problem with the CKFG solution, and that is that outflowing charges cross regions of alternating charge sign. Please notice that crossing a region of opposite corotating charge, for an outflowing current, is not a problem per se; the outflowing current might be provided in fact by charges of like sign, as those found in the corotating regions but flowing in the opposite direction. In fact, suitable charges might be sucked up from infinity by the favorable potential difference or generated at the star surface via a mechanism we will present shortly (section 8.4). The problem arises when regions containing charges of *both* signs are crossed one after the other by the outflowing currents, because there is no way then to arrange charges so that currents are carried by same-sign charges as those being crossed. Ideally, we would like the solution to have a curve that is both a magnetic surface $f = $ constant and the $\rho = 0$ surface. In this way, field-aligned currents would cross charges of the same sign everywhere; but this is *not* a property of the CKFG solution.

Where have we gone wrong? Nobody knows exactly, at this point. One possibility is that equation 8.17 fails close to the pulsar surface. That this equation cannot hold everywhere through space is clear enough. From it we immediately deduce

$$\vec{E} \cdot \vec{B} = 0 \qquad (8.76)$$

that is, magnetic field lines are equipotentials. But we already know that as the magnetic field becomes weak further out in radius, the Larmor radius of outflowing (or inflowing!) particles increases and carries them to regions of significantly different electrostatic potential; currents are not *field aligned* any more, and the equipotentiality of magnetic lines is spoiled. We tacitly assumed that this would happen way out in radius, where this effect would not matter, but, given the impossibility to obtain a self-consistent solution for a charge-separated plasma, this assumption appears now dubious at best.

Another possibility widely discussed in the literature is that there exist *vacuum gaps*: namely, gaps between corotating positive and negative charges, which are completely empty (vacuum) of charges (Holloway 1973). Here the key idea is that adjoining regions of opposite charge sign are kept apart by an electric field; stray charges crossing the vacuum might be accelerated, giving rise to a pair cascade. The electric field then gives rise to a current that appears to have carried positive charges through a negatively charged region (or viceversa), without this having really taken place. The problem of this explanation is that no one has ever managed to calculate the form and dimensions of these regions, nor to describe their topology, much less to demonstrate that they are stable rather than unstable, as some authors suspect. Therefore, though attractive, this solution is, for the moment, wholly hypothetical.

Another possibility, that has been discussed is that aligned rotators are not pulsars at all; only when the rotation is inclined with respect to the magnetic moment by an angle

$\alpha \neq 0$, does the rotating star display the typical features of pulsars.

8.3 The Blandford-Znajek Model

The Blandford-Znajek model is the exact equivalent for black holes with accretion disks of the model for pulsar magnetospheres, which we just discussed. The new difficulty is that this problem can no longer be discussed in flat Euclidean space; it requires a full general-relativistic discussion. In their original work, Blandford and Znajek (1977) derived the equivalent of the pulsar equation (eq. 8.61) using the exact Boyer-Lundquist metric. Before proceeding, however, it is necessary to discuss, in qualitative terms, the nature of the magnetic field of a black hole.

8.3.1 The Magnetic Field of a Black Hole

It is well known (Hawking and Ellis 1973) that an isolated black hole without net electric charge cannot have a magnetic field; only electrically charged, rotating black holes have a nonzero magnetic field. As discussed already in this book, it is extremely unlikely that a black hole may acquire a dynamically significant amount of electric charge. This, however, does not imply that an astrophysical black hole has no magnetic field, because the hole is not isolated. We must consider the possibility that the magnetic field is generated by currents carried by matter accreting onto the black hole. For example, consider a fluid element that having reached the inner radius of the accretion disk, falls on the black hole. Because of the frozen magnetic flux, the fluid particle drags with itself the magnetic field line to which it is attached. When the particle crosses the event horizon, the black hole seems to possess two magnetic field lines, with opposite polarities, stretching from the black hole to the accretion disk. Naturally enough, the currents generating the field must be

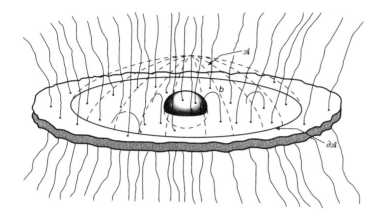

Figure 8.5. The structure of the magnetic field around a black hole surrounded by an accretion disk. From MacDonald and Thorne 1982.

located outside the event horizon, since each physical phenomenon inside the black hole is causally disconnected from the outside.

It is therefore reasonable to wonder which configuration will be assumed by the magnetic field. We must assume axial symmetry and stationarity, because the black hole is surrounded by a stationary accretion disk. Let us assume furthermore that the disk already lies on the equatorial plane of the black hole, as required by viscous dissipation of the Lense-Thirring precession (section 6.9).

Let us follow MacDonald and Suen (1985) in this discussion, and let us consider the surface $\mathcal{A}$ of figure 8.5 and apply the law of induction:

$$\frac{d}{dt} \int_{\mathcal{A}} \vec{B} \cdot d\vec{\mathcal{A}} = -c \oint \alpha \vec{E} \cdot d\vec{l} \qquad (8.77)$$

The quantity α is a function of position, which appears when we write the Maxwell equations in non-Euclidean geometry; it will be determined in the next section, since it is wholly irrelevant to our present aims. The integral on the right-hand side extends to a line lying in the accretion disk, where the condition of ideal magnetohydrodynamics holds

$$\vec{E} + \frac{\vec{v}}{c} \wedge \vec{B} = 0 \qquad (8.78)$$

which can be replaced in the above equation to obtain

$$\frac{d}{dt} \int_A \vec{B} \cdot d\vec{A} = \oint \alpha \vec{v} \wedge \vec{B} \cdot d\vec{l} \qquad (8.79)$$

Since $\vec{v}$ has radial and tangential components only, it follows that the variation of the magnetic flux through the surface A can only be due to the magnetic flux crossing the disk perimeter by the accreting matter; in other words, the total flux is conserved.

This result allows us to understand two important things. First, the magnetic field line attached to matter falling on the black hole cannot fly away toward infinity, because if it did so, it would violate equation 8.79. This line has three alternatives: It may reenter the disk, through an instability called *exchange instability*, which is the magnetohydrodynamic analogue of the Rayleigh-Taylor instability.[4] Alternatively, it may be pushed to cross the event horizon by the magnetic pressure of the surrounding regions, which simply drives it toward the black hole. Finally, it may be completely annihilated; a closed line, freed from the plasma to which it was attached, is subject to the magnetic tension, but not to matter's inertia (which is no longer there, since it fell into the hole). In this case, the line becomes smaller and smaller, until the magnetic field gradient becomes so large that we can no longer neglect currents; it is then destroyed by ohmic dissipation. With reference to figure 8.6, this is the fate of the line L in the top panel.

Second, let us consider the lines extending to a large distance from the black hole, with their feet close to the poles

---

[4] The Rayleigh-Taylor instability, in a hydrodynamics, takes place when a dense fluid presses on a light one and leads to the exchange of positions of the two fluids.

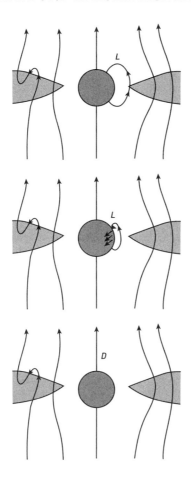

Figure 8.6. Fate of the magnetic field line for disk accretion onto a black hole. Adapted from Thorne, Price, and MacDonald 1986.

(the lines D in the bottom panel of fig. 8.6); these lines extend perhaps to infinity or may reconnect to the disk at large distances, but, in any case, they pierce the surface $\mathcal{A}$ of figure 8.5. Thanks to equation 8.79, we know for sure that these lines cannot be destroyed and must be conserved in time.

We must remark that the magnetic flux through the surface $\mathcal{A}$ is not strictly constant and may vary in direction; according to the polarity of the accreted matter, it may either increase or decrease. Like all stochastic phenomena, the magnetic field that is built in this way is unlikely to vanish; the black hole has acquired a magnetic configuration, supported by currents located on the right side of the event horizon.

The magnetic field lines extending to large distances do not stay in their place of their own accord; they are kept in their position by the magnetic pressure of the lines anchored to the accretion disk. If the disk is exhausted, the confining magnetic pressure also disappears, and the magnetic lines D fly away from the hole at the Alfvén speed, which, because of the very low density of the matter, is very close to the speed of light. The black hole would then remain without a magnetic field, as required by the theory for an isolated, electrically neutral black hole.

As a further consequence of this confining pressure, we note that the lines D may be sometimes compressed by a pressure excess in the lines anchored to the disk, but their magnetic pressure will lead to reexpansion, a purely sonic mechanism. When this occurs, the appropriate expansion speed is, as usual, the Alfvén speed, $v_A \approx c$, so that the lines D will stretch once again on the time scale $\approx GM/c^3 \approx 10^3$ s$(M/10^8 M_\odot)$.

In summary, the magnetic field configuration near a black hole divides into two zones: the zone closer to the equatorial plane, dominated by the magnetic field anchored to the accretion disk, where transient phenomena like exchange instability, magnetic reconnection, and dissipation abound; and a zone closer to the rotation axis, in which the field is more orderly. For the sake of simplicity, in the following, we shall completely neglect the transient phenomena of the first zone and shall treat the whole magnetosphere as if it were filled with a stationary, axially symmetric field. It goes without saying that this description is correct only in a time-averaged sense.

Once again, we shall assume that there are enough electric charges to make the magnetic field lines equipotential. The origin of these charges will be discussed in section 8.4, but we can anticipate that the mechanism is similar to that close to pulsars. As a consequence, the condition of equilibrium of the magnetospheres of black holes is, once again, the so-called *degeneracy condition*:

$$\rho \vec{E} + \frac{\vec{j}}{c} \wedge \vec{B} = 0 \qquad (8.80)$$

The magnetic fields around black holes are much weaker than the ones around pulsars; as we have just said, the black hole does not have a magnetic field of its own but acquires the disk's field. From equation 6.71, for $M = 10^8 \, M_\odot$, corresponding to $m = 10^8$, and using $v_\phi \approx c$ for the disk's innermost orbits, we see that the equipartition magnetic field $(B^2/8\pi = \rho v_\phi^2/2)$ is rather small, $B \approx 10^4$ G. The magnetic field is, *at most*, of these dimensions; various processes, such as reconnection of opposite, adjacent polarities, and exchange instability tend to make it weaker than the limit just derived.

This small magnetic field explains why charge separation probably does not hold in these environments. In discussing pulsars, we have seen that it was impossible keep charges of one sign above the star surface. The only possible mechanism to contrast gravitational and electric forces is the pressure gradient, but the required high temperature could not be maintained because of the considerable radiative losses in the pulsar's strong magnetic field. Now, however, the magnetic field is 8 to 10 orders of magnitude smaller, and the radiative losses, which scale like B^2, are 16 to 20 orders of magnitude smaller! Therefore, it is wholly reasonable that the losses may be neglected, and that a solution without separation of charge is acceptable. In fact, the whole reason for discussing pulsars' magnetospheres, despite our inability to build a self-consistent model for them, is that immediate generalization to black holes (which have no magnetic field of their own) is perfectly acceptable.

However, the potential differences generated by massive black holes are much higher than those of pulsars. In order of magnitude, consider a black hole with mass $M = 10^8 \, M_\odot$, specific angular momentum GM/c, and field $B \approx 10^4$ G. From equation 8.27 we find $V \approx 10^{20}$ eV. This number is so close to the highest energies of the cosmic rays observed so far, that we can seriously take into consideration the possibility that they are generated in this environment (Boldt and Ghosh 1999).

Notice also that magnetic D-type lines carrying current to infinity does not conflict with the elementary (and correct!) idea that near the event horizon, matter falls into the black hole. For example, a positive current propagating to infinity, which seems to originate from inside the black hole, may well be carried by negative charges *falling* inside the event horizon.

8.3.2 The Black Hole Equation

Just like with pulsars, it is possible to exploit the conditions of stationarity and axial symmetry of magnetospheres around rotating black holes to derive a single equation for Euler's potential, thus simplifying the problem. The derivation is nearly identical to that for pulsars, the only difference being that the computation must be carried out in a non-Euclidean geometry. In this section, we shall concentrate on the details of the computation, since the physics of the problem is identical to that pulsars.

A Few Useful Formulae from General Relativity

The notions of general relativity necessary for this problem are surprisingly limited. In this section we shall assume units such that $G = c = 1$, and we shall express the angular momentum of the black hole through the formula $J = aM$, where a is the specific (i.e., per mass unit) angular momentum; the quantity a varies between $0 \leq a \leq M$, or, in physical units, $0 \leq a \leq GM/c$. The nonrotating, or Schwarzschild, black hole corresponds to the case $a = 0$. Moreover, we shall not use an explicitly covariant description, but the one called the $3 + 1$ split of general relativity (Thorne, Price, and Mac-Donald 1986), which allows us to build on the physical understanding we have developed in the Newtonian case.

The metric we choose is the Boyer-Lindquist one. We can visualize this metric as follows (Thorne, Price, and MacDonald 1986). Space-time is full of observers who are as still as possible with respect to the black hole. If the black hole were of the Schwarzschild kind, observers could be exactly at rest; obviously, it would be necessary to assume that they are on a rocket providing the boost that counters the black hole gravitational attraction. However, apart from this, there is no conceptual difficulty in the notion of an observer at rest with respect with a Schwarzschild black hole. If, on the other hand, we consider a rotating black hole, this is *never* possible: the black hole drags space-time in its rotation, so that no observer can ever avoid rotating in the same direction as the black hole. In other words, there is no rocket capable of keeping the observers at rest with respect to the black hole. As a consequence, we are forced to consider observers rotating with speed $\equiv -\vec{\beta} = -\beta_\phi \hat{e}_\phi$.

The metric is given by

$$ds^2 = -\alpha^2 dt^2 + g_{jk}(dx^j + \beta^j dt)(dx^k + \beta^k dt) \quad (8.81)$$

where α is called the *lapse function*, since it describes the fact that the observers' time $d\tau = \imath ds$, is slower than the coordinate time dt, sometimes also called *universal time*.

In order to write the metric explicitly, it is useful to introduce the following definitions[5]:

$$\triangle^2 \equiv r^2 + a^2 - 2Mr, \quad \rho^2 \equiv r^2 + a^2 \cos^2\theta$$
$$\Sigma^2 \equiv (r^2 + a^2)^2 - a^2 \triangle^2 \sin^2\theta, \quad \varpi = \frac{\Sigma}{\rho}\sin\theta \quad (8.82)$$

---

[5] Notice that the definition of $\triangle$ differs from the one usually present in the literature.

thanks to which the metric coefficients become

$$\alpha = \frac{\rho}{\Sigma} \triangle \tag{8.83}$$

$$\beta^r = \beta^\theta = 0, \quad \beta^\phi = -\frac{2aMr}{\Sigma^2} \tag{8.84}$$

$$g_{\rm rr} = \frac{\rho^2}{\triangle^2}, \quad g_{\theta\theta} = \rho^2 \quad g_{\phi\phi} = \varpi^2 \tag{8.85}$$

and all the other coefficients of $g_{\rm ij}$ vanish.

All coefficients are independent of t and ϕ, as required by stationarity and axial symmetry. The quantity $-\vec{\beta}$ is the velocity of our observers, which reduces to $\vec{\beta} = 0$ for $a = 0$, the Schwarzschild limit; the function $\vec{\beta}$ is often called the *shift function*.

Notice also that the circumference of a circle, with its center on the axis of symmetry, is given by $2\pi\sqrt{g_{\phi\phi}} = 2\pi\varpi$, which will prove useful shortly. The simplest way to prove this is to consider the circumference of the circle in figure 8.7 in parametric form:

$$
\begin{aligned}
r &= r_0 \\
\theta &= \theta_0 \\
\phi &= \xi, \quad 0 \le \xi \le 2\pi
\end{aligned}
\tag{8.86}
$$

In this way, the (square) distance between two infinitesimal points becomes only a function of the parameter ξ:

$$ds^2 = g_{\rm jk}dx^j\,dx^k = g_{\rm rr}dr^2 + g_{\theta\theta}d\theta^2 + g_{\phi\phi}d\phi^2$$

$$= g_{\phi\phi}\left(\frac{d\phi}{d\xi}\right)^2 d\xi^2 = \varpi^2 d\xi^2 \tag{8.87}$$

Therefore, the circle's circumference is

$$C = \int_0^{2\pi} ds = \int_0^{2\pi} \varpi\,d\xi = 2\pi\varpi \tag{8.88}$$

as was to be proved.

Maxwell's Equations

From now on, we shall assume that all measurements of the electromagnetic field are made by local observers (local with respect to the point where we want to measure $\vec{E}, \vec{B}$). This requires two corrections to Maxwell's equations. First, each observer will express time not in the above-defined universal time t, but in his/her own time τ. In order to avoid this nuisance, we shall rewrite Maxwell's equations using t as the time coordinate, not the more natural τ. The two are connected by the obvious relation

$$\frac{d\tau}{dt} = \alpha \qquad (8.89)$$

Therefore, we should expect that the shift function α appears in the Maxwell equations.

Second, when we calculate the derivatives of the electromagnetic field, since we have defined E and B as quantities measured by rotating observers at the *exact point of measurement*, we find ourselves comparing values of E and B defined by two different observers who *are in relative motion*! As a consequence, Maxwell's equations are different from those we know for Euclidean space, where the various observers who measure the fields at different points are all at rest in the same reference frame.

The exact computation (MacDonald and Thorne 1982) gives

$$\nabla \cdot \vec{E} = 4\pi\rho_{\mathrm{e}} \qquad (8.90)$$

$$\nabla \cdot \vec{B} = 0 \qquad (8.91)$$

$$\nabla \wedge (\alpha\vec{E}) = -\frac{\partial\vec{B}}{\partial t} - \omega \pounds_{\mathrm{m}}\vec{B} + (\vec{B} \cdot \nabla\omega)\vec{m} \qquad (8.92)$$

$$\nabla \wedge (\alpha\vec{B}) = 4\pi\alpha\vec{j} + \frac{\partial\vec{E}}{\partial t} + \omega \pounds_{\mathrm{m}}\vec{E} - (\vec{E} \cdot \nabla\omega)\vec{m} \qquad (8.93)$$

where the density of charge is ρ_{e}, to distinguish it from the metric coefficient ρ. In this equation, ω is the observers' angular velocity, $\omega = -\beta_\phi/\varpi$, $\vec{m} = \varpi \hat{e}_\phi$, and

$$\pounds_m \vec{X} \equiv (\vec{m} \cdot \nabla)\vec{X} - (\vec{X} \cdot \nabla)\vec{m} \qquad (8.94)$$

is the so-called *Lie derivative* of $\vec{X}$ along $\vec{m}$.

These equations simplify considerably in the case of a stationary problem with axial symmetry. The axial symmetry, indeed, gives us $\pounds_m \vec{X} = 0$, and stationarity gives $\partial/\partial t = 0$. Therefore, in our case, Maxwell's equations are simply given by

$$\nabla \cdot \vec{E} = 4\pi \rho_e \qquad (8.95)$$
$$\nabla \cdot \vec{B} = 0 \qquad (8.96)$$
$$\nabla \wedge (\alpha \vec{E}) = (\vec{B} \cdot \nabla \omega)\vec{m} \qquad (8.97)$$
$$\nabla \wedge (\alpha \vec{B}) = 4\pi \alpha \vec{j} - (\vec{E} \cdot \nabla \omega)\vec{m} \qquad (8.98)$$
$$\vec{m} = \varpi \hat{e}_\phi \qquad (8.99)$$

It is useful also to have the integral formulation for the Faraday-Neumann-Lenz equation (MacDonald and Thorne 1982):

$$\oint_{C(t)} \alpha(\vec{E} + \vec{v} \wedge \vec{B}) \cdot d\vec{l} = -\frac{d}{dt} \int_{A(t)} \vec{B} \cdot d\vec{A} \qquad (8.100)$$

This is the generalization in the Boyer-Lindquist metric of the equation

$$\oint_C (\vec{E} + \frac{\vec{v}}{c} \wedge \vec{B}) \cdot d\vec{l} = -\frac{1}{c}\frac{d\Phi_B}{dt} \qquad (8.101)$$

This equation, probably known from elementary courses (see, for example, Landau and Lifshitz 1984, section 63), is the integral formulation of the Faraday-Neumann-Lenz equation (eq. 8.41), when the circuit C on which the integral is to be computed is *not* stationary in the reference frame in which we measure $\vec{E}$ and $\vec{B}$. In this case, indeed, $\vec{v}$ is the velocity of the points belonging to the contour C. In this case, the time derivative of Φ_B, the magnetic field flux, includes the contributions due to the time variation of $\vec{B}$ and to the circuit change with time.

A Few Comments on Geometry

We have adopted, for the description of physical phenomena in a gravitational field, the $3+1$ split because the description is very similar to the one in Euclidean geometry, as shown by the simple Maxwell's equations we have just described. However, there is something new: although the coordinates are spherical, r, θ, ϕ, since the three-dimensional metric tensor is not the one known from elementary courses, but the more complex one, g_{jk}, described above, the expressions for the differential operators are different from the ones in spherical but Euclidean coordinates. We derive here adequate expressions.

Let us consider arbitrary curvilinear coordinates, u, v, w, and unitary, mutually orthogonal vectors $\hat{e}_u, \hat{e}_v, \hat{e}_w$, parallel to coordinate lines, taken in the direction of increasing coordinates. Then, the vector distance of two arbitrarily close points is given by

$$\delta \vec{x} = h_1 \delta u \hat{e}_u + h_2 \delta v \hat{e}_v + h_3 \delta w \hat{e}_w \qquad (8.102)$$

where, obviously, $h_1 = |\partial \vec{x}/\partial u|$, and so on. Since the unit vectors have been supposed to be orthonormal, we see that the square modulus of the infinitesimal distance between these two points is given by

$$dx^2 = (h_1 \delta u)^2 + (h_2 \delta v)^2 + (h_3 \delta w)^2 \qquad (8.103)$$

whence we see that the coefficients h_1^2, h_2^2, h_3^3 correspond to the coefficients of the metric tensor g_{jk}, provided the latter is diagonal.

It is easy to obtain expressions for the gradient,

$$\nabla \psi = \frac{1}{h_1} \frac{\partial \psi}{\partial u} \hat{e}_u + \frac{1}{h_2} \frac{\partial \psi}{\partial v} \hat{e}_v + \frac{1}{h_3} \frac{\partial \psi}{\partial w} \hat{e}_w \qquad (8.104)$$

for the divergence,

$$\nabla \cdot \vec{F} = \frac{1}{h_1 h_2 h_3} \left(\frac{\partial}{\partial u} (h_2 h_3 F_u) + \frac{\partial}{\partial v} (h_1 h_3 F_v) + \frac{\partial}{\partial w} (h_1 h_2 F_w) \right)$$

$$(8.105)$$

and for the curl,

$$\nabla \wedge \vec{F} = \frac{1}{h_1 h_2 h_3} \begin{vmatrix} h_1 \hat{e}_u & h_2 \hat{e}_v & h_3 \hat{e}_w \\ \frac{\partial}{\partial u} & \frac{\partial}{\partial v} & \frac{\partial}{\partial w} \\ h_1 F_u & h_2 F_v & h_3 F_w \end{vmatrix} \qquad (8.106)$$

Specializing to our case, we find

$$u = r, \quad v = \theta, \quad w = \phi \qquad (8.107)$$

$$h_1 = \frac{\rho}{\triangle}, \quad h_2 = \rho, \quad h_3 = \varpi \qquad (8.108)$$

$$\nabla \psi = \frac{\triangle}{\rho} \frac{\partial \psi}{\partial r} \hat{e}_r + \frac{1}{\rho} \frac{\partial \psi}{\partial \theta} \hat{e}_\theta + \frac{1}{\varpi} \frac{\partial \psi}{\partial \phi} \hat{e}_\phi \qquad (8.109)$$

$$\nabla \cdot \vec{F} = \frac{\triangle}{\rho^2 \varpi} \left(\frac{\partial}{\partial r} (\rho \varpi F_r) + \frac{\partial}{\partial \theta} \left(\frac{\rho \varpi F_\theta}{\triangle} \right) + \frac{\partial}{\partial \phi} \left(\frac{\rho^2 F_\phi}{\triangle} \right) \right) \qquad (8.110)$$

$$\nabla \wedge \vec{F} = \frac{\triangle}{\rho^2 \varpi} \left(\frac{\rho}{\triangle} \hat{e}_r \left(\frac{\partial}{\partial \theta} (\varpi F_\phi) - \frac{\partial}{\partial \phi} (\rho F_\phi) \right) \right.$$
$$+ \rho \hat{e}_\theta \left(\frac{\partial}{\partial \phi} \frac{\rho F_r}{\triangle} - \frac{\partial}{\partial r} (\varpi F_\phi) \right)$$
$$\left. + \varpi \hat{e}_\phi \left(\frac{\partial}{\partial r} (\rho F_\theta) - \frac{\partial}{\partial \theta} \frac{\rho F_r}{\triangle} \right) \right) \qquad (8.111)$$

The Equation

We are now ready to deduce the black hole equation, equivalent to the pulsar equation 8.61. We note once again that axial symmetry and stationarity of the process imply

$$E_\phi = 0 \qquad (8.112)$$

In fact, we apply equation 8.97 by integrating it on the surface in figure 8.7, recalling that $\vec{m} \propto \hat{e}_\phi$, thus it does not contribute to the integral.

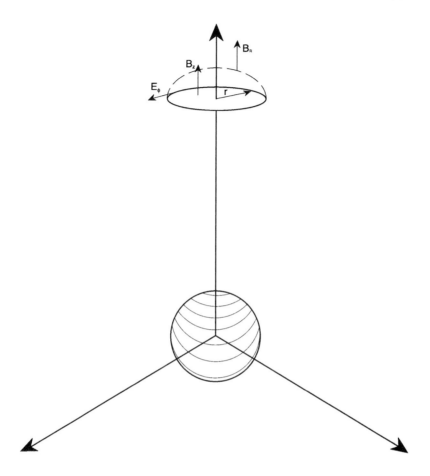

Figure 8.7. Computation for Black Hole.

In the same way, from equation 8.98, we easily find, integrating on the surface of the usual circle of figure 8.7, with circumference C,

$$\int (\nabla \wedge (\alpha \vec{B})) \cdot d\vec{A} = \oint_C \alpha \vec{B} \cdot d\vec{l} = \oint_C \alpha B_\phi ds = 2\pi \alpha B_\phi \varpi = 4\pi I$$

(8.113)

where B_ϕ cannot depend on ϕ because of the assumption of axial symmetry, and I is, as usual, the current flowing through the circle

$$I \equiv \int \alpha \vec{j} \cdot d\vec{A} \tag{8.114}$$

At first sight, this definition of I may seem unusual compared to the traditional $I = \int \vec{j} \cdot d\vec{A}$, but one may note that $\vec{j}$ is the flux of charge per unit surface and *proper time τ*, whereas $\alpha \vec{j}$ is the flux of charge per unit surface and *universal time t*. We therefore find

$$B_\phi = \frac{2I}{\alpha \varpi} \tag{8.115}$$

Notice that this time, we have integrated on the spherical cap, instead of the usual circle; this makes no difference whatsoever because of $\nabla \cdot \vec{B} = 0$.

In this way, we have once again identified the field's toroidal component, and we must find the poloidal ones. As far as the poloidal parts of $\vec{E}$ are concerned, we note once again that the condition of degeneracy, $\vec{E} \cdot \vec{B} = 0$, which automatically follows from the vanishing of the Lorentz force, equation 8.17, and the fact that $E_\phi = 0$ tell us once again that

$$\vec{E} = -h\hat{e}_\phi \wedge \vec{B} \tag{8.116}$$

just as in the case of pulsars. We can now use equation 8.100 to find

$$\vec{E} = -v\hat{e}_\phi \wedge \vec{B} \tag{8.117}$$

Here v is a velocity of rotation to be specified. However, even before determining v, we know that this is a velocity with respect to our observers. Therefore,

$$v = \frac{\varpi \Omega + \beta_\phi}{\alpha} \equiv \frac{\varpi}{\alpha}(\Omega - \omega) \tag{8.118}$$

where Ω is the velocity of rotation with respect to infinity, not to local observers. $\omega \equiv -\beta_\phi/\varpi$ is the observers' angular velocity. Here the shift function appears again, because we want to express the angular velocities Ω and ω in absolute time t, whereas in Maxwell's equations and in the Lorentz force we see speed with respect to the proper time, τ; we recall that $d\tau/dt = \alpha$. We still have to specify to which object the angular velocity Ω refers, which we will do shortly.

Once again, as in the case of pulsars, exploiting the axial symmetry of the problem, we know that B_ϕ does not depend on ϕ, so that when it is introduced into Maxwell's equation $\nabla \cdot \vec{B} = 0$ (eq. 8.110), the term containing B_ϕ disappears. We see then that the poloidal component of $\vec{B}$, $\vec{B}_{\mathrm{p}}$, satisfies $\nabla \cdot \vec{B}_{\mathrm{p}} = 0$. We can therefore write the poloidal part of the magnetic field $\vec{B}_{\mathrm{p}}$ as

$$\vec{B}_{\mathrm{p}} = \frac{\nabla f \wedge \hat{e}_\phi}{\varpi} \qquad (8.119)$$

We easily realize that this leads to the solution of $\nabla \cdot \vec{B}_{\mathrm{p}} = 0$. Using equation 8.109, we find

$$\vec{B}_{\mathrm{p}} = -\frac{\triangle}{\varpi\rho} \frac{\partial f}{\partial r} \hat{e}_\theta + \frac{1}{\rho\varpi} \frac{\partial f}{\partial \theta} \hat{e}_r \qquad (8.120)$$

We can now introduce this expression in the equation $\nabla \cdot \vec{B}_{\mathrm{p}} = 0$, using the equation 8.110; we these find

$$\nabla \cdot \vec{B}_{\mathrm{p}} = \frac{\partial^2 f}{\partial\theta\partial r} - \frac{\partial^2 f}{\partial r\partial\theta} = 0 \qquad (8.121)$$

Euler's potential f is once again the magnetic flux (apart from the usual factor 2π). Indeed, we can obtain the flux through the circle in figure 8.7 by integrating on the spherical cap, since $\nabla \cdot \vec{B} = 0$. For the surface element, we have

$$d\vec{A} = \sqrt{g_{\phi\phi}g_{\theta\theta}}\hat{e}_\phi \wedge \hat{e}_\theta = \rho\varpi d\theta d\phi\hat{e}_r \qquad (8.122)$$

and thus,

$$\int \vec{B} \cdot d\vec{A} = \int B_{\mathrm{r}}\rho\varpi d\theta d\phi = 2\pi \int \frac{\partial f}{\partial\theta}d\theta = 2\pi f \qquad (8.123)$$

where I used equation 8.120. This shows in fact that f is the magnetic flux, apart from a constant factor.

Introducing equation 8.119 into equation 8.117, we obtain

$$\vec{E} = -\frac{\Omega - \omega}{\alpha}\nabla f \qquad (8.124)$$

We can now prove that Ω is a function only of f. This, as we shall soon see, is the condition to satisfy equation 8.100, which is the mathematical equivalent of Maxwell's equation in differential form, equation 8.97. In order to go on, let us apply equation 8.100 to a circuit C, which is fixed in absolute space. Locally, this circuit moves, relative to local observers, with a velocity $\vec{v}$, obviously opposite to that of the observers relative to the absolute space; therefore, $\vec{v} = \beta_\phi \hat{e}_\phi$. Notice, however, that the right-hand side of equation 8.100 also vanishes, thanks to the problem's axial symmetry, so that

$$\oint_C \left(\vec{E} + \frac{\beta_\phi}{\alpha}\hat{e}_\phi \wedge \vec{B}\right) \cdot \alpha d\vec{l} = 0 \qquad (8.125)$$

Using equation 8.124 and recalling that $\beta_\phi = -\omega\varpi$, this can be rewritten as

$$-\oint_C \Omega \nabla f \cdot d\vec{l} = 0 \qquad (8.126)$$

which must hold for *any circuit* C; therefore, there must be a potential ψ such that

$$\nabla\psi = \Omega\nabla f \qquad (8.127)$$

Whence we see that $\nabla\psi$ and ∇f are always parallel, and that the level curves of ψ and f are identical. Thus also Ω has the same level curves as f:

$$\Omega = \Omega(f) \qquad (8.128)$$

Ω is constant on magnetic surfaces, that is, the surfaces identified by the equation $f = $ constant (see fig. 8.8).

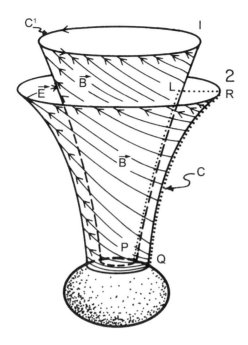

Figure 8.8. On each of these magnetic surfaces Ω = constant.

This is the right time to discuss the physical interpretation of the angular velocity Ω. In the case of pulsars, where observers were at rest ($\omega = 0$), Ω was simply the star's rotation velocity. However, the black hole has no definite rotation speed. Since the event horizon is not a physical place, but a mere geometric place, it cannot hold a luminous spot rotating with a definite frequency. Nor, on the other hand, are field lines dragged by the star's rotation, since we cannot suppose that the black hole's surface is a perfect conductor, just like the pulsar, *also* because this surface does not exist. Therefore, Ω is only the angular velocity of the magnetic surface; it is not the black hole's angular speed (which does not exist), and it is not even constant from surface to surface. However, it corresponds to something physical, namely, the angular velocity of matter in the disk, which obviously is an excellent conductor. The magnetic field will be discontinuous there because of the presence of toroidal disk currents. However, the total electric field in the comoving reference frame (which is the same for matter orbiting in the disk and for the observers moving with velocity $-\vec{\beta}$) must vanish because of the assumption of high conductivity for the disk matter. Therefore, $\Omega(f)$ is parametrically given by the rotation velocity of matter in the disk, and by the potential f at the same point.

Since equation 8.117 holds, the condition for the cancelation of the Lorentz force,

$$\rho_e \vec{E} + \vec{j} \wedge \vec{B} = 0 \qquad (8.129)$$

may be rewritten, eliminating $\vec{E}$, as

$$\left(\vec{j} - \frac{\rho_e \varpi}{\alpha}(\Omega - \omega)\hat{e}_\phi\right) \wedge \vec{B} = 0 \qquad (8.130)$$

which gives

$$\vec{j} = \frac{\rho_e \varpi}{\alpha}(\Omega - \omega)\hat{e}_\phi + \kappa\vec{B} \qquad (8.131)$$

An argument absolutely identical to the one for the pulsars tells us that

$$\kappa = \frac{1}{2\pi\alpha}\frac{dI}{df} \qquad (8.132)$$

and therefore,

$$\vec{j} = \frac{\rho_e \varpi}{\alpha}(\Omega - \omega)\hat{e}_\phi + \frac{dI}{df}\frac{\vec{B}}{2\pi\alpha} \qquad (8.133)$$

We now use the component ϕ of equation 8.98 to obtain the black hole equation. Introducing equation 8.120 into equation 8.111, we find

$$(\nabla \wedge \alpha\vec{B})_\phi = -\frac{\triangle}{\rho^2}\left(\frac{\partial}{\partial r}\left(\frac{\triangle\alpha}{\varpi}\frac{\partial f}{\partial r}\right) + \frac{\partial}{\partial\theta}\left(\frac{\alpha}{\triangle\varpi}\frac{\partial f}{\partial\theta}\right)\right)$$

$$= -\varpi\nabla \cdot \left(\frac{\alpha\nabla f}{\varpi^2}\right) \qquad (8.134)$$

The right-hand side of equation 8.98 is

$$4\pi\alpha j_\phi = (\nabla \cdot \vec{E})(\Omega - w)\varpi + \frac{dI}{df}\frac{4I}{\alpha\varpi} \tag{8.135}$$

The divergence of $\vec{E}$ is obtained using equation 8.124 in equation 8.110

$$\nabla \cdot \vec{E} = -\nabla \cdot \left(\frac{\Omega - w}{\alpha} \nabla f \right) \tag{8.136}$$

If we use the two preceding equations in equation 8.98, we find the equation for f, in the form of MacDonald and Thorne (1982), not in that of Blandford and Znajek (1977):

$$\nabla \cdot \left(\frac{\alpha}{\varpi^2} \left(1 - \frac{(\Omega - w)^2 \varpi^2}{\alpha^2} \right) \nabla f \right) + \frac{\Omega - w}{\alpha} \frac{d\Omega}{df} (\nabla f)^2$$

$$+ \frac{4}{\alpha\varpi^2} I \frac{dI}{df} = 0 \tag{8.137}$$

This is, once again, a second-order elliptic equation that replaces equation 8.61.

Boundary Conditions

The boundary conditions deriving from the symmetry of the problem for the pulsars' equation continue to hold. Thus we have

$$f_\theta = 0, \quad \theta = 0 \tag{8.138}$$
$$f_\theta = 0, \quad \theta = \pi/2 \tag{8.139}$$

Moreover, again in analogy with pulsars, at large distances we want to insure that f becomes a function of the angle θ only, in other words, that the magnetic field becomes radial. Therefore we require

$$\frac{\partial f}{\partial r} \to 0 \text{ as } r \to \infty \tag{8.140}$$

As for the equation of pulsars, the coefficients of the second-order terms may vanish on critical surfaces. There are two different kinds of surfaces, but the more easily identifiable one is where

$$\triangle = \alpha = 0 \tag{8.141}$$

This surface is just the event horizon for the black hole in question, namely, the surface (in geometric units, $G = c = 1$)

$$r = r_{\rm H} \equiv M + \sqrt{M^2 - a^2} \tag{8.142}$$

The other surface on which this happens, as can be easily shown, is the one on which

$$D^2 \equiv \alpha^2 - (\Omega - \omega)^2 \varpi^2 = 0 \tag{8.143}$$

which is the generalization of the light cylinder. In fact, in the Newtonian case, $\alpha \rightarrow 1$, $\omega \rightarrow 0$, $\varpi \rightarrow r \sin \theta \equiv R$, the radius in cylindrical coordinates, and, finally, $\Omega \rightarrow \Omega_{\rm pulsar}$, the pulsar's angular velocity.

The solution of equation 8.137 is still far in the future, so we shall not even try to present an in-depth discussion.

8.3.3 The Transport of Energy and of Angular Momentum

Just like in the Newtonian case, the solution for black holes transports energy and angular momentum to infinity. If the magnetic field lines never crossed the accretion disk, we could say that the process of Blandford-Znajek would manage to extract energy and angular momentum from the black hole. However, since there may also be configurations of the type drawn in figure 8.9, we might discover that the exact solution of equation 8.137 (which, we remind you, is still far in the future) removes them from the disk. Currently, we cannot tell exactly which fraction will come from the disk, and which from the black hole.

We can easily realize that each local observer sees an energy flux in a radial direction, given by the appropriate component of the Poynting vector, $\vec{F}_{\rm E} = c\vec{E} \wedge \vec{B}/4\pi$, and a flux

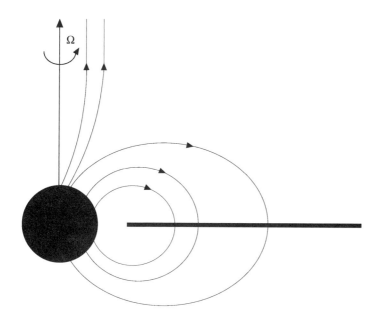

Figure 8.9. Qualitative scheme representing a possible config-uration for the magnetic field in a system constituted by an accretion disk and a black hole.

of angular momentum, given by $r \sin \theta M_{\mathrm{R}\phi}$, where M is, as usual, the Maxwell tensor, equation 6.72; and, still in analogy with pulsars, this flux is not zero, even though $E_\phi = 0$, be-cause $B_\phi \neq 0$. However, these fluxes are computed in terms of the coordinates of each local observer, who moves relative to the observer at infinity (where observations are made). We can define a new vector for energy flux and angular momen-tum from the modified Maxwell's equations (eq. 8.90), which we gave above. Thus we find (MacDonald and Thorne 1982)

$$\vec{F}_{\mathrm{E}} = \frac{1}{4\pi} \left(\alpha \vec{E} \wedge \vec{B} - \omega(\vec{E} \cdot \vec{m})\vec{E} - \omega(\vec{B} \cdot \vec{m})\vec{B} + \frac{\omega}{2}(E^2 + B^2)\vec{m} \right)$$
(8.144)

$$\vec{F}_{\mathrm{L}} = \frac{1}{4\pi} \left((-\vec{E} \cdot \vec{m})\vec{E} - (\vec{B} \cdot \vec{m})\vec{B} + \frac{1}{2}(E^2 + B^2)\vec{m} \right)$$
(8.145)

where I defined $\vec{F}_L$ as the angular-momentum flux vector, per unit absolute surface and time; the radial component of $\vec{F}_L$ is the one we have called $r \sin \theta M_{R\phi}$, in Euclidean space. We also recall that $\vec{m} = \varpi \hat{e}_\phi$, and that $\omega = -\beta_\phi/\varpi$.

Again these two expressions simplify considerably in the case of axial symmetry, since $\vec{E} \cdot \vec{m} = \varpi E_\phi = 0$. Now we consider a spherical surface large enough to contain the black hole and the accretion disk, and we compute the energy and angular momentum fluxes through this sphere. Since $d\vec{A} \propto \hat{e}_r$, the terms along $\vec{m} = \varpi \hat{e}_\phi$, in $\vec{F}_E$ and $\vec{F}_L$, disappear, leaving

$$F_{Er} = \frac{1}{4\pi}((\alpha \vec{E} \wedge \vec{B})_r - \omega \varpi B_\phi B_r) \qquad (8.146)$$

$$F_{Lr} = -\frac{1}{4\pi} \varpi B_\phi B_r \qquad (8.147)$$

which can now be simplified thanks to equations 8.109, 8.120, and 8.124:

$$F_{Er} = -\frac{\Omega \varpi}{4\pi} B_\phi B_r \qquad (8.148)$$

$$F_{Lr} = -\frac{\varpi}{4\pi} B_\phi B_r \qquad (8.149)$$

The important thing to notice here is that the two fluxes differ by a factor Ω. You may remember that in the case of pulsars, the two fluxes differ by a factor $\Omega_\star$, the star's angular velocity; this confirms our interpretation of the quantity Ω as the angular velocity of the whole magnetic surface.

We get the total fluxes by integrating the above equation on the sphere's surface; as we have already demonstrated, $d\vec{A} = \rho \varpi \hat{e}_r$, so the total fluxes are

$$\frac{dE}{dt} = -\int I(f)\Omega(f)df \qquad (8.150)$$

$$\frac{dL_z}{dt} = -\int I(f)df \qquad (8.151)$$

where I used equations 8.120 and 8.115. Notice that here a factor α disappeared because the integrals must be calculated for $r \to \infty$ (since, in this limit, f becomes independent from r), in which case, however, $\alpha \to 1$.

8.3.4 A Qualitative Discussion

We present here an analogy between the rotating black hole and a battery; we wish to show that a particle, starting from infinity and moving toward the hole to its surface and then moving back to infinity feels a nonzero electromotive force.

Our starting point is a reformulation of the Faraday-Neumann-Lenz law, in integral form

$$\oint_{\mathcal{C}} \alpha \left(\vec{E} + \frac{\vec{v}}{c} \wedge \vec{B} \right) \cdot d\vec{l} = -\frac{1}{c} \frac{d}{dt} \int_{\mathcal{S}(t) \vec{B} \cdot d\vec{A}} \equiv -\frac{1}{c} \frac{d\Phi_{\mathrm{B}}}{dt} \quad (8.152)$$

Here $\mathcal{C}$ is an arbitrary closed path and $\vec{v}$ the velocity of its perimeter, and Φ_{B} is the magnetic field flux through $\mathcal{S}$, the area enclosed by the path $\mathcal{C}$. Notice that this formulation is slightly different from the usual one because of the term $\vec{v} \wedge \vec{B}$. The latter appears because we have *not* assumed that the path $\mathcal{C}$ is stationary but assume instead that it may depend on time. You can find the derivation of this expression in any elementary text; Landau and Lifshitz (1984) discuss it in their section 63. It differs from the traditional (Newtonian) formula,

$$\oint_{\mathcal{C}} \vec{E} \cdot d\vec{l} = -\frac{1}{c} \frac{d\Phi_{\mathrm{B}}}{dt} \quad (8.153)$$

essentially because electric and magnetic fields are measured in a reference frame in which the path is moving.

We now apply this expression to the path $PQRL$ in figure 8.8; the line segments QR and LP lie on magnetic surfaces, $f = $ constant, the segment PQ on the hole surface, and RL at infinity. Notice that we are using magnetic surfaces reaching infinity, neglecting the corotating magnetosphere. It is convenient to consider the path as stationary with respect to the local observers in whose frame we measure the electromagnetic field. In this case,

$$\frac{d\Phi_{\mathrm{B}}}{dt} = 0 \quad (8.154)$$

because the magnetic field does not vary with time. However, with respect to the star-fixed coordinates, the curve rotates (differentially), so that $\vec{v} \neq 0$. As a consequence,

$$\oint_C \alpha \vec{E} \cdot d\vec{l} \equiv \mathcal{E} = -\oint_C \alpha \frac{\vec{v}}{c} \wedge \vec{B} \cdot d\vec{l} \qquad (8.155)$$

where $\mathcal{E}$ is the *electromotive force* along the circuit. We shall now show that $\mathcal{E} \neq 0$. Let us start with QR and LP. We notice that $\vec{v} \propto \Omega r \hat{e}_\phi$, and therefore $\vec{v} \wedge \vec{B} \propto \hat{e}_\phi \wedge \vec{B}_{\mathrm{p}}$ is a poloidal vector, but orthogonal to $\vec{B}_{\mathrm{p}}$. Since $\vec{B}_{\mathrm{p}}$ is contained within the magnetic surface $f = $ constant by definition of the magnetic surface itself, $\vec{v} \wedge \vec{B} \cdot d\vec{l} = 0$. Therefore, the part of the integral along the magnetic surfaces vanishes.

However, the part at infinity, which crosses from one magnetic surface to the other, also vanishes. It is in fact easy to show that for $r \to \infty$, the integrand is proportional to ω, the rotation speed of the local observers, but this $\omega \to 0$ as $r \to \infty$. Therefore,

$$\int_{\mathrm{R}}^{L} \frac{\vec{v}}{c} \wedge \vec{B} \cdot d\vec{l} = 0 \qquad (8.156)$$

On the other hand, the contribution along PQ is not zero. As a consequence,

$$\mathcal{E} = -\frac{\Omega_{\mathrm{H}}}{c} df \qquad (8.157)$$

There is a net electromotive force along the path, where, obviously, Ω_{H} is the value of $\alpha v_\phi / \varpi$ at the event horizon.

We learn in elementary courses that a battery loses energy because there is a resistance, which, however, is absent in this case. However, we have an energy loss even in the case in which the resistance (i.e., collisions) is strictly zero. Imagine an electron covering this circuit N times; after N iterations, its energy is $N\mathcal{E}$, and this energy must have been lost by the battery. It makes no difference whether the electron radiates away this energy or not; the battery has lost energy. In the same way, it makes no difference whether the circuit is not geometrically closed; at infinity, the magnetic field has become so weak that particles may drift from one magnetic field line to another, and it is therefore *as if* the circuit were geometrically closed. Again, it makes no difference whether the very same electrons are not crossing the circuit several times (why?), nor if, instead of electrons crossing LP, there are instead positrons crossing the same path in the opposite direction (once again, the explanation is left to the reader). When the particles reach infinity, the potential in which they find themselves is zero by definition, and they must realize that they come from zones of nonzero potential. The only thing that counts is that there is a nonzero electromotive force $\mathcal{E}$, which generates a current density. Therefore, there is energy loss.

What loses energy? The only possible source of energy is the rotational energy of the star. Please notice also that since for a rotator the energy loss $I\Omega\dot{\Omega}$ gives origin also to a loss of angular momentum $I\dot{\Omega}$, the mechanism of the battery must necessarily be capable of losing angular momentum as well.

8.3.5 A Simplified Discussion of Total Energetics

Despite our ignorance of the functions $I(f)$ and $\Omega(f)$, it is useful to have at least one approximate formula for the total loss of energy by the black hole. An order of magnitude for the loss of angular momentum can then be obtained from the formula

$$\frac{dE}{dt} \approx \Omega_{\mathrm{H}} \frac{dL_z}{dt} \equiv \frac{a}{2Mr_{\mathrm{H}}} \frac{dL_z}{dt} \qquad (8.158)$$

Here Ω_{H} is the so-called angular velocity of the black hole, which does not correspond to the angular velocity of anything on the surface of the black hole, as discussed above; it is simply defined as $\alpha v_\phi / \varpi$ at the event horizon.

Unfortunately, a simple estimate of the energy loss cannot be made on just a dimensional basis. In fact, if we use *only* dimensional arguments, we find $dE/dt \approx cB^2 r_{\mathrm{H}}^2$, which does not even depend on the rotation frequency of the black hole! As a consequence, we must adopt a different strategy (Thorne, Price, and MacDonald 1986).

Let us start by demonstrating that the surface of the black hole (or, better said, its event horizon) behaves as if it were endowed with surface resistivity[6] $R_H = 4\pi/c = 377$ ohm. The boundary conditions at the event horizon are easy to specify[7]; indeed, they must give fields with a finite amplitude and must correspond to waves that only *enter* the horizon, whereas *nothing* gets out of it. If we call $\vec{n}$ the local outward-directed normal to the event horizon, the outgoing waves satisfy

$$\vec{B} = \vec{n} \wedge \vec{E} \qquad (8.159)$$

There is a simple way to determine these conditions, provided by Znajek (1978): we imagine that on the event horizon, a (fictitious) current density $\vec{J}$ cancels the field components directed *outside* the event horizon. We are not saying that this distribution of currents really exists, just that outside the hole, the electromagnetic situation is identical to one where this distribution exists. The current density necessary to cancel a surface magnetic field $\vec{B}$ is well known from elementary courses:

$$\vec{B}_+ - \vec{B}_- = \frac{4\pi}{c} \vec{J} \wedge \vec{n} \qquad (8.160)$$

where we have momentarily abandoned our convention on units, $c = 1$. However, from equation 8.159, we see that

$$\vec{E} = \frac{4\pi}{c} \vec{J} \qquad (8.161)$$

which is essentially Ohm's law per unit surface

$$\vec{J} = \frac{c}{4\pi} \vec{E} = \frac{1}{R_H} \vec{E} \qquad (8.162)$$

---

[6] But careful: this is not the usual resistivity, linking the electric field to the current density per unit of volume.

[7] Strictly speaking, these conditions are not exactly valid at the event horizon, but an infinitesimal distance above it, because of the singularity of the metric at the event horizon.

Here, $R_{\rm H}$ is therefore the surface resistivity of the black hole. In other words, the condition that at the event horizon, there are only incoming electromagnetic waves, makes the fields behave as if the black hole surface had a surface resistance given by $R_{\rm H} = 4\pi/c = 377$ ohms.

The existence of this resistivity allows us to estimate the emitted power. Let us integrate equation 8.100 on the circuit $\mathcal{C}$ $PQRL$ in figure 8.8; the circuit is assumed to be at rest in the reference frame at infinity, and we assume that the segment RL is located infinitely far away. We find

$$dV \equiv \oint_{\mathcal{C}} \alpha \vec{E} \cdot d\vec{l} = - \oint_{\mathcal{C}} \vec{\beta} \wedge \vec{B} \cdot d\vec{l} \qquad (8.163)$$

Since $\vec{\beta}$ has only a toroidal component, it follows that $\vec{\beta} \wedge \vec{B} = \vec{\beta} \wedge \vec{B}_{\rm p}$ is a poloidal vector, perpendicular to $\vec{B}_{\rm p}$; on the other hand, $d\vec{l}$ is a poloidal vector parallel to $\vec{B}_{\rm p}$ by definition of magnetic surface. Therefore $(\vec{\beta} \wedge \vec{B}_{\rm p}) \cdot d\vec{l} = 0$, so the two integrals along QR and LP vanish. If we use equation 8.124, we can easily calculate the integrals along PQ, obtaining

$$dV = \Omega_{\rm H} df \qquad (8.164)$$

where I used the fact that $\omega \to \Omega_{\rm H}$ when $r \to r_{\rm H}$. This potential difference acts on a surface resistivity $dR_{\rm H}$ given by

$$dR_{\rm H} = R_{\rm H}\frac{dl}{2\pi\varpi} = R_{\rm H}\frac{df}{2\pi\varpi^2 B_{\rm r}} \qquad (8.165)$$

where, once again, I used equation 8.120 in order to eliminate dl

$$dl = \sqrt{g_{\theta\theta}d\theta^2} = \rho d\theta \qquad (8.166)$$

$$\varpi B_{\rm r}\rho d\theta = df \qquad (8.167)$$

However, apart from the resistance $R_{\rm H}$, there will be a resistance also along RL, which we shall call R_∞. This resistance produces the energy transport to infinity. Indeed, the resistance R_H dissipates in place, and cannot count as energy transport to infinity. In order to calculate it, we note that from equation 8.124, we find the potential drop to infinity,

$$dV_\infty = -\int_R^L \alpha \vec{E} \cdot d\vec{l} = \Omega df \qquad (8.168)$$

and therefore we find the potential difference at the ends of PQ as

$$dV - dV_\infty = (\Omega_H - \Omega)df \qquad (8.169)$$

Since this potential difference passes through the surface resistance dR_H given above, we find the current as

$$IdR_H = IR_H \frac{df}{2\pi\varpi^2 B_r} = dV - dV_\infty \qquad (8.170)$$

As usual, the dissipated power is given by

$$dP = I^2 R_\infty = IdV_\infty = \frac{\Omega(\Omega_H - \Omega)}{2}\varpi^2 B_r df \qquad (8.171)$$

which is the expression we were looking for. We cannot proceed any further without specifying $\Omega(f)$.

In order to estimate the total power lost, we use the following approximations: $\Omega \approx \Omega_H/2$, $\varpi^2 = r_H^2/2$, and $\int df \approx 2B_r r_H^2$:

$$P \approx 10^{45} \text{ erg s}^{-1} \left(\frac{a}{M} \frac{M}{10^9 M_\odot} \frac{B_r}{10^4 \, G} \right)^2 \qquad (8.172)$$

whereas the total potential drop is

$$\triangle V = \Omega_H f \approx 10^{20} \text{ volt} \frac{a}{M} \frac{M}{10^9 M_\odot} \frac{B_r}{10^4 G} \qquad (8.173)$$

At this point a comment is in order. This description of the energetics shows that there is a battery associated with the black hole. In fact, the integral of the electric field along a closed circuit, equation 8.164, is not zero. We can therefore say that the rotating black hole can be seen like a battery, with an internal resistance $R_{\mathrm{H}} = 377$ ohms, which generates electric currents. This analogy is useful for understanding where the energy of the configuration comes from; it is less useful for computing the dissipation rate.

8.4 The Generation of Charges

We have so far postponed the question of how electric charges outside pulsars and black holes are generated; we just assumed that a sufficiently large number of charges can be produced. There is a mechanism, originally proposed by Ruderman and Sutherland (1975), that shows that charges of both signs can easily be created outside pulsars. They suggest that when charges are globally lacking, there is a small area of thickness h with an underabundance of charges with respect to those required for a stationary solution; the major consequence of the local underabundance is the impossibility of canceling the electric field $E_{\parallel}$ parallel to the magnetic field. An electric charge situated in this region will then be accelerated, radiating photons of energy $>m_{\mathrm{e}}c^2$ because of curvature radiation. These photons can produce pairs in the pulsar's magnetic field. Even though, initially, they move parallel to the magnetic field, after a while, because of the field's curvature, they will perceive a component of the magnetic field perpendicular to their direction of motion, and in that moment, they can form pairs against the magnetic field. The newly formed pairs are later reaccelerated, thus producing a pair cascade. This process is called *vacuum breakdown* and takes place in a *spark gap*.

Let us now determine the conditions for the gap to work. We shall do this roughly, simply determining the width h of the gap, for each charge inside to produce at least two more charges. For the condition of degeneracy to be realized, we must have, in order of magnitude,

$$E \approx \frac{\Omega R}{c} B \qquad (8.174)$$

where Ω and R are the pulsar's angular velocity and radius, respectively. Once again in order of magnitude, the density of charge necessary to realize this field is

$$\rho = \frac{1}{4\pi} \nabla \cdot \vec{E} \approx \frac{\Omega}{4\pi c} B \qquad (8.175)$$

If this charge density leaves a void with a height h, a potential difference builds up of magnitude

$$V \approx \frac{\Omega}{8\pi c} B h^2 \qquad (8.176)$$

which can accelerate electrons up to a Lorentz factor

$$\gamma \approx \frac{\Omega \omega_L h^2}{8\pi c^2} \qquad (8.177)$$

where $\omega_L = eB/m_e c$ is the electron's Larmor frequency. An electron reaching these energies can emit through curvature radiation (section 3.4.7) photons with energy

$$\epsilon_\gamma \approx \hbar \omega \approx \gamma^3 \frac{\hbar c}{R} \qquad (8.178)$$

where we have assumed that the spark gap is located near the pulsar's surface, and therefore that the magnetic field lines have curvature radius $\rho_c \approx R$.

At this point, we should impose three conditions: first of all, that emitted photons have energy $\gtrsim 2m_e c^2$, in order to make sure that they can create pairs; second, that the total number of photons emitted within the gap is >1; third, that these photons have a mean free path against pair creation in the field B shorter than h. The first condition is

$$\gamma^3 \frac{\hbar c}{R} \gtrsim 2m_e c^2 \qquad (8.179)$$

The second one is derived by recalling the power from curvature radiation (eq. 3.93) and assuming that it is emitted exclusively in photons with an energy given by equation 8.178.

Computing from this the total number of photons emitted in the gap n_γ, we find

$$n_\gamma \approx \frac{Ph}{\epsilon_\gamma c} \approx \frac{e^2}{\hbar c} \gamma \frac{h}{R} \qquad (8.180)$$

We now impose the condition $n_\gamma \gtrsim 1$, namely,

$$\frac{e^2}{\hbar c} \gamma \frac{h}{R} \gtrsim 1 \qquad (8.181)$$

Finally, the condition that these photons have the time to form pairs inside the gap is

$$\frac{h^2}{2R} \gtrsim \frac{1}{\alpha} \qquad (8.182)$$

where α is the coefficient of attenuation, given in equation 3.100, and $h^2/2R$ is the photon path perpendicular to the magnetic field. The coefficient of attenuation can be rewritten in the limit $\chi \ll 1$, which holds here, as

$$\frac{h^2}{2R} \gtrsim \frac{4\hbar c}{e^2} \frac{c}{\omega_L} \exp\left(\frac{8}{3} \frac{m_e c^2}{\epsilon_\gamma} \frac{m_e c^2}{\hbar \omega_L}\right) \qquad (8.183)$$

It is easy to see that these three conditions are easily satisfied, since we have $h \gtrsim 10^4$ cm $\ll R$, for typical pulsar parameters ($R = 10^6$ cm, $\Omega = 10$ s^{-1}, $B = 10^{12}$ G); this gives typical energies for accelerated electrons

$$\gamma m_e c^2 \approx 1 \text{ TeV} \qquad (8.184)$$

As a consequence, it is easy to generate through pair creation all the electric charges necessary to maintain the condition of degeneracy in the pulsar's atmosphere.

Around black holes, the argument is absolutely identical, apart from the fact that the magnetic field and angular momentum of the black hole are unknown, unlike for pulsars. In this case, we should assume $h \lesssim r_{Sch} = 2GM/c^2$ and deduce constraints on the magnetic field around the black hole, as well as on its total angular momentum J. We find (Blandford and Znajek 1977)

$$B \gtrsim 2 \times 10^4 \text{ G}, \quad \frac{M}{\text{M}_\odot} \gtrsim 10^{-11} \left(\frac{Jc}{GM^2} \right)^{1/2} \quad (8.185)$$

Once again, these rather modest constraints persuade us that it is possible to build a spark gap around a black hole.

Of course, the exact physics of the spark gap can be more complex. For example, other radiative processes such as inverse Compton of photons on energetic particles can reduce the estimate of the particle Lorentz factor γ just presented. However, in this case, the photons produced by inverse Compton are often capable of producing pairs, and this leads, in any case, to a vacuum breakdown. The exact details of the spark gap must therefore be studied in the various situations we encounter.

The only true difference between pulsars and black holes, in this respect, is where the spark gap is located. In the case of pulsars, there is an obvious site: immediately outside the pulsar's surface. In the case of black holes, the situation is more complex because the surface of the black hole is not a physical place. Therefore, there is even a possibility that the gap is not static at all.

8.5 Disk-Jet Coupling

It is well known that sources showing a jet of matter emitted from a compact object also display an accretion disk; each accreting source uses a part of the gravitational energy liberated to accelerate a wind. Furthermore the wind is often collimated, sometimes to a few degrees. These observational facts have led to the obvious speculation that the two facts are intimately connected. The model that most clearly shows this connection has been proposed by Blandford and colleagues: the jet is used to remove angular momentum from the disk by means of a strong toroidal magnetic field, which also forces the jet collimation.

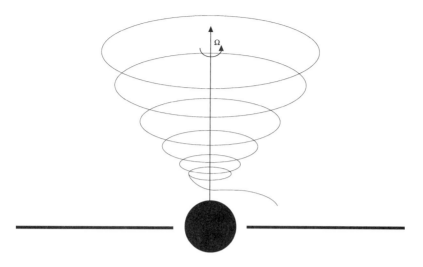

Figure 8.10. How magnetic tension collimates a jet.

We have already seen, in a qualitative manner in chapter 7 (section 7.4.2), how the torque due to the magnetic field can transfer angular momentum to the wind, thus removing it from the disk. We still need to understand how the magnetic field can collimate the jet. The idea is very simple: The magnetic field lines, which have their feet in the disk, are forced to corotate with the disk, by the flux-freezing theorem. The matter leaving the disk drags with itself the magnetic field lines, again because of its large conduction coefficient; in doing this, it wraps the magnetic field around the rotation axis. We know from chapter 2 (fig. 2.1) that the magnetic field lines have a tension that tries to shorten them; in other words, this is a radial force directed toward the rotation axis. This force compresses the jet of matter, just like a string tying a salami; see figure 8.10, which is obviously idealized but descriptive. The magnetic tension is therefore responsible for the jet collimation. These are called *hoop stresses*, that is, stresses deriving from a circular configuration of the magnetic field.

In the following, we shall introduce two special models for this coupling. In the first one (Blandford 1976; Lovelace 1976), the plasma emitted is assumed massless; this simplification allows us to solve the structure of the magnetosphere that extends over the disk. The second model (Blandford and Payne 1982), on the other hand, includes the inertial effects of the wind; to solve it, we have to make other approximations. In any case, these two highly idealized models are, at the moment, the best quantitative explanations of the disk-jet connection we have.

8.5.1   The Lovelace-Blandford Model

In the 1970s (Blandford 1976; Lovelace 1976) a mechanism was proposed for the removal of angular momentum from disks through the Maxwell tensor, mutuated from that of pulsars. It was immediately clear that the very same mechanism could also be responsible for the acceleration of plasma jets, which are universally associated with accretion disks from T Tauri stars to AGNs. Here we will discuss the connection between a rotating object and the jet.

In order to do this, let us consider an infinitely thin disk in the plane $z = 0$, which can carry currents. Since the disk rotates, a situation similar to that of pulsars ensues, namely, there must be an electric field outside the disk, and therefore a charge distribution satisfying, in equilibrium, the condition of *degeneracy*,

$$\rho \vec{E} + \frac{\vec{j}}{c} \wedge \vec{B} = 0 \qquad (8.186)$$

The boundary conditions are obviously different from those for pulsars, first because the disk does not have just one rotation velocity (it rotates differentially), and second because the magnetic field is not purely dipolar.

However, before presenting the computation, we want to remark that this solution is free from the criticism levied against the pulsar solution; in fact a quasi-neutral plasma is acceptable in this case. There are two reasons for this. First of all, the magnetic fields in accretion disks around black holes of galactic dimensions (i.e., those powering AGNs) are of the order 10^4 G, 8 orders of magnitude lower than those around pulsars. Since the rate of energy loss scales like the square of the magnetic field, this implies cooling rates 16 orders of magnitude smaller than in pulsars' magnetospheres. Second, since there is an accretion disk, plasma in the magnetosphere may be heated by the disk's emission. Therefore the existence of a quasi-neutral plasma, supported by a pressure gradient, is certainly not unrealistic.

8.5.2 A Special Solution

We now derive the properties of a specific solution (Blandford 1976), with the aim of demonstrating the disk-jet coupling. Stated otherwise, we are not trying to find all possible solutions for the equations to written, but just one, which displays the coupling in simple, fully analytic terms. This is why at some point we shall make a simplifying hypothesis leading us to a specific solution.

Because of axial symmetry and stationarity, all the relationships derived for pulsars on the basis of these properties still hold. In particular, we know that

$$E_\phi = 0, \quad B_\phi = \frac{2I}{cr} \tag{8.187}$$

for the fields' toroidal components. Moreover, the electric field is

$$\vec{E} = -\frac{\omega r}{c}\hat{e}_\phi \wedge \vec{B} \tag{8.188}$$

where, however, we can no longer have $\omega = \Omega$ because the disk does not have just one rotation velocity. Therefore, in general, ω will be a function, still to be determined, of r and z (not of ϕ because of axial symmetry), which reduces to $\omega = \omega_0(r)$ when $z = 0$. In other words, on the disk plane, the continuity of (tangential and normal) components of the (respectively, electric and magnetic) fields obviously requires that the rotation velocity be that of the disk, $\omega_0(r)$, which we, of course, suppose as known.

We also have again that

$$\vec{j} = \rho\omega r \hat{e}_\phi + \kappa \vec{B} \tag{8.189}$$

and, of course,

$$\vec{B} = \frac{1}{r}\nabla f \wedge \hat{e}_\phi \tag{8.190}$$

which implies

$$\vec{E} = -\frac{\omega}{c}\nabla f \tag{8.191}$$

Moreover

$$\kappa = \frac{1}{2\pi}\frac{dI}{df} \tag{8.192}$$

The Faraday equation, $\nabla \wedge \vec{E} = 0$, immediately tells us that

$$\nabla\omega \wedge \nabla f = 0 \tag{8.193}$$

and therefore also ω, in addition to I and κ, is constant on the surfaces $f = $ constant. Therefore, we can also write

$$\omega = \omega(f) \tag{8.194}$$

We can now derive a new version of equation 8.61; from equation 8.191, we find

$$\nabla \cdot \vec{E} = 4\pi\rho = -\frac{\omega}{c}\nabla^2 f - \frac{1}{c}\nabla\omega \cdot \nabla f \tag{8.195}$$

therefore, from the toroidal component of $\nabla \wedge \vec{B} = 4\pi\vec{j}/c$ we find, just as we did in the case of pulsars,

$$f_{zz} + f_{rr} - \frac{1}{r}\frac{c^2 + \omega^2 r^2}{c^2 - \omega^2 r^2}f_r = -\frac{4}{c^2 - \omega^2 r^2}I\frac{dI}{df}$$

$$+ \frac{\omega r^2}{c^2 - \omega^2 r^2}(f_z\omega_z + f_r\omega_r) \tag{8.196}$$

The anticipated simplification consists in specifying a priori the form of equipotential surfaces, $f = $ constant. It is convenient to assume a very simple form for $f(r, z)$, and Blandford (1976) suggests parabolae,

$$|z| = \frac{r^2 - r_0^2}{2r_0} \tag{8.197}$$

where the modulus takes into account symmetry with respect to the plane $z = 0$, and the parameter r_0 is the distance from the origin when $z = 0$, that is, the distance from the origin when the parabola intersects the disk. The same equation can also be rewritten as

$$\sqrt{r^2 + z^2} - |z| = R - |z| = r_0 \tag{8.198}$$

from which we have

$$f = f(R - |z|) = f(r_0) \tag{8.199}$$

Notice that at this point, we do not know how f depends on $R - |z|$; we know only that f does not depend separately on R and z, but only on their combination $R - |z|$.

Despite this ignorance, we know the disk angular velocity of rotation, ω. In fact, we know that ω is constant on the same surfaces on which f is constant, so that $\omega = \omega(R - |z|)$. Besides, we know that at the surface of the disk, ω coincides with the angular velocity of the disk itself, $\omega_0(r_0)$. Therefore, we find

$$\omega = \omega_0(r_0) = \omega_0(R - |z|) \tag{8.200}$$

Now we can introduce this expression into equation 8.196, together with our assumption, equation 8.199. It proves convenient to use as independent variables $r_0 = R - z$ and z, so that $r^2 = r_0^2 + 2r_0 z$. We obtain, with the notation $\dot{f} \equiv df/dr_0$ and so on

$$\left(c^2 - \omega^2 r_0^2\right)\left(2r_0\ddot{f} + \dot{f}\right) - \left(c^2 + \omega^2 r_0^2\right)\dot{f} + 4r_0 I \frac{dI}{df} - 2\omega\dot{\omega}r_0^3\dot{f}$$

$$= 2zr_0\left(-\omega^2(2r_0\ddot{f} + \dot{f}) - \omega^2\dot{f} + \frac{2IdI/df}{r_0} - 2\omega\dot{\omega}r_0\dot{f}\right) \tag{8.201}$$

whence we see that the left-hand side and the coefficient of z on the left-hand side are functions of r_0 only. For the equality to hold, it is therefore necessary that both vanish:

$$\left(c^2 - \omega^2 r_0^2\right)\left(2r_0\ddot{f} + \dot{f}\right) - \left(c^2 + \omega^2 r_0^2\right)\dot{f} + 4r_0 I\frac{dI}{df} - 2\omega\dot{\omega}r_0^3\dot{f} = 0$$
(8.202)

$$-\omega^2\left(2r_0\ddot{f} + \dot{f}\right) - \omega^2\dot{f} + \frac{2I\,dI/df}{r_0} - 2\omega\dot{\omega}r_0\dot{f} = 0 \qquad (8.203)$$

Solving with respect to I, we find

$$I\frac{dI}{df} = \omega\dot{\omega}\dot{f}r_0^2 + \omega^2\dot{f}r_0 + \omega^2 r_0^2\ddot{f} \qquad (8.204)$$

$$\frac{\ddot{f}}{\dot{f}} = -\frac{\omega\dot{\omega}r_0^2 + \omega^2 r_0}{c^2 + \omega^2 r_0^2} \qquad (8.205)$$

The last equation can be integrated easily

$$f(r_0) = \frac{K}{2}\int_0^{r_0}\frac{dx}{(c^2 + \omega^2(x)x^2)^{1/2}} \qquad (8.206)$$

where K is an arbitrary constant. If we introduce this solution in the equation for I (eq. 8.204) we find

$$I^2 = \frac{\omega^2 r_0^2}{c^2 + \omega^2 r_0^2}K^2 \qquad (8.207)$$

The constant of integration in the equation for I has been chosen so as to avoid currents to infinity. At this point, we know f and I as functions of r_0, and we have therefore solved our problem. The separability of equation 8.196 does not hold for all shapes of magnetic surfaces; it is an exceptional property that holds only under specific conditions (Blandford 1976). But this need not concern us here; our aim is to find *one* explicit solution, not all of them.

This solution arises for a specific distribution of currents in the disk. Indeed, we know from elementary courses that since the disk carries electric currents, the component of $\vec{B}$ parallel to the plane of the disk is discontinuous across the plane, while the normal component (B_z) is continuous. Integrating the component r of Ampère's law on an infinitesimal interval, from $z = -\eta$ and $z = +\eta$, and letting $\eta \to 0$, we find

$$-2B_\phi = \frac{4\pi}{c} J_r \qquad (8.208)$$

where J_r is the surface current flowing *in the disk*, not in the magnetosphere. Comparing this with equation 8.187, we find

$$J_r = -\frac{I(r,0)}{\pi r} \qquad (8.209)$$

We can do the same for the component ϕ of Ampère's law, thus finding

$$2B_r = \frac{4\pi}{c} J_\phi \qquad (8.210)$$

which yields the ϕ component of the surface current in the plane of the disk. These forms of J_r and J_ϕ are the only ones compatible with the form of f we have assumed in equation 8.199. In doing this exercise, we have not solved our problem with boundary conditions. Rather, we have found the only boundary condition compatible with the form assumed for the surfaces $f =$ constant, which have been chosen only to make the problem tractable.

It is now possible to identify all the physical quantities of the problem. It is simpler to use r_0, not z; the two are connected by equation 8.197. We shall give the solution directly, leaving out the computations. Therefore, we remind you that ω is always considered as a function of r_0 in the following formulae. The components (r, z, ϕ) of the various vectors are

$$\vec{B} = \frac{K r_0}{r \left(r^2 + r_0^2\right) \left(1 + \omega^2 r_0^2/c^2\right)^{1/2}} \left(r_0, r, -\omega \left(r^2 + r_0^2\right)/c\right)$$

$$(8.211)$$

$$\vec{E} = \frac{K r_0 \omega}{c \left(r^2 + r_0^2\right) \left(1 + \omega^2 r_0^2/c^2\right)^{1/2}} (-r, r_0, 0) \quad (8.212)$$

$$\frac{4\pi \vec{j}}{c} = \frac{2 K r_0}{cr \left(r^2 + r_0^2\right) \left(1 + \omega^2 r_0^2/c^2\right)^{3/2}} \left(\frac{d}{dr_0} r_0 \omega\right) \left(-r_0, -r, \omega r_0^2/c\right)$$

$$(8.213)$$

$$4\pi \rho = \frac{-2 K r_0}{c \left(r^2 + r_0^2\right) \left(1 + \omega^2 r_0^2/c^2\right)^{3/2}} \left(\frac{d}{dr_0} r_0 \omega\right) \quad (8.214)$$

Here K is an arbitrary constant. It is also possible to identify the surface currents in the disk explicitly:

$$\frac{4\pi J_r}{c} = \frac{2 K \omega}{c\sqrt{1 + \omega^2 r_0^2/c^2}} \quad (8.215)$$

$$\frac{4\pi J_\phi}{c} = \frac{K}{r_0 \sqrt{1 + \omega^2 r_0^2/c^2}} \quad (8.216)$$

8.5.3   Discussion

This solution is absolutely complete on both sides of the light surface.[8] The latter solution is continuous, and electric and magnetic fields are continuous and differentiable everywhere; it is therefore a physically acceptable solution.

The structure of the solution is illustrated in figure 8.11, where a Keplerian angular velocity is assumed, $\omega(r) \propto r^{-3/2}$.

---

[8] Notice that since the disk does not have just one angular velocity, there no longer is a light cylinder, which is replaced by the so-called *light surface*.

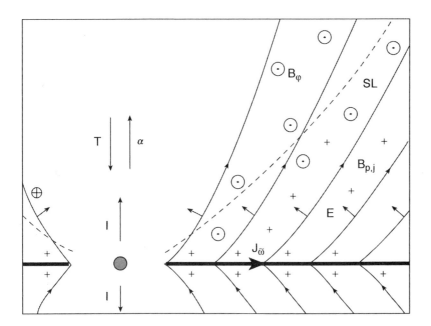

Figure 8.11. Diagram representing the solution for a disk-jet system. Blandford 1976.

It is easy to see from the above equations that charge conservation holds

$$\frac{1}{r_0}\frac{\partial}{\partial r_0}(r_0 J_{\rm r}) = -2j_{\rm z} \qquad (8.217)$$

where the factor 2 is due to the fact that the disk loses the same amount of current $j_{\rm z}$ from the sides, that is, above and below.

In this model, there is no charge separation. The easiest way to see this is to calculate the relation $j/\rho c$ close to the disk. If we use equations 8.214 and 8.213, specialized to the case $r = r_0$, we find

$$\frac{j}{\rho c} = \sqrt{2} > 1 \qquad (8.218)$$

As discussed in the section about pulsars, $v = j/\rho > c$ only in the absence of charge separation.

However, at large distances $(r \gg r_0)$, we find for the poloidal and toroidal components of the current

$$
\frac{j_{\mathrm{p}}}{c} = \rho \frac{\sqrt{r^2 + r_0^2}}{r} \to 1
$$

$$
\frac{j_\phi}{c} = -\rho \frac{\omega r_0^2}{r} \to 0 \tag{8.219}
$$

This shows that the toroidal component of the current is associated with a decreasing speed $(v = j_\phi/\rho \to 0)$, whereas the asymptotic particles' velocity at infinity, given by j_{p}/ρ, tends to the speed of light. This is important because it shows that at infinity, there is charge separation: infinity is reached only by the particles emitted by the disk, and since each magnetic line carries just one sign of current, charge separation ensues.

The fact that $v \to c$ at large distances from the disk implies that at least potentially, the mechanism can accelerate particles up to very high energies. Obviously, several loss mechanisms limit the maximum attainable energy, such as radiative losses, generation of Alfvén waves, and so on.

A torque acts on the disk. The best way to calculate it is to recall that the disk plasma, which carries currents, is subject to a force given by $\vec{J} \wedge \vec{B}$, therefore by a torque density (per unit surface)

$$
C_z = \left(\vec{r} \wedge (\vec{J} \wedge \vec{B}) \right)_z = -\frac{K^2 \vec{\omega}}{4\pi c \left(c^2 + \omega^2 r_0^2 \right)} \tag{8.220}
$$

There is also a loss of rotational kinetic energy, due to this loss of angular momentum, given by

$$
P = -C_z \omega = \frac{K^2 \omega^2}{4\pi c \left(c^2 + \omega^2 r_0^2 \right)} \tag{8.221}
$$

The total losses (integrated on the disk surface) that can be obtained from these formulae are obviously identical to those obtained by integrating Maxwell's stress tensor on a surface at a large distance. Therefore, the disk losses are exclusively due to electromagnetic energy removal to infinity.

Notice that asymptotically, the magnetic field has the behavior

$$\frac{B_p}{B_\phi} \to -\frac{c}{\omega r} \to 0 \tag{8.222}$$

Therefore, the field becomes essentially toroidal, but the poloidal component is necessary to carry away angular momentum. In fact, the flux of angular momentum along the z axis is given by the component $R\phi$ of the Maxwell tensor, which is $\propto B_R B_\phi$. Therefore, if $B_R = 0$, there is no transport of angular momentum to infinity.

The most remarkable aspect of the model is its capacity to focus the electromagnetic wind. In order to see this, we use the fact that the energy flux is given by the Poynting vector, $c\vec{E} \wedge \vec{B}/4\pi$, which is obviously perpendicular to $\vec{B}$ and is thus parallel to the magnetic surfaces, $f = $ constant. These are given by equation 8.197, from which we see that the angle θ between the Poynting vector and the z axis is given by

$$\cos\theta = \frac{z}{R} = \frac{z}{r_0 + z} \to 1 \tag{8.223}$$

where the limit is realized at large distances ($z \gg r_0$) from the disk. Therefore $\theta \to 0$, and all the energy is focused along the z axis with aperture angle θ, which tends asymptotically to zero. Another way to see it consists in computing the distribution of the energy flux on the surface of a sphere of radius R as a function of the angle θ. In this case, we can show (see problem 6) that the distribution in angle, when $R \to \infty$, tends to $\delta(\cos\theta - 1)$, where $\delta(x)$ is Dirac's delta. In this limit, the jet is perfectly collimated.

Essentially, this is the reason why this simple case has been studied: although idealized, it shows how, removing angular momentum along the z axis, it is at the same time possible to power an electromagnetic wind (or jet, if you prefer) along the same axis and to collimate it. The reason why this remains a toy model, is that there is no obvious reason why

the disk should carry such well-ordered currents, with coherence length $\approx r$. In our study of accretion disks, we saw
that the magnetic field is locally amplified and has a coherence length $\approx H \ll r$, which is also, necessarily, the coherence
length of the currents on the plane of the disk.

8.5.4   A Model Including Inertial Effects

The model discussed above neglects altogether the plasma inertial effects, and therefore we cannot use it to study the wind
acceleration from corotation with the disk to a highly super-
Alfvénic motion at large distances from the origin. In order
to follow this transition, and confirm the intuition that the
disk-jet coupling allows acceleration of the observed (even, occasionally, hyperrelativistic!) jets, we must take into account
the wind mass. This is the goal of the model by Blandford
and Payne (1982).

Also in this case it is impossible to obtain a solution of the
magnetohydrodynamics (MHD) equations for an arbitrary
configuration of the magnetic field. And also in this case,
we introduce a simplification leading to a particular solution,
rather than the most general solution. The advantage of this
approach is to demonstrate explicitly the disk-jet connection
and the wind acceleration from the surface of the accretion
disk.

The need to include a massive wind leads us to an approximation different from equation 8.17. Indeed, exactly because
we assume that the wind has a high density, it will be made
of many free charged particles, in which case we know, from
Chapter 2, that the most suitable approximation is that of
ideal magnetohydrodynamics. Therefore, the equation

$$\vec{E} + \frac{\vec{v}}{c} \wedge \vec{B} = 0 \qquad (8.224)$$

replaces equation 8.17; please remember that here $\vec{v}$ indicates
the average velocity of the fluid, not of individual particles.

Together with this equation, we know that the equations of magnetohydrodynamics assume the form

$$\frac{\partial}{\partial t}(\rho v_i) + \frac{\partial}{\partial x_j}(R_{ij} + M_{ij}) = -\rho \frac{\partial \psi}{\partial x_i} \qquad (8.225)$$

where ψ is the gravitational potential of the compact object, to avoid confusion with the coordinate ϕ. Moreover, since we are treating this problem in the limit of ideal magnetohydrodynamics, the Maxwell tensor (equation 2.23) contains only the terms due to the magnetic field, for the reason given in section 6.7.

We use the equation for energy conservation, equation 2.25

$$\frac{\partial}{\partial t}\left(\frac{\rho v^2}{2} + \rho \epsilon + \frac{B^2}{8\pi}\right)$$

$$= -\nabla \cdot \left(\rho \vec{v}\left(\frac{v^2}{2} + w + \psi\right) + \frac{1}{4\pi}\vec{B} \wedge (\vec{v} \wedge \vec{B})\right) (8.226)$$

At this point, we should perhaps note that the solution we are going to find is purely Newtonian, unlike the one in the previous paragraph. Indeed, in order to deduce the equations of ideal magnetohydrodynamics, we have neglected all the terms of order $\mathcal{O}(v^2/c^2)$, keeping only those of order v/c.

We shall use cylindrical coordinates r, ϕ, z. The system we have in mind is, once again, an infinitely thin disk of matter in the plane $z = 0$, which gives origin to a wind expanding to infinity. The system will be assumed to be stationary, with axial symmetry; thus, for any physical quantity X, $\partial X/\partial t = \partial X/\partial \phi = 0$.

A General but Incomplete Solution

Equations 8.224, 8.225, and 8.226 may be solved in part, in a wholly general manner, under the obvious hypotheses of stationarity and axial symmetry. Let us proceed as follows.

We can show that because of axial symmetry, we must have $E_\phi = 0$. Integrating one of Maxwell's equations,

$$\nabla \wedge \vec{E} = -\frac{1}{c}\frac{\partial \vec{B}}{\partial t} \tag{8.227}$$

on the circle of figure 8.2, we find, for the left-hand side,

$$\int \nabla \wedge \vec{E} \cdot d\vec{A} = \int_C \vec{E} \cdot d\vec{l} = \int_0^{2\pi} E_\phi r d\phi = 2\pi r E_\phi \tag{8.228}$$

The integral of the right-hand side of equation 8.227 on the same circle vanishes because of the problem's stationarity. It follows that

$$E_\phi = 0 \tag{8.229}$$

The condition of ideal magnetohydrodynamics, equation 8.224, implies that $\vec{E} \cdot \vec{B} = 0$; therefore, $\vec{E}$ is perpendicular both to $\hat{e}_\phi$ and to $\vec{B}$. Such a vector may always be written

$$\vec{E} = -h\hat{e}_\phi \wedge \vec{B} \tag{8.230}$$

where the quantity h, so far unknown, is obviously dimensionless. That is why it may also be written as

$$\vec{E} = -\frac{\omega r}{c}\hat{e}_\phi \wedge \vec{B} \tag{8.231}$$

where ω is a function to be determined. The only thing we know at the moment is that it does not depend on ϕ, because of the assumption of axial symmetry.

On the other hand, if we proceed just like for pulsars, we know that

$$\nabla \cdot \vec{B} = \frac{1}{r}\frac{\partial}{\partial r}(rB_r) + \frac{1}{r}\frac{\partial B_\phi}{\partial \phi} + \frac{\partial B_z}{\partial z}$$

$$= \frac{1}{r}\frac{\partial}{\partial r}(rB_r) + \frac{\partial B_z}{\partial z} = \nabla \cdot \vec{B}_p = 0 \tag{8.232}$$

where we see that thanks to the hypothesis of axial symmetry ($\partial X/\partial \phi = 0$), the divergence of the poloidal component of the magnetic field also vanishes. We can therefore set

$$\vec{B}_p = \frac{1}{r}\nabla f \wedge \hat{e}_\phi \tag{8.233}$$

where, as usual, f is the so-called Euler's potential, which, however, is still to be determined. It is now easy to see that

$$\nabla \cdot \vec{B}_{\mathrm{p}} = \frac{1}{r}\left(-\frac{\partial^2 f}{\partial z \partial r} + \frac{\partial^2 f}{\partial r \partial z}\right) \equiv 0 \qquad (8.234)$$

which shows why we have introduced Euler's potential: it automatically ensures that $\vec{B}$ satisfies the divergence equation. When we introduce equation 8.233 into equation 8.231, we find

$$\vec{E} = -\frac{\omega}{c}\nabla f \qquad (8.235)$$

If $\nabla \wedge \vec{E} = 0$ is to hold, we must have

$$\nabla \omega \wedge \nabla f = 0 \qquad (8.236)$$

which shows that the poloidal vector $\nabla \omega$ is always parallel to ∇f, and therefore always perpendicular to $\vec{B}$ (eq. 8.233)

$$(\vec{B} \cdot \nabla)\omega = 0 \qquad (8.237)$$

which is the equation of propagation for ω, which we were looking for.

Now, if we introduce equation 8.231 into 8.224, we find

$$(\vec{v} - \omega r \hat{e}_\phi) \wedge \vec{B} = 0 \qquad (8.238)$$

from which

$$\vec{v} = \frac{\kappa}{4\pi\rho}\vec{B} + \omega r \hat{e}_\phi \qquad (8.239)$$

This is identical to equation 8.34, except for the factor $4\pi\rho$, which will prove, however, very useful. It is possible to determine immediately the equation obeyed by κ, using the equation for mass conservation, $\nabla \cdot (\rho\vec{v}) = 0$. From the above equation we obtain

$$0 = \nabla \cdot (\kappa\vec{B}) + 4\pi\nabla \cdot (\rho\omega r \hat{e}_\phi) \qquad (8.240)$$

We now notice that

$$\nabla \cdot (\kappa \vec{B}) = (\vec{B} \cdot \nabla)\kappa + \kappa \nabla \cdot \vec{B} = (\vec{B} \cdot \nabla)\kappa \qquad (8.241)$$

thanks to Maxwell's equation, $\nabla \cdot \vec{B} = 0$. Furthermore,

$$\nabla \cdot (\rho \omega r \hat{e}_\phi) = \frac{1}{r}\frac{\partial}{\partial \phi}(\rho \omega r) = 0 \qquad (8.242)$$

thanks to the identity C.15 (see appendix C), and to the hypothesis of axial symmetry. Thus we obtain

$$(\vec{B} \cdot \nabla)\kappa = 0 \qquad (8.243)$$

another transport equation, identical to the one for ω.

It is now time to consider the equations of motion, which we can specialize to the stationary case. Mass conservation yields

$$\nabla \cdot (\rho \vec{v}) = 0 \qquad (8.244)$$

It is now possible to find two useful integrals of motion. Let us consider the component ϕ of the equations of motion

$$\frac{\partial}{\partial x_i}(T_{\phi i} + M_{\phi i}) = -\frac{\rho}{r}\frac{\partial \psi}{\partial \phi} = 0 \qquad (8.245)$$

This may be made explicit[9] in cylindrical coordinates

$$\frac{\partial}{\partial x_i}\left(\rho r v_\phi v_i + r p g_{\phi i} - \frac{1}{4\pi}\left(r B_\phi B_i - \frac{r}{2}g_{\phi i}B^2\right)\right) = 0 \quad (8.246)$$

---

[9] We remark that when the coordinates are not Cartesian, the correct generalization of this equation is $(T^{\phi k} + M^{\phi k})_{;k} = 0$, where ; indicates the covariant derivative. This may be written as

$$(T^{ik} + M^{ik})_{;k} = \frac{1}{\sqrt{g}}\frac{\partial}{\partial x_i}(\sqrt{g}(T^{ik} + M^{ik})) - \frac{\partial g_{kl}}{\partial \phi}(T^{kl} + M^{kl}) = 0$$

where g is the determinant of the metric, $g = r^2$. However, in axial symmetry, the coefficients of the metric do not depend on ϕ, since they are measurable physical quantities. As a consequence, the equation written in the text is correct.

We note the disappearance of two terms of the type

$$\nabla \cdot (X g_{\phi i}) \propto \frac{\partial X}{\partial \phi} = 0 \qquad (8.247)$$

once again because of the assumption of axial symmetry. The remaining terms give

$$
\begin{aligned}
0 &= r v_\phi \nabla \cdot (\rho \vec{v}) + \rho (\vec{v} \cdot \nabla) r v_\phi - \frac{1}{4\pi}(r B_\phi \nabla \cdot \vec{B} + (\vec{B} \cdot \nabla) r B_\phi) \\
&= \rho (\vec{v}_{\mathrm{p}} \cdot \nabla) r v_\phi - \frac{1}{4\pi}(\vec{B} \cdot \nabla) r B_\phi \\
&= \frac{1}{4\pi}\left((\kappa \vec{B}_{\mathrm{p}} \cdot \nabla) r v_\phi - (\vec{B}_{\mathrm{p}} \cdot \nabla) r B_\phi \right) \qquad (8.248)
\end{aligned}
$$

where I repeatedly used Maxwell's equation, $\nabla \cdot \vec{B} = 0$, as well as axial symmetry. If we now recall equation 8.237, we find

$$(\vec{B} \cdot \nabla)\left(r v_\phi - \frac{r B_\phi}{\kappa} \right) \equiv (\vec{B} \cdot \nabla) l = 0 \qquad (8.249)$$

The quantity l is an integral of motion in the sense that it remains constant along the magnetic field lines. Physically, it obviously corresponds to the angular momentum per unit mass of the system, taking into account the contribution of the magnetic field.

The equation for energy conservation gives an important integral of motion; equation 2.25 can be rewritten as

$$\nabla \cdot \vec{f}_{\mathrm{E}} = 0 \qquad (8.250)$$

As usual, the toroidal component of $\vec{f}_{\mathrm{E}}$ does not depend on ϕ and therefore disappears from the equation (see appendix C, eq. C.15), which reduces to

$$\nabla \cdot \vec{f}_{\mathrm{Ep}} = 0 \qquad (8.251)$$

It is easy to calculate the poloidal component of $\vec{f}_{\mathrm{E}}$. We have, from equation 8.239,

$$\rho \vec{v}_{\mathrm{p}} = \frac{\kappa \vec{B}_{\mathrm{p}}}{4\pi} \qquad (8.252)$$

and therefore

$$\vec{f}_{\rm Ep} = \frac{1}{4\pi}\left(\kappa\vec{B}_{\rm p}\left(\frac{v^2}{2}+w+\psi\right)-wrB_\phi\vec{B}_{\rm p}\right) \quad (8.253)$$

The equation of energy conservation then becomes

$$\nabla\cdot(\kappa\epsilon\vec{B}_{\rm p}) = 0 \quad (8.254)$$

where ϵ is the energy per unit mass, including the work done by the magnetic torque:

$$\epsilon \equiv \frac{v^2}{2}+w+\psi-\frac{wrB_\phi}{\kappa} \quad (8.255)$$

Using Maxwell's equation, $\nabla\cdot\vec{B}_{\rm p} = 0$, and equation 8.243, namely, energy conservation, the above simplifies to

$$(\vec{B}\cdot\nabla)\epsilon = 0 \quad (8.256)$$

So far, we have obtained equations 8.239 and 8.244, and equation 8.237, 8.243, 8.249, and 8.256: eight equations in the nine unknown quantities $\vec{v}, \vec{B}, w, w, \kappa$. We need only one more equation, the z component of the equation of motion

$$\rho(\vec{v}\cdot\nabla)v_z = -\frac{\partial p}{\partial z}-\rho\frac{\partial\psi}{\partial z}-\frac{1}{8\pi}(\vec{B}\cdot\nabla)B_z \quad (8.257)$$

The remaining equations, like that for the magnetic field in the ideal magnetohydrodynamic limit (eq. 2.14), or the radial component of the equation 8.225; are all automatically satisfied (see problem 7).

8.5.5 A Special Solution

So far, no specific hypothesis has been made about the solution we are looking for, so the results obtained this far apply to any stationary problem with axial symmetry. If we arbitrarily choose a configuration for the magnetic field for the

disk, we would have an extraordinarily difficult problem to
solve. On the other hand, as in the case without inertial effects, we can look for a specific solution with peculiar properties, chosen so as to simplify the computational side of our
work. Although, obviously, the solution thus identified lacks
generality, it will, however, offer the advantage of displaying
explicitly the wind acceleration from the disk and its collimation by the magnetic field.

Let us consider the equation for a generic field line inter-
secting the plane of the disk $(z = 0)$ at the point $r = r_0, \phi =
\phi_0$. In parametric form, if we call t the parameter varying
along the line, the equation that describes this field line is

$$
\begin{aligned}
r &= r(r_0, \phi_0, t) \\
\phi &= \phi(r_0, \phi_0, t) \\
z &= z(r_0, \phi_0, t)
\end{aligned}
\tag{8.258}
$$

The matter's velocity along the same line is

$$
\begin{aligned}
v_r &= v_r(r_0, \phi_0, t) \\
v_\phi &= v_\phi(r_0, \phi_0, t) \\
v_z &= v_z(r_0, \phi_0, t)
\end{aligned}
\tag{8.259}
$$

We have explicitly exhibited the dependence on r_0 and ϕ_0, the
two constants that identify the various field lines, to remind
us that all quantities may vary from one line to another.
At this point we remember that in the hypothesis of axial
symmetry, there can be no dependence on ϕ_0, apart from ϕ.
Thus we have

$$
\begin{aligned}
r &= r(r_0, t) \\
\phi &= \phi_0 + \phi(t) \\
z &= z(r_0, t)
\end{aligned}
\tag{8.260}
$$

$$
\begin{aligned}
v_r &= v_r(r_0, t) \\
v_\phi &= v_\phi(r_0, t) \\
v_z &= v_z(r_0, t)
\end{aligned}
\tag{8.261}
$$

What we have done so far is true in absolute generality; but Blandford and Payne (1982) consider a particularly simple solution in which the dependencies on r_0 are given by

$$r = r_0 \xi(\chi)$$
$$\phi = \phi_0 + h(\xi)$$
$$z = r_0 \chi \tag{8.262}$$

$$v_r = \left(\frac{GM}{r_0}\right)^{1/2} \xi'(\chi) f(\chi)$$

$$v_\phi = \left(\frac{GM}{r_0}\right)^{1/2} g(\chi)$$

$$v_z = \left(\frac{GM}{r_0}\right)^{1/2} f(\chi) \tag{8.263}$$

In other words, they fix a priori the dependence of the solution on r_0, that is, the distance from the center at which the magnetic field line intersects the plane of the disk; the way in which they do this is of course dimensionally correct, but it is certainly not the only possible way. Therefore, the solution we are going to find is a special solution, *not* the most general solution possible. This is done to simplify the mathematical treatment. Indeed, we see that at this point, while the variables $\vec{v}$ were functions of two arguments, r_0 and t, the new variables are functions of *one* argument only. It is therefore possible to transform the system of partial differential equations into a system of ordinary differential equations, which is a major simplification indeed.

To which physical approximation does this mathematical approximation correspond? Notice from equation 8.262 that if we fix the angle $\tan \theta \equiv z/r = \chi/\xi(\chi)$, we also fix the quantity χ; and, from equation 8.263, we see that along a given direction, all physical quantities scale with the spherical radius; therefore, these are self-similar solutions.

We only have to show why we have taken $v_r \propto \xi'(\chi)$ $f(\chi) \equiv d\xi/d\chi f(\chi)$. The direction of the magnetic field is obviously given by the derivative of equation 8.262, by definition of the magnetic field line itself

$$\vec{B} \propto (r_0\xi'(\chi), \phi'(\chi), r_0) \tag{8.264}$$

At the same time, we know that the velocity satisfies equation 8.239. As a consequence, $v_r \propto \xi'(\chi)f(\chi)$.

8.5.6 Results

Before showing the solutions obtained numerically for the angular dependence of physical quantities, we should notice that as in all self-similar solutions, the relationship between quantities with the same dimensions may depend only on the self-similarity variable (χ in our case). Therefore, the Alfvén velocity $\approx B/\rho^{1/2}$ and the velocity along the z axis of the wind, v_z, scale like the rotation velocity in the plane of the disk, $\propto r^{-1/2}$. Moreover, the energy and the angular momentum per unit mass must scale like the corresponding quantities in the accretion disk, $\propto r^{-1}$ and $r^{1/2}$, respectively. The mass loss from the disk may be calculated by considering a cylinder of infinitesimal height h and a radius r around the disk (fig. 6.2). Therefore, all the mass loss caused by the wind is due to the flux through the two bases of the cylinder and is therefore given by

$$\dot{M} = 2\pi r^2 v_z \rho \tag{8.265}$$

since, in stationary conditions, $\dot{M}$ must be independent from the radius, we must have $\rho \propto r^{-3/2}$. From the scaling of the Alfvén velocity, we find that $B \propto r^{-5/4}$. This obviously applies everywhere and, in particular, also to the disk surface. As a consequence, the solution of Blandford and Payne is applicable only to disks with a magnetic field scaling like $r^{-5/4}$. Moreover, since the sound speed must scale like $r^{-1/2}$, the gas pressure scales like $p \propto r^{-5/2}$; we thus obtain that $p \propto \rho^{5/3}$. Therefore, the self-similar solution exists only for fluids with $\gamma = 5/3$, in which case entropy is conserved (which is fine since no dissipative effect has been included). However, it is also possible that the sound speed is $\ll v_A$, the Alfvén velocity, in which case the solutions can also be applied to fluids with $\gamma \neq 5/3$.

From now on, the solution of equations 8.239, 8.244, 8.237, 8.243, 8.249, 8.256, and 8.257, subject to the conditions of self-similarity, equations 8.262 and 8.263, does not present further difficulties. There are only algebraic manipulations that allow us to reduce the problem to the solution of one ordinary differential equation in terms of the constants of motion

$$e \equiv \frac{\epsilon r_0}{GM} , \quad \lambda \equiv \frac{l}{\sqrt{GMr_0}} \qquad (8.266)$$

There are two other quantities that define the properties of the solutions. The first one is $\xi_0' \equiv d\xi/d\chi(\chi = 0)$, which defines the angle θ between the magnetic field and the plane of the disk (see eq. 8.264), $\xi_0' = \cot\theta$; you probably remember from the preceding chapter (section 7.4.2) that in a purely ballistic treatment of the problem, we have found instability for $\xi_0' > 1/\sqrt{3}$, and therefore we now expect something similar, although not exactly the same because the description of the fluid is no longer ballistic, but magnetohydrodynamic.

Notice that the solutions must pass through an Alfvén-sonic point (which we shall simply call sonic from now on), namely, a point where the poloidal velocity equals the Alfvén velocity, also due to the poloidal component only. Here, $4\pi\rho(v_r^2 + v_z^2) = (B_r^2 + B_z^2)$. This point presents the usual mathematical difficulty typical of sonic points (see chapter 5). In this case, we can choose the quantity ξ_0' to avoid divergences; as a consequence, ξ_0' *is not* a free parameter related to the boundary conditions, but must assume a precise value that allows a stationary solution. The existence of the sonic point corresponds, physically, to a change in the properties of the solutions. In fact, inside the sonic point, where the magnetic field contribution dominates, matter is forced to corotate with the field; this is the region in which we can apply the simple ballistic solution, discussed in section 7.4.2. Beyond the sonic point, matter drags the magnetic field through flux freezing. Moreover, at the sonic point, it is no longer possible to force matter to corotate with the region of the disk from which it originates; the wind lags behind, with respect to corotation, and dragging the magnetic field generates a strong toroidal magnetic field B_ϕ. Although this field also tends to become weaker and weaker, it becomes nonetheless dominant with respect to the poloidal components (exactly as with pulsars) and cannot vanish altogether because of the necessity to transport to infinity the angular momentum subtracted from the disk, through Maxwell's stress tensor. In fact, from the solution at large radii we can ascertain that

$$\frac{4\pi\rho v_\phi^2}{B_\phi^2} \to 0 \text{ as } \chi \to \infty \qquad (8.267)$$

Also the quantity κ can be dedimensionalized

$$k = \kappa\sqrt{1 + \xi_0'^2}\left(\frac{GM}{r_0 B(r_0)}\right)^{1/2} \qquad (8.268)$$

This number defines the relative importance of the Reynolds' and Maxwell's tensors, that is, whether the wind is dominated by the magnetic field or by the matter's inertia.

The solutions we are looking for are those that at large radii, are super-Alfvénic and collimated. Notice that by collimated solutions we mean the ones for which

$$v_{\rm r} = \left(\frac{GM}{r_0}\right)^{1/2}\xi'(\chi)f(\chi) \to 0 \quad \text{as } \chi \to \infty \quad (8.269)$$

This implies that at large distances, the matter's motion is directed along the z axis.

Not all initial conditions, e, λ, k, lead to this kind of solution. In fact, this is a rather delicate balance. On one hand, too strong magnetic fields may block the wind's acceleration to super-Alfvénic velocities; on the other hand, too weak magnetic fields will be incapable of exerting those hoop stresses

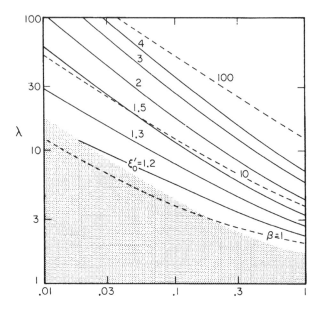

Figure 8.12. The diagram indicates the zone, in the space of the parameters λ, k, which leads to acceptable solutions, namely, super-Alfvénic and collimated winds. The broken-line zone, defined by the curve $k\lambda(2\lambda - 3)^{1/2} = 1$, indicates the flows that never become super-Alfvénic. For $k \gtrsim 1$, the solutions are dominated by the inertia of particles, and the weak magnetic fields are incapable of collimating the jet. Finally, the light zone indicates the space of initial parameters, leading to super-Alfvénic solutions; we can see the curves $\xi_0' =$ constant (note that $\xi_0' > 1/\sqrt{3}$). The condition of collimation is $\beta \equiv k(2\lambda - 3)^{3/2} > 3$; the curves $\beta =$ constant are also shown. From Blandford and Payne 1982.

described above, which lead to jet collimation. In the presence of a particular solution, such as the one by Blandford and Payne, our main interest lies in demonstrating that there is a region in the space of parameters e, λ, k leading to solutions with the features we are looking for.

This space is illustrated in figure 8.12. Notice that for $k\lambda\sqrt{2\lambda - 3} < 1$ the solutions are sub-Alfvénic, for $k \gtrsim 1$ they are excessively dominated by matter, so that the magnetic field cannot collimate them. Finally, the condition of collimation, which can be shown to be

$$\beta \equiv k(2\lambda - 3)^{3/2} > 3 \qquad (8.270)$$

which is obtained by means of an asymptotic analysis of the solutions for $\chi \to \infty$, holds in most of the light zone of the figure.

The solutions display the interesting transition discussed above; before the sonic point, it is the magnetic field that transports most energy flux:

$$\frac{F_B}{F_{\text{matter}}} \leq 2(\lambda - 1) \qquad (8.271)$$

where the equality is exactly valid on the disk, whereas, at large distances (and therefore far beyond the sonic point)

$$\frac{F_B}{F_{\text{matter}}} \to \frac{2}{\mathcal{M}^2 - 2} \qquad (8.272)$$

where the Mach number $\mathcal{M}$ is defined by $\mathcal{M}^2 \equiv 4\pi\rho(v_z^2 + v_r^2)/B^2$. In all the solutions discussed here, $\mathcal{M} \gg 1$, and therefore $F_B/F_{\text{matter}} \ll 1$. At large distances, the centrifugal acceleration of matter has been so efficient that it manages to carry away all energy. Note that this is in definite contrast with all the other solutions studied so far for pulsars, black holes, or disks in a nonmagnetohydrodynamic regime, where the contribution of matter has always been neglected.

Finally, it is also possible to estimate the total luminosity of the jet. We obtain

$$L = \left[\frac{k(\lambda - 3/2)}{1 + \xi_0'^2}\right] B^2(r_{\text{m}}) r_{\text{m}}^2 \left(\frac{GM}{r_{\text{m}}}\right)^{1/2} \qquad (8.273)$$

where r_{m} is the minimum radius of the disk. The first term, inside the square parentheses, is specific to this model of outflow. In this sense, it is not very important. The dimensional term can be estimated using $B = 10^4$ G, typical for an equipartition field around a black hole in an AGN, $r_{\rm m} = 5r_{\rm Sch} = 10GM/c^2$, and $M = 10^8$ M$_\odot$. We find

$$L \approx 10^{46} \text{ erg s}^{-1} \left(\frac{M}{10^8 \text{ M}_\odot} \right)^2 \qquad (8.274)$$

compatible with the AGN luminosities.

The main shortcoming of this solution consists in the assumption of the existence of open magnetic lines; to be more exact, what is difficult to understand is how a dynamically significant quantity of open magnetic field can be generated. First of all, the field will not have the same polarity everywhere; for example, in the case of the solar wind, it has a sectorial structure. The accreting matter carries, in the accretion process, its magnetic field, which will constitute the seed from which the magnetorotational instability process (section 6.7) produces the amplification. Inevitably, this magnetic field will have a coherence length comparable to the thickness of the disk, $H \ll r$, and therefore the lines will be closed. The most plausible way to generate open lines is through magnetic flares, as in the interaction between disk and corona. In this case, indeed, the zones that are going to reconnect (Galeev, Rosner, and Vaiana, 1979) may be pushed away by magnetic buoyancy. If they pass the sonic point before reconnecting, the dominant force pushing them to reconnect fails because matter (which, at this point, has become dynamically dominant) tends to expand. These lines may then be considered, in all respects, open and contribute to establishing the solution with open lines, which we have just discussed.

8.5.7 A Brief Summary

In this book, we have discussed various mechanisms for jet acceleration and their possible connection with accretion disks. The ubiquity of jets and their equally constant connection with accretion disks seems to suggest that there is a fundamental reason for their existence, but unfortunately, this does not seem to be the case. While a blind observer shut inside a room could predict the existence of stars and the lower and upper limits on their masses from first principles, nothing of the sort is possible for jets. Moreover, the mystery shrouding the ubiquity of jets has thickened since the discovery of the magnetorotational instability (chapter 6) persuaded most astronomers that the removal of the angular momentum to the outer disk zones can be accomplished without invoking mechanisms that simultaneously generate a wind (Blandford 1976; Lovelace 1976; Blandford and Payne 1982).

The first mechanism for the acceleration and collimation of jets obviously is the pressure of an ordinary gas that confines them (Blandford and Rees 1974); a similar idea was revived by Woosley (1993) for the generation of gamma ray bursts. This idea is essentially based on the existence of the de Laval nozzle, of the kind described in chapter 1. Unfortunately, later observations showed that jets acquire their collimation at very short distances from the central black hole, the shortest ever investigated ($\approx 10^3 r_{\text{Sch}}$) by means of the very long baseline interferometry (VLBI). Collimation on these scales by a thermal gas requires such high temperatures and densities that it would be impossible to keep the gas hot for a long time; moreover, the gas luminosity turns out to be much larger than Eddington's, making its observational detection very easy, also thanks to its thermal spectrum.

A variation on this theme is given by the advection-dominated accretion flow model of Narayan and Yi (1994) (see chapter 7), according to which the accretion flux has a Bernoulli number >0, so that matter is essentially unbound. This fact may obviously be helpful for accelerating a jet, but the details of the process and the reasons for the collimation are not contained in the model.

The next idea was radiation pressure, and, in particular, a great interest has been raised by the existence of funnels in models of thick accretion disks. Hopes were quickly dashed, however. On one hand, the high radiation density inside the funnel can only partially be used to accelerate matter, since its contribution is soon transformed in inverse Compton braking, as discussed in chapter 7. On the other hand, the violent dynamic instabilities to which all thick disks are subject make them unsuitable as models for steady sources.

The third idea uses the magnetohydrodynamic phenomena discussed above. In short, this is a centrifugal acceleration. Inside the Alfvén point, matter is forced to rotate with the disk by the magnetic field, which controls the dynamics. The acceleration phase terminates at the Alfvén point, but hoop stresses, due to the toroidal component of the magnetic field, which always exceed the centrifugal force (eq. 8.267), ensure the jet collimation.

The weak point of these models is the large coherence-length of the necessary magnetic field, which exceeds that of magnetorotational instability by several orders of magnitude. These doubts lead to a different question, namely, whether a configuration with a more turbulent and less orderly magnetic field can reproduce, at least qualitatively, the results of Blandford and Payne. The beginning of an answer comes from the simulations of Proga and Begelman (2003), which show that in magnetohydrodynamic conditions, the wind is more powerful and remains more collimated than in purely hydrodynamic conditions. However, the details of this jet acceleration in a turbulent environment are at the moment unknown.

We also totally ignore (apart from the idealized model of Blandford and Payne) the loss rates of mass, angular momentum, momentum, and energy due to the presence of the jet.

8.6 Problems

1. Show that if a pulsar is in vacuum, it has a surface charge

$$\sigma = -\frac{B_{\mathrm{p}}\Omega R_{\mathrm{p}}}{4\pi c}\cos^2\theta \qquad (8.275)$$

Hint: Remember that in a conductor, the discontinuity of the normal component of the electric field is linked to the density of surface charges by the relation $E_\perp^{(\mathrm{out})} - E_\perp^{(\mathrm{in})} = 4\pi\sigma$.

2. In deriving the equation of pulsars, we have used only two of the three vector components of Ampère's law, equation 8.46, namely, the components ϕ and z. Show that the component r of this equation gives, once again, equation 8.51.

3. Using the radiation energy-momentum tensor (Landau-Lifshitz 1987a), prove that for rotating pulsars, there is also an energy flux to infinity, and that the pulsar's energy loss rate is $dE/dt = \Omega dL_z/dt < 0$.

4. Derive the general version of equation 8.19, namely, the one that reduces to this for $\kappa = 0$:

$$4\pi\rho = \frac{4\pi\Omega}{c^2}rj_\phi - \frac{2\Omega}{c}B_z \qquad (8.276)$$

or:

$$\rho = \frac{1}{1 - \Omega^2 r^2}c^2\left(\frac{\Omega r\kappa B_\phi}{c^2} - \frac{\Omega B_z}{2\pi c}\right) \qquad (8.277)$$

Hint: Use equation 8.43.

5. Determine explicitly the light surface for the disk studied in section 8.2, under the hypothesis that the rotation is Keplerian, $\omega(r) \propto r^{-3/2}$.

6. For an electromagnetic wind from a Newtonian disk, compute the energy flux through the surface of a large sphere with radius R, as a function of the angle θ. Show that this distribution tends, when $R \to \infty$, to $\sin\theta\delta(\cos\theta-1)d\theta$, where $\delta(x)$ is Dirac's delta.

7. In section 8.5.4, we have neither considered equation 2.14, nor the radial component of equation 8.225. Can you show that they are already verified? (Hints: For the first one, use stationarity and equations 8.239, 8.236; for the second one, assume entropy conservation, equation 2.25, and the toroidal and z components of equation 8.225; demonstrate that it is possible to infer from these the radial component.)

Appendix A

Propagation of Electromagnetic Waves

The propagation of electromagnetic waves in the universe does not take place in a void, but inside the interstellar or intergalactic media. This gives origin to some interesting phenomena, which we describe here, namely, absorption below the plasma frequency, plasma dispersion, and Faraday rotation.

In the text, we have treated these media as fluids, a correct approximation when all the time scales for the phenomena in question are long with respect to the time between collisions t_c. The collision time t_c (Spitzer 1962) is given by

$$t_c = \frac{m^{1/2}(3kT)^{3/2}}{5.7\pi n e^4 \ln \Lambda} = 0.27 \frac{T^{3/2}}{n \ln \Lambda} \text{ s} \qquad (A.1)$$

for the electrons. Here T is the temperature in degrees kelvin, and $\ln \Lambda \approx 30$ is an approximate factor, called the Coulomb factor. Since the typical electromagnetic waves observed on Earth have oscillation periods $\nu^{-1} \ll t_c$, their propagation cannot be described by magnetohydrodynamics; it is necessary to consider plasma effects.

When an electromagnetic wave propagates in a dielectric medium, electrons are pushed by the wave's fields and emit new electromagnetic waves. The final wave is therefore given by the superposition of the original wave plus the sum of the contributions of all the electrons present. Two circumstances greatly simplify the analysis. On one hand, since the densities are low, the collision times are long; collisions with other electrons or protons may then be neglected. Furthermore, since in high-energy astrophysics temperatures are relatively high, plasmas are essentially ionized, and electrons may be considered totally free. Let us therefore consider an electromagnetic wave of the type

$$\vec{E} = \vec{E}_0 e^{i(\vec{k}\cdot\vec{x} - \omega t)} \tag{A.2}$$

The microscopic Maxwell equations (namely, the ones explicitly including charges and currents present in the dielectric field) for this wave are

$$i\vec{k}\cdot\vec{E} = 4\pi\rho \qquad i\vec{k}\cdot\vec{B} = 0$$
$$i\vec{k}\wedge\vec{E} = i\frac{\omega}{c}\vec{B} \qquad i\vec{k}\wedge\vec{B} = \frac{4\pi}{c}\vec{j} - i\frac{\omega}{c}\vec{E} \tag{A.3}$$

We need to specify charge densities and currents. To this aim, we note that each electron is subject to the equation of motion,

$$m_{\rm e}\frac{d\vec{v}}{dt} = -e\vec{E} \tag{A.4}$$

where the electron's charge is $-e$, and, in the Lorentz force, since $E_0 = B_0$ for the electromagnetic wave, the term due to the magnetic field has been neglected because it is proportional to $v/c \ll 1$. Its solution is easy,

$$\vec{v} = -\frac{ie}{\omega m_{\rm e}}\vec{E} \tag{A.5}$$

and the current density $\vec{j} = -en\vec{v}$ is also proportional to the electric field,

$$\vec{j} = \sigma\vec{E} \qquad \sigma = \frac{ine^2}{\omega m_{\rm e}} \tag{A.6}$$

where σ is the plasma conductivity. It is also useful to have an expression for the *perturbed* charge density in terms of $\vec{E}$. From the equation for charge conservation we find

$$-\imath\omega\rho + \imath\vec{k}\cdot\vec{j} = 0 \tag{A.7}$$

which immediately yields

$$\rho = \frac{\sigma}{\omega}\vec{k}\cdot\vec{E} \tag{A.8}$$

We can now use these expressions for $\vec{j}$ and ρ in Maxwell's equations and obtain

$$\imath\vec{k}\cdot\epsilon\vec{E} = 0 \qquad \imath\vec{k}\cdot\vec{B} = 0$$
$$\imath\vec{k}\wedge\vec{E} = \imath\frac{\omega}{c}\vec{B} \qquad \imath\vec{k}\wedge\vec{B} = -\imath\frac{\omega}{c}\epsilon\vec{E} \tag{A.9}$$

where we have introduced the definition

$$\epsilon = 1 - \frac{4\pi\sigma}{\imath\omega} = 1 - \frac{4\pi n e^2}{m_e\omega^2} \equiv 1 - \frac{\omega_p^2}{\omega^2} \tag{A.10}$$

The quantity ω_p is called the plasma frequency.

Maxwell's equations are just like those in *vacuum*, except that

$$\frac{\omega^2}{k^2} = \frac{c^2}{\epsilon} = \frac{c^2}{n^2} \tag{A.11}$$

where n obviously is the plasma refraction index. It is useful to rewrite this equation as

$$k^2 c^2 = \omega^2 - \omega_p^2 \tag{A.12}$$

From this, we see the first interesting phenomenon: if the wave frequency is lower than the plasma frequency, $k^2 < 0$. From equation A.2, we see that this corresponds to a spatial absorption (the solution that grows exponentially exists only formally, of course, and has no physical meaning).

Besides, we see that the apparent velocity of propagation $c/\epsilon^{1/2}$ exceeds the speed of light in *vacuum*, c, since $\epsilon < 1$.

However, this happens because we are discussing the phase velocity, which has no physical meaning. We are interested in the group velocity,

$$v_{\mathrm{g}} = \frac{\partial \omega}{\partial k} = c\sqrt{1 - \frac{\omega_{\mathrm{p}}^2}{\omega^2}} < c \qquad (A.13)$$

This equation also illustrates the second interesting physical phenomenon: the velocity of propagation of electromagnetic waves in a plasma depends on its frequency. This phenomenon is called *dispersion*.

The last phenomenon we discuss is Faraday rotation, which takes place when a circularly polarized wave propagates in a magnetized plasma. For the sake of simplicity, we consider only the case of propagation along the magnetic field, which we take as $\vec{B} = B_0 \hat{z}$. Let us now consider a circularly polarized wave:

$$\vec{E} = E_0(\hat{\epsilon}_1 \pm \hat{\epsilon}_2)e^{\imath(\vec{k}\cdot\vec{r} - \omega t)} \qquad (A.14)$$

Maxwell's equations (eq. A.3), obviously do not change, whereas the equation of motion changes, because we are forced to introduce the Lorentz force due to the constant magnetic field B_0, which is supposed to be much larger than the wave's. Thus we have

$$m_{\mathrm{e}}\frac{d\vec{v}}{dt} = -e\vec{E} - e\frac{\vec{v}}{c} \wedge \vec{B} \qquad (A.15)$$

which has the solution

$$\vec{v} = -\frac{\imath e}{m_{\mathrm{e}}(\omega \pm \omega_{\mathrm{c}})}\vec{E} \qquad (A.16)$$

where ω_{c} is the cyclotron, or Larmor frequency,

$$\omega_{\mathrm{c}} \equiv \frac{eB_0}{m_{\mathrm{e}}c} \qquad (A.17)$$

From the solution for $\vec{v}$ we can see that there is a difference between right-handed and left-handed polarized waves, because the electron has its own direction of rotation through the magnetic field. Following the same steps leading to equation A.10, we find

$$\epsilon_{\pm} = 1 - \frac{\omega_p^2}{\omega(\omega \pm \omega_c)} \tag{A.18}$$

which shows that there is a difference between the wave numbers for the two polarizations. In fact we have

$$k_{\pm} = \frac{\omega}{c}\sqrt{\epsilon_{\pm}} \tag{A.19}$$

As a consequence, in a magnetized medium, if we decompose a linearly polarized wave into two waves of opposite circular polarization, the two circularly polarized waves propagate differently, depending upon whether they are left- or right-handed.

Appendix B

Orbits Around Black Holes

In this appendix, we derive a few elementary results on circular orbits around black holes. All these results are well known and are discussed in detail in any modern text on general relativity. We assume knowledge of the Schwarzschild and Kerr solutions of the field equations of general relativity.

We can derive the properties we need from those of the metric. Initially, we concentrate on the metric for the Schwarzschild, nonrotating black hole,

$$ds^2 = \left(1 - \frac{2GM}{c^2 r}\right) dt^2 - \frac{dr^2}{1 - \frac{2GM}{c^2 r}} - r^2 d\theta^2 - r^2 \sin^2\theta d\phi^2 \tag{B.1}$$

which describes exactly the metric *outside* any massive object with spherical symmetry. Among these, the black hole is the only one for which this metric holds for every $r > 2GM/c^2 \equiv r_s$. For any other object, this metric does not hold for $r \leq R_\star$, the star's radius. From this, we can immediately write the Lagrangian of motion:

$$\mathcal{L} = \left(1 - \frac{r_s}{r}\right) \dot{t}^2 - \frac{\dot{r}^2}{1 - \frac{r_s}{r}} - r^2 \dot{\theta}^2 - r^2 \sin^2\theta \dot{\phi}^2 \tag{B.2}$$

Here derivatives are taken with respect to any parameter (which we call λ) parameterizing the orbit.

Since the Lagrangian is cyclic in ϕ (i.e., does not depend on ϕ), its conserved conjugate momentum is

$$p_\phi \equiv \frac{\partial \mathcal{L}}{\partial \dot{\phi}} = -r^2 \sin^2 \theta \dot{\phi} \equiv l_z \qquad \text{(B.3)}$$

where l_z is the specific angular momentum around the z axis, measured locally at infinity, and is a constant of motion.

However, the Lagrangian is cyclic also in t, therefore,

$$p_t \equiv \frac{\partial \mathcal{L}}{\partial \dot{t}} = \left(1 - \frac{r_s}{r}\right)\dot{t} \equiv E_\infty \qquad \text{(B.4)}$$

where E_∞, the energy measured at infinity, is also a constant.

The Lagrangian is not cyclic in θ. The component of the equation of motion along θ turns out to be

$$\frac{d}{d\lambda}(-r^2 \dot{\theta}) = r^2 \sin\theta \cos\theta \dot{\phi}^2 \qquad \text{(B.5)}$$

Since the metric is spherically symmetric, we can choose its equatorial plane as we like. Calling $\vec{r}_o$ and $\vec{v}_o$ the initial position and velocity of the particle, respectively, their cross product $\vec{r}_o \wedge \vec{v}_o$ singles out one plane that contains both. Let us therefore choose the axes of the system of coordinates such that $\theta = \pi/2$ is exactly this plane. In this way, since the particle's initial velocity has no component outside this plane, $\dot{\theta}_o = 0$. As a consequence, at the initial instant $\ddot{\theta} = 0$, and the above equation admits only the solution $\theta = \pi/2$ at all times.

The equation of motion in the direction r can be immediately obtained if we recall that in relativity, the modulus of the four-velocity is a constant, q: $q = 1$ for massive particles, and $q = 0$ for photons. Thus we have $g_{\mu\nu}\dot{x}^\mu \dot{x}^\nu = q$, and therefore,

$$\frac{E_\infty^2}{2} = \frac{\dot{r}^2}{2} + \frac{1}{2}\left(q + \frac{l_z^2}{r^2}\right)\left(1 - \frac{r_s}{r}\right) \equiv \frac{\dot{r}^2}{2} + V_{\text{eff}}(r) \qquad \text{(B.6)}$$

We have deliberately written this equation in this form to emphasize the analogy with the Newtonian case:

$$E = \frac{\dot{r}^2}{2} + \frac{l_z^2}{2r^2} - \frac{GM}{r} = \frac{\dot{r}^2}{2} + \phi_{\text{eff}} \qquad (B.7)$$

Thus the general relativity case behaves just like the Newtonian case, except that the two effective potentials differ. For particles with a nonzero mass, we have, indeed,

$$V_{\text{eff}} - \phi_{\text{eff}} = \frac{1}{2} - \frac{GM l_z^2}{c^2 r^3} \qquad (B.8)$$

Therefore, Newtonian and general relativity treatments differ for a constant term (which does not influence dynamics) and for the term $-1/r^3$, which, however, has serious consequences.

The constant term is present because of the different definition of energy in Newtonian dynamics and in general relativity, where energy also includes the rest mass. It is easy to see that a particle at rest at infinity has $E_\infty = 1$, whereas, in Newtonian physics, its total energy vanishes.

Circular orbits are, by definition, those with $\dot{r} = \ddot{r} = 0$. These conditions lead to

$$\frac{\partial V_{\text{eff}}}{\partial r} = 0 \qquad (B.9)$$

This defines the relation between the orbital radius and the specific angular momentum. If we introduce these values into equation B.6, we find the total energy for a circular orbit, as a function of orbital radius or of specific angular momentum.

However, the most unusual feature of circular orbits in general relativity emerges when we consider the stability of the orbit, which is, of course, determined by the sign of $\partial^2 V_{\text{eff}}/\partial r^2$; this is to be evaluated for circular orbits, in other words, using equation B.9 after having taken the derivative. When $\partial^2 V_{\text{eff}}/\partial r^2 > 0$, the orbit is stable; the effective energy has a minimum, and orbits with slightly higher energies are in any case confined to a small region around the equilibrium radius. On the other hand, if $\partial^2 V_{\text{eff}}/\partial r^2 < 0$, the circular orbit corresponds to a maximum of the effective potential. Therefore, orbits with energies slightly different from that of the circular orbit are not confined to a small neighborhood of the circular orbit.

We find that V_{eff} has a maximum and a minimum for radii:

$$r_{\text{e}} = \frac{r_{\text{s}}}{4}\left(\frac{l_z c}{GM}\right)^2\left(1 \pm \sqrt{1 - 12\left(\frac{GM}{l_z c}\right)^2}\right) \qquad \text{(B.10)}$$

Therefore, when $l_z < \sqrt{12}GM/c$, the effective potential does not even have an extremum, and there can be no circular orbits. When $l_z = \sqrt{12}GM/c$, the first inflexion point appears. This is the minimum value of l_z, which allows a circular orbit, at a radius

$$r_{\text{mso}} = 3r_{\text{s}} = \frac{6GM}{c^2} \qquad \text{(B.11)}$$

This orbit is called marginally stable, for obvious reasons. We can now calculate the energy of this orbit by introducing its radius, and $l_z = \sqrt{12}GM/c$ in equation B.6; thus

$$E_\infty = \sqrt{8/9} \qquad \text{(B.12)}$$

We have already mentioned the fact that the orbit at rest at infinity has $E_\infty = 1$, which is $>\sqrt{8/9}$. Where did the energy difference go? In our theory of accretion disks, we show that the missing energy must have been radiated; thus, the radiative efficiency of disk accretion on a nonrotating black hole is

$$\eta = 1 - \sqrt{8/9} \approx 0.06 \qquad \text{(B.13)}$$

In other words, the quantity of radiated energy, if the black hole swallows a quantity of mass m, is $\approx 0.06 \; mc^2$, much higher than that of nuclear combustion, which is (for hydrogen burning) only $\eta_{\text{H}} \approx 0.007$.

An important note concerns the so-called Paczynski-Wiita potential (1980). This is a Newtonian potential of the form

$$\phi_{\text{eff}} = -\frac{GM}{r - r_{\text{s}}} + \frac{l_z^2}{2r^2} \tag{B.14}$$

which, despite its use in Newtonian dynamics, has some similarities with that of general relativity, V_{eff}. In particular, it has a marginally stable orbit, located exactly at $r_{\text{mso}} = 6GM/c^2$, and a radiative efficiency $\eta = 1/16$, only marginally larger than the exact one. For these reasons, this potential is often used in analytic calculations and in entirely Newtonian numerical simulations to approximate a fully general relativistic treatment of the problem.

In the Kerr metric the analysis is more complex. We state only the results for circular orbits in the equatorial plane of the black hole. We recall that the various solutions are parameterized by the specific angular momentum of the black hole $l_z = Ga/c$; we know that a varies in the interval $0 \leq a \leq M$. However, for realistic configurations, a is probably limited to the value $a \approx 0.998M$ (Thorne 1974), because among the photons emitted near the black hole, some will be emitted in orbits with a negative angular momentum. These are caught by the hole more easily than those emitted in orbits with a positive angular momentum, thus contributing to limit the maximum attainable angular momentum. Although this is a small difference from unity, it does have a certain importance, as we shall soon see.

Even around a rotating black hole, there are circular orbits. The orbits' properties are very different, depending on whether they are corotating (i.e., they rotate in the same direction as the black hole) or counterrotating. The reason for this difference between the two rotations is illustrated in problem 1. The radius of the marginally stable orbit is

$$r_{\text{mso}} = \frac{GM}{c^2}\left(3 + q_2 \mp \sqrt{(3 - q_1)(3 + q_1 + 2q_2)}\right) \tag{B.15}$$

where

$$q_1 \equiv 1 + \left(1 - \frac{a}{M}\right)^{1/3} \left(\left(1 + \frac{a}{M}\right)^{1/3} + \left(1 - \frac{a}{M}\right)^{1/3}\right) \quad \text{(B.16)}$$

and

$$q_2 \equiv \left(3\frac{a^2}{M^2} + q_1^2\right)^{1/2} \quad \text{(B.17)}$$

The minus sign refers to corotating orbits, and the plus sign to counterrotating orbits. For a maximally rotating black hole, $a = M$, we find

$$r_{\text{mso}} = \begin{cases} \frac{GM}{c^2} & \text{corotating} \\ \frac{9GM}{c^2} & \text{counterrotating} \end{cases} \quad \text{(B.18)}$$

We can calculate the energy of marginally stable orbits, and from this the radiative efficiency. For corotating orbits and a maximally rotating black hole, it is

$$\eta = 1 - 1/\sqrt{3} \approx 0.42 \quad \text{(B.19)}$$

The much higher efficiency is obviously due to the fact that marginally stable, corotating orbits have radius GM/c^2, a factor of 6 smaller than the nonrotating case. However, the radiative efficiency is a very steep function of a, and when a is limited to $a = 0.998M$, we find that $\eta \approx 0.30$, thus confirming the importance of the Thorne constraint (1974).

Rotating black holes also have a peculiar property, namely, their rotational energy can be lost to infinity (and the hole slowed down) through processes that take place in a special region, immediately outside their event horizon, called the *ergosphere*. These processes will not be reviewed here, and the reader is invited to study the beautiful book by Wald (1984) for further details.

There is one further effect we need to discuss for later reference: Lense-Thirring precession. The easiest way to do this is through the 3+1 interpretation of gravity presented by

Thorne, Price, and McDonald (1986), specialized to the weak-field case. Here, gravity is interpreted as a weak perturbation, which can be derived from a quadripotential $\phi, \vec{A}$ (why not a scalar potential?), which is related to the metric tensor via the formulae

$$\phi = \frac{1 + g_{00}}{2} \; , \quad A_i = -g_{0i} \qquad (\text{B.20})$$

The equations to which $\phi, \vec{A}$ are subject can obviously be derived from the linearization of the Einstein equations and must obviously be similar to Maxwell's equations, because these are the relativistically covariant equations for a quadripotential, and we certainly cannot derive anything that is *not* relativistically covariant from Einstein's equations. Calling, as usual,

$$\vec{g} \equiv -\nabla\phi - \frac{1}{2c}\frac{\partial \vec{A}}{\partial t} \; , \quad \vec{B} \equiv \frac{1}{2}\nabla \wedge \vec{A} \qquad (\text{B.21})$$

we find the Maxwell-like equations

$$\nabla \cdot \vec{g} = -4\pi G\rho$$

$$\nabla \cdot \vec{B} = 0$$

$$\nabla \wedge \vec{g} = -\frac{1}{c}\frac{\partial \vec{B}}{\partial t}$$

$$\nabla \wedge \vec{B} = -\left(4\pi G\rho\frac{\vec{v}}{c} - \frac{1}{c}\frac{\partial \vec{g}}{\partial t}\right) \qquad (\text{B.22})$$

This set of equations is of course very similar to Maxwell's, except that all signs are wrong because gravity is a purely attractive force and thus it behaves as if only *negative* charges exist.

These Maxwell-like equations must be supplemented with a Lorentz-like force,

$$\vec{f} = m\left(\vec{g} + \frac{\vec{v}}{c} \wedge 2\vec{B}\right) \qquad (\text{B.23})$$

since this is, notoriously, not included in Maxwell's equations.

The extra factor of 2 is due to a more subtle effect, which is that gravity is a tensor force, unlike electromagnetism, which is a vector force. This is often expressed in quantum-mechanical terms by saying that gravity is mediated by a spin 2 particle (the *gravitino*), while electromagnetism is mediated by the spin 1 photon. However, this extra factor does not preclude mass conservation, which can easily be inferred from the equations above.

From the equations above we see that mass motion is a source of gravity, not just mass, in perfect analogy with electromagnetism, where currents generate forces even when the total charge vanishes. Thus, consider a test particle orbiting a rotating black hole on a plane slightly inclined to the equatorial plane. Since the hole rotates, it produces not just a gravity force $\vec{g}$, which causes the rotation of the test particle, but also a gravitomagnetic field $\vec{B}$, and a gravitomagnetic force $\propto \vec{v} \wedge \vec{B}$, which induces observable effects on the test particle motion. To see what these effects are, we simply note that in electromagnetism, dipole fields due to a magnetic moment

$$\vec{M} = \frac{1}{2c} \int \vec{x}' \wedge \vec{j}(\vec{x}')d^3x' = \frac{e\vec{L}}{2mc} \qquad \text{(B.24)}$$

where $\vec{j}$ is the current density and $\vec{L}$ the object's angular momentum, are given by

$$\vec{B} = \frac{3\hat{n}(\hat{n} \cdot \vec{M}) - \vec{M}}{r^3} \qquad \text{(B.25)}$$

If we take into account the different factor $-Gm/e$, which links mass current to the gravitomagnetic field $\vec{B}$ in the Maxwell-like equations above, we can set

$$\vec{M} = -\frac{\vec{L}}{2c} = -\frac{aM_{\text{BH}}}{2c}\hat{z} \qquad \text{(B.26)}$$

where $\vec{L}$, the hole angular momentum, is oriented along the z axis and is given by M_{BH}, the hole specific angular momentum a times the hole's mass M_{BH}.

From the pseudo-Lorentz force, equation B.23, we find the force

$$\vec{f} = m\frac{\vec{v}}{c} \wedge 2B \tag{B.27}$$

which leads to a torque

$$\vec{N} = \vec{r} \wedge \vec{f} = -\frac{1}{c}\vec{l} \wedge 2\vec{B} \tag{B.28}$$

where $\vec{l}$ is the particle's angular momentum. $\vec{B}$ is computed at the instantaneous position of the particle; we find the precession equation,

$$\vec{N} = -\frac{2\vec{B}}{c} \wedge \vec{l} \tag{B.29}$$

which shows that there will be no precession for a particle in the hole equatorial plane. We can now identify the Lense-Thirring precession frequency as

$$\omega_{\text{LT}} \equiv \frac{2B}{c} \tag{B.30}$$

B.1 Problems

1. In order to illustrate the difference between corotating and counterrotating orbits, consider the following thought experiment. Two identical spheres, with the same mass M and the same electric charge Q, are placed at rest at a relative distance D. Charge and mass are chosen so that $Q^2 = GM^2$. An observer at rest with respect to the spheres deduces that the two forces are always equal, and therefore that the two spheres are in an equilibrium. Now the observer starts moving relative to the spheres, in a direction perpendicular to their separation. The gravitational force increases because the apparent masses increase by a factor γ due to the motion, while the separation D is unaffected; the electrostatic force does not

change since the electric charge is a Lorentz invariant. But a magnetic force now appears since our motion causes the appearance of two parallel currents, and the ensuing magnetic force is attractive. The total electromagnetic force weakens, whereas the gravitational force strengthens; the observer in motion infers that the two spheres attract each other. How can we solve this paradox? (Suggestion: Nothing we have said so far is wrong, and there is no mistake in the Lorentz transformations. But we are missing something, whose sign explains why corotating orbits are closer to the black hole than the counterrotating ones. For a different hint, read the preface.).

2. Can you derive eq. 6.101 from eq. B.30 (Wilkins, D.C., 1972, *Phys. Rev. D*, 5, 814)?

Appendix C

Useful Formulae

C.1 Vector Identities

$$(\vec{a} \wedge \vec{b}) \wedge \vec{c} = (\vec{a} \cdot \vec{c})\vec{b} - (\vec{b} \cdot \vec{c})\vec{a} \tag{C.1}$$

$$\nabla \wedge \nabla \phi = 0 \ , \quad \nabla \cdot (\nabla \wedge \vec{F}) = 0 \tag{C.2}$$

$$\nabla \cdot (\phi \vec{F}) = \phi \nabla \cdot \vec{F} + \vec{F} \cdot \nabla \phi \tag{C.3}$$

$$\nabla \wedge (\phi \vec{F}) = \phi \nabla \wedge \vec{F} + (\nabla \phi) \wedge \vec{F} \tag{C.4}$$

$$\nabla \wedge (\vec{F} \wedge \vec{G}) = (\vec{G} \cdot \nabla)\vec{F} - (\vec{F} \cdot \nabla)\vec{G} + \vec{F}(\nabla \cdot \vec{G}) - \vec{G}(\nabla \cdot \vec{F}) \tag{C.5}$$

$$\nabla \cdot (\vec{F} \wedge \vec{G}) = \vec{G} \cdot (\nabla \wedge \vec{F}) - \vec{F} \cdot (\nabla \wedge \vec{G}) \tag{C.6}$$

$$\nabla(\vec{F} \cdot \vec{G}) = \vec{F} \wedge (\nabla \wedge \vec{G}) + \vec{G} \wedge (\nabla \wedge \vec{F}) + (\vec{F} \cdot \nabla)\vec{G} + (\vec{G} \cdot \nabla)\vec{F} \tag{C.7}$$

$$(\vec{F} \cdot \nabla)\vec{F} = (\nabla \wedge \vec{F}) \wedge \vec{F} + \frac{1}{2}\nabla(\vec{F} \cdot \vec{F}) \tag{C.8}$$

$$\nabla^2 \vec{F} = \nabla(\nabla \cdot \vec{F}) - \nabla \wedge (\nabla \wedge \vec{F}) \tag{C.9}$$

C.2 Cylindrical Coordinates

$$x = r \cos \phi \ , \quad y = r \sin \phi \ , \quad z = z \tag{C.10}$$

$$\hat{e}_r = \cos \phi \hat{e}_x + \sin \phi \hat{e}_y \ , \quad \hat{e}_\phi = -\cos \phi \hat{e}_x + \cos \phi \hat{e}_y \ , \quad \hat{e}_z = \hat{e}_z \tag{C.11}$$

$$\hat{e}_r \wedge \hat{e}_\phi = \hat{e}_z \ , \quad \hat{e}_\phi \wedge \hat{e}_z = \hat{e}_r \ , \quad \hat{e}_z \wedge \hat{e}_r = \hat{e}_\phi \tag{C.12}$$

529

The unit vectors do not depend on r or z, but do depend on the angle ϕ:

$$\frac{\partial \hat{e}_r}{\partial \phi} = \hat{e}_\phi \ , \quad \frac{\partial \hat{e}_\phi}{\partial \phi} = -\hat{e}_r \ , \quad \frac{\partial \hat{e}_z}{\partial \phi} = 0 \qquad \text{(C.13)}$$

$$\nabla \psi = \hat{e}_r \frac{\partial \psi}{\partial r} + \hat{e}_\phi \frac{1}{r} \frac{\partial \psi}{\partial \phi} + \hat{e}_z \frac{\partial \psi}{\partial z} \qquad \text{(C.14)}$$

$$\nabla \cdot \vec{a} = \frac{1}{r} \frac{\partial}{\partial r}(r a_r) + \frac{1}{r} \frac{\partial a_\phi}{\partial \phi} + \frac{\partial a_z}{\partial z} \qquad \text{(C.15)}$$

$$\nabla \wedge \vec{a} = \hat{e}_r \left(\frac{1}{r} \frac{\partial a_z}{\partial \phi} - \frac{\partial a_\phi}{\partial z} \right) + \hat{e}_\phi \left(\frac{\partial a_r}{\partial z} - \frac{\partial a_z}{\partial r} \right)$$

$$+ \hat{e}_z \frac{1}{r} \left(\frac{\partial}{\partial r}(r a_\phi) - \frac{\partial a_r}{\partial \phi} \right) \qquad \text{(C.16)}$$

$$\nabla^2 \psi = \frac{1}{r} \frac{\partial}{\partial r} \left(r \frac{\partial \psi}{\partial r} \right) + \frac{1}{r^2} \frac{\partial^2 \psi}{\partial \phi^2} + \frac{\partial^2 \psi}{\partial z^2} \qquad \text{(C.17)}$$

$$\vec{a} \cdot \nabla = a_r \frac{\partial}{\partial r} + \frac{a_\phi}{r} \frac{\partial}{\partial \phi} + a_z \frac{\partial}{\partial z} \qquad \text{(C.18)}$$

$$(\vec{v} \cdot \nabla)\vec{v} = \hat{e}_r \left((\vec{v} \cdot \nabla) v_r - \frac{v_\phi^2}{r} \right) + \hat{e}_\phi \left((\vec{v} \cdot \nabla) v_\psi + \frac{v_\phi v_r}{r} \right)$$

$$+ \hat{e}_z (\vec{v} \cdot \nabla) v_z \qquad \text{(C.19)}$$

C.3   Spherical Coordinates

$$x = r \sin \theta \cos \phi \ , \quad y = r \sin \theta \sin \phi \ , \quad z = r \cos \theta \qquad \text{(C.20)}$$

$$\hat{e}_r = \sin \theta \cos \phi \hat{e}_x + \sin \theta \sin \phi \hat{e}_y + \cos \theta \hat{e}_z \ ,$$
$$\hat{e}_\phi = -\sin \phi \hat{e}_x + \cos \phi \hat{e}_y$$
$$\hat{e}_\theta = \cos \theta \cos \phi \hat{e}_x + \cos \theta \sin \phi \hat{e}_y - \sin \theta \hat{e}_z \qquad \text{(C.21)}$$

$$\hat{e}_r \wedge \hat{e}_\phi = \hat{e}_\theta \ , \quad \hat{e}_\phi \wedge \hat{e}_\theta = \hat{e}_r \ , \quad \hat{e}_\theta \wedge \hat{e}_r = \hat{e}_\phi \qquad \text{(C.22)}$$

The unit vectors do not depend on r, but do depend on the angles ϕ and θ:

$$\frac{\partial \hat{e}_r}{\partial \phi} = \sin\theta \hat{e}_\phi \; , \quad \frac{\partial \hat{e}_\phi}{\partial \phi} = -\sin\theta \hat{e}_r - \cos\theta \hat{e}_\theta \; , \quad \frac{\partial \hat{e}_\theta}{\partial \phi} = \cos\theta \hat{e}_\phi$$

(C.23)

$$\frac{\partial \hat{e}_r}{\partial \theta} = \hat{e}_\theta \; , \quad \frac{\partial \hat{e}_\theta}{\partial \theta} = -\hat{e}_r \; , \quad \frac{\partial \hat{e}_\phi}{\partial \theta} = 0 \tag{C.24}$$

$$\nabla\psi = \hat{e}_r \frac{\partial \psi}{\partial r} + \hat{e}_\theta \frac{1}{r}\frac{\partial \psi}{\partial \theta} + \hat{e}_\phi \frac{1}{r\sin\theta}\frac{\partial \psi}{\partial \phi} \tag{C.25}$$

$$\nabla\cdot\vec{a} = \frac{1}{r^2}\frac{\partial}{\partial r}(r^2 a_r) + \frac{1}{r\sin\theta}\frac{\partial a_\phi}{\partial \phi} + \frac{1}{r\sin\theta}\frac{\partial}{\partial \theta}(\sin\theta a_\theta) \tag{C.26}$$

$$\nabla\wedge\vec{a} = \hat{e}_r \frac{1}{r\sin\theta}\left(\frac{\partial}{\partial \theta}(\sin\theta a_\phi) - \frac{\partial a_\theta}{\partial \phi}\right)$$

$$+\hat{e}_\theta \left(\frac{1}{r\sin\theta}\frac{\partial a_r}{\partial \phi} - \frac{1}{r}\frac{\partial}{\partial r}(r a_\phi)\right) + \hat{e}_\phi \frac{1}{r}\left(\frac{\partial}{\partial r}(r a_\theta) - \frac{\partial a_r}{\partial \theta}\right)$$

(C.27)

$$\nabla^2\psi = \frac{1}{r^2}\frac{\partial}{\partial r}\left(r^2 \frac{\partial \psi}{\partial r}\right) + \frac{1}{r^2\sin^2\theta}\frac{\partial^2 \psi}{\partial \phi^2} + \frac{1}{r^2\sin\theta}\frac{\partial}{\partial \theta}\left(\sin\theta \frac{\partial \psi}{\partial \theta}\right)$$

(C.28)

$$\vec{a}\cdot\nabla = a_r \frac{\partial}{\partial r} + \frac{a_\phi}{r\sin\theta}\frac{\partial}{\partial \phi} + \frac{a_\theta}{r}\frac{\partial}{\partial \theta} \tag{C.29}$$

$$(\vec{v}\cdot\nabla)\vec{v} = \hat{e}_r\left((\vec{v}\cdot\nabla)v_r - \frac{v_\phi^2 + v_\theta^2}{r}\right)$$

$$+\hat{e}_\phi\left((\vec{v}\cdot\nabla)v_\phi + \frac{v_r v_\phi + v_\phi v_\theta \cot\theta}{r}\right)$$

$$+\hat{e}_\theta\left((\vec{v}\cdot\nabla)v_\theta + \frac{v_r v_\theta - v_\phi^2 \cot\theta}{r}\right) \tag{C.30}$$

Bibliography

Abramowicz, M.A., Calvani, M., Nobili, L., 1980, *Astrophys. J.*, 242, 772.

Abramowicz, M.A., Jaroszynski, M., Sikora, M., 1978, *Astron. and Astrophys.*, 63, 221.

Abramowicz, M.A., Lanza, A., Percival, M.J., 1997, *Astrophys. J.*, 479, 179.

Alfvén, H., 1942, *Ark. Mat. Astron. Fys.*, 28A, 6; 29B, 2.

Balbus, S.A., 2003, *Annu. Rev. Astron. Astrophys.*, 41, 555.

Balbus, S.A., Hawley, J.F., 1991, *Astrophys. J.*, 376, 214.

Ballard, K.R., Heavens, A.F., 1992, *Monthly Not. R. Astron. Soc.*, 259, 89.

Bardeen, J.M., Petterson, J.A., 1975, *Astrophys. J. Lett.*, 195, L65.

Barenblatt, G.I., 1996, *Scaling, self-similarity, and intermediate asymptotics: dimensional analysis and intermediate asymptotics*, Cambridge University Press, Cambridge.

Beck, R., Brandenburg, A., Moss, D., Shukurov, A., Sokoloff, D., 1996, *Annu. Rev. Astron. Astrophys.*, 34, 155.

Begelman, M.C., Chiueh, T., 1988, *Astrophys. J.*, 332, 872.

Bell, A.R., 1978, *Monthly Not. R. Astron. Soc.*, 182, 147.

Bender, C.M., Orszag, S.A., 1978, *Mathematical methods for scientists and engineers*, McGraw Hill, New York.

Biermann, L., 1950, *Z. Naturforsch.*, 5A, 65.

Binney, J., Tremaine, S., 1987, *Galactic dynamics*, Princeton Univ. Press, Princeton, N.J.

Bisnovatyi-Kogan, G.S., Lovelace, R.V.E., 1997, *Astrophys. J. Lett.*, 486, L43.

Bisnovatyi-Kogan, G.S., Zel'dovich, Ya.B., Sunyaev, R.A., 1971, *Sov. Astron.*, 15, 17.

Blackman, E.G., 1999, *Monthly Not. R. Astron. Soc.*, 302, 723.

Blaes, O., 1987, *Monthly Not. R. Astron. Soc.*, 227, 975.

Blandford, R.D., 1976, *Monthly Not. R. Astron. Soc.*, 176, 465.

Blandford, R.D., Begelman, M.C., 1999, *Monthly Not. R. Astron. Soc. Lett.*, 303, L1.

Blandford, R.D., McKee, C.F., 1976, *Phys. Fluids*, 19, 1130.

Blandford, R.D., Ostriker, J.P., 1978, *Astrophys. J. Lett.*, 221, L29.

Blandford, R.D., Payne, D.G., 1982, *Monthly Not. R. Astron. Soc.*, 199, 883.

Blandford, R.D., Rees, M.J., 1974, *Monthly Not. R. Astron. Soc.*, 169, 395.

Blandford, R.D., Znajek, R.L., 1977, *Monthly Not. R. Astron. Soc.*, 179, 433.

Blasi, P., Vietri, M., 2005, *Astrophys. J.*, 626, 877.

Blumenthal, G.R., Gould R.J., 1970, *Rev. Mod. Phys.*, 42, 237.

Boldt, E., Ghosh, P., 1999, *Monthly Not. R. Astron. Soc.*, 307, 491.

Bondi, H., 1952, *Monthly Not. R. Astron. Soc.*, 112, 195.

Bondi, H., Hoyle, F., 1944, *Monthly Not. R. Astron. Soc.*, 104, 273.

Cavallo, G., Rees, M.J., 1978, *Monthly Not. R. Astron. Soc.*, 125, 52.

Chandrasekhar, S., 1981, *Hydrodynamic and hydromagnetic stability*, Dover, New York.

Chevalier, R.A., 1974, *Astrophys. J.*, 188, 501.

Chevalier, R.A., 1990, *Astrophys. J. Lett.*, 359, 463.

Chevalier, R.A., Theys, J.C., 1975, *Astrophys. J.*, 195, 53.

Contopoulos, I., Kazanas, D., Fendt, C., 1999, *Astrophys. J.*, 511, 351.

Cowie, L.L., 1977, *Monthly Not. R. Astron. Soc.*, 180, 491.

Cowling, T.G., 1934, *Monthly Not. R. Astron. Soc.*, 94, 39.

Cowling, T.G., 1981, *Annu. Rev. Astron. Astrophys.*, 19, 115.

Dalgarno, A., McCray, R.A., 1972, *Annu. Rev. Astron. Astrophys.*, 10, 375.

Davidson, K., 1973, *Nature*, 246, 1.

Dendy, R.O., 1994, *Plasma dynamics*, Oxford Univ. Press, Oxford.

D'yakov, S.P., 1954, *Zhurnal eksperimental'noi i teoreticheskoi fiziki*, 27, 288.

Ellison, D.C., Reynolds, S.P., Jones, F.C., 1990, *Astrophys. J.*, 360, 702.

Erber, T., 1966, *Rev. Mod. Phys.*, 38, 626.

Fermi, E., 1954, *Astrophys. J.*, 119, 1.

Ferreira, J., 1997, *Astron. Astrophys.* 319, 340.

Field, G.B., 1965, *Astrophys. J.*, 142, 531.

Galeev, A.A., Rosner, R., Vaiana, G.S., 1979, *Astrophys. J.*, 229, 318.

Ghosh, P., Lamb, F.K., 1978, *Astrophys. J. Lett.*, 223, L83.

Ghosh, P., Lamb, F.K., 1979, *Astrophys. J.*, 232, 259; 234, 296.

Goldreich, P.J., Julian, W.H., 1969, *Astrophys. J.*, 157, 869.

Goldreich, P.J., Sridhar, S., 1995, *Astrophys. J.*, 438, 763.

Gould, R.J., 1982, *Astrophys. J.*, 263, 879.

Greisen, K., 1967, *Phys. Rev. Lett.*, 16, 748.

Gruzinov, A.V., 1998, *Astrophys. J.*, 501, 787.

Gruzinov, A.V., 2005, *Phys. Rev. Lett.*, 94, 2101.

Haardt, F., Maraschi, L., 1991, *Astrophys. J. Lett.*, 380, L51.

Hawking, S.W., Ellis, G.F.R., 1973, *The large scale structure of space-time*, Cambridge Univ. Press, Cambridge.

Holloway, N.J., 1973, *Nature*, 24, 6.

Illarionov, A.F., Sunyaev, R.A., 1975, *Astron. Astrophys.*, 39, 185.

Jokipii, J.R., 1987, *Astrophys. J.*, 313, 842.

Jones, T.W., O'Dell, S.L., Stein, W.A., 1974, *Astrophys. J.*, 188, 353.

Karzas, W.J., Latter, R., 1961, *Astrophys. J. Suppl.*, 6, 167.

Kobayashi, S., Piran, T., Sari, R., 1999, *Astrophys. J.*, 513, 669.

Kraichnan, R.H., 1965, *Phys. Fluids*, 8, 1385.

Kraus, U., Zahn, C., Weth, C., Ruder, H., 2003, *Astrophys. J.*, 590, 424.

Krause, F., Radler, K.H., Rüdiger, G., 1993, *The cosmic dynamo: Proceedings of the 157th symposium of the IAU in Potsdam*, Kluwer, Dordrecht.

Krolik, J.H., 1998, *Astrophys. J. Lett.*, 498, L13.

Krolik, J.H., 1999, *Active galactic nuclei*, Princeton Univ. Press, Princeton, N.J.

Kulsrud, R.M., Cen, R., Ostriker, J.P., Ryu, D., 1997, *Astrophys. J.*, 480, 481.

Landau, L.D., Lifshitz, E.M., 1981a, *Physical kinetics*, Pergamon Press, Oxford.

Landau, L.D., Lifshitz, E.M., 1981b, *Quantum electrodynamics*, Pergamon Press, Oxford.

Landau, L.D., Lifshitz, E.M., 1984, *Electrodynamics of continuous media*, Pergamon Press, Oxford.

Landau, L.D., Lifshitz, E.M., 1987a, *The classical theory of fields*, Pergamon Press, Oxford.

Landau, L.D., Lifshitz, E.M., 1987b, *Fluid mechanics*, Pergamon Press, Oxford.

Lemoine, M., Pelletier, G., 2003, *Astrophys. J. Lett.*, 589, L73.

Lightman, A.P., Eardley, D.M., 1974, *Astrophys. J. Lett.*, 187, L1.

Loeb, A., Laor, A., 1992, *Astrophys. J.*, 384, 115.

Loeb, A., McKee, C.F., Lahav, O., 1991, *Astrophys. J. Lett.*, 374, L44.

Longair, M.S., 1999, *High energy astrophysics*, Cambridge Univ. Press, Cambridge.

Lovelace, R.V.E., 1976, *Nature*, 262, 649.

Mansfield, V.N., Salpeter, E.E., 1974, *Astrophys. J.*, 190, 305.

MacDonald, D.A., Suen, W.-M., 1985, *Phys. Rev. D*, 32, 848.

MacDonald, D.A., Thorne, K.S., 1982, *Monthly Not. R. Astron. Soc.*, 198, 345.

Meszaros, P., Laguna, P., Rees, M.J., 1993, *Astrophys. J.*, 415, 181.

Mezzaappa, A., 2005, *Annu. Rev. Nucl. Part. Sci.*, 55, 467.

Michel, F.C., 1972, *Astrophys. Space Sci.*, 15, 153.

Michel, F.C., 1973, *Astrophys. J. Lett.*, 180, L133.

Miller, K.A., Stone, J.M., 1997, *Astrophys. J.*, 489, 890.

Narayan, R., 1992, *R. Soc. Philos. Trans. A*, 341, 151.

Narayan, R., Yi, I., 1994, *Astrophys. J. Lett.*, 428, L13.

Narayan, R., Yi, I., 1995, *Astrophys. J.*, 444, 231.

Osterbrock, D., Ferland, G., 2005, Astrophysics of Gaseous nebulae and active galactic nuclei, University Science Books, Sausalito.

Ostriker, J.P., 1977, *Annu. N.Y. Acad. Sci.*, 302, 2290.

Pacini, F., 1967, *Nature*, 216, 567.

Paczynski, B., 1978, *Acta Astron.*, 28, 111; 28, 241.

Paczynski, B., Wiita, P.J., 1980, *Astron. Astrophys.*, 88, 23.

Page, D.N., Thorne, K.S., 1974, *Astrophys. J.*, 191, 499.

Papaloizou, J.C., Pringle, J.E., 1983, *Monthly Not. R. Astron. Soc.*, 202, 1181.

Papaloizou, J.C., Pringle, J.E., 1984, *Monthly Not. R. Astron. Soc.*, 208, 721.

Parker, E.N., 1955, *Astrophys. J.*, 121, 491; 122, 293.

Parker, E.N., 1979, *Cosmic magnetic fields*, Oxford Univ. Press, Oxford.

Perna, R., Vietri, M., 2002, *Astrophys. J. Lett.*, 569, L47.

Petschek, H.E., 1964, *Physics of solar flares*, in *AAS-NASA Symposium on the Physics of Solar Flares*, ed. W.N. Ness, NASA, Washington, D.C.

Pietrini, P., Krolik, J.H., 1995, *Astrophys. J.*, 447, 526.

Piran, T., 1978, *Astrophys. J.*, 221, 652.

Piran, T., 2005, *Rev. Mod. Phys.*, 76, 1143.

Poutanen, J., Svensson, R., 1996, *Astrophys. J.*, 470, 249.

Pozd'nyakov, L.A., Sobol, I.M., Sunyaev, R.A., 1977, *Sov. Astron.*, 21, 708.

Pozd'nyakov, L.A., Sobol, I.M., Sunyaev, R.A., 1983, *Astrophys. Space Rev.*, 2, 189.

Pringle, J.E., 1976, *Monthly Not. R. Astron. Soc.*, 177, 65.

Pringle, J.E., 1981, *Annu. Rev. Astron. Astrophys.*, 19, 137.

Proga, D., Begelman, M.C., 2003, *Astrophys. J.*, 582, 69; 592, 767.

Proga, D., Macfayden, A.I., Armitage, P.J., Begelman, M.C., 2003, *Astrophys. J. Lett.*, 599, L5.

Protheroe, R.J., Johnson, P.A., 1996, *Astropart. Phys.*, 4, 253.

Quataert, E., 1998, *Astrophys. J.*, 500, 978.

Rees, M.J., 1966, *Nature*, 211, 468.

Ross, R.R., Weaver, R., McCray, R.M., 1978, *Astrophys. J.*, 219, 292.

Ruderman, M.A., Sutherland, P.G., 1975, *Astroph. J.*, 196, 51.

Rybicki, G.B., Lightman, A.P., 1979, *Radiative processes in astrophysics*, J. Wiley and Sons, New York.

Ryu, D. Vishniac, E.T., 1987, *Astrophys. J.*, 313, 820.

Sakurai, J.J., 1967, *Advanced quantum mechanics*, Addison-Wesley, Reading, Mass.

Sari, R., Piran, T., 1995, *Astrophys. J. Lett.*, 455, L143.

Scharlemann, E.T., Wagoner, R.V., 1973, *Astrophys. J.*, 182, 951.

Schlickeiser, R., 2001, *Cosmic ray astrophysics*, Springer-Verlag, Berlin.

Sedov, L.I., 1946, *Prikl. Mat. Mekh.*, 10, 241.

Shakura, N.I., Sunyaev, R.A., 1973, *Astron. Astrophys.*, 24, 337.

Shapiro, S.L., 1973, *Astrophys. J.*, 180, 531.

Shapiro, S.L., Lightman, A.P., Eardley, D.M., 1976, *Astrophys. J.*, 204, 187.

Shapiro, S.L., Teukolsky, S.A., 1984, *Black holes, white dwarfs, and neutron stars*, John Wiley and Sons, New York.

Shima, E., Matsuda, T., Takeda, H., Sawada, K., 1985, *Monthly Not. R. Astron. Soc.*, 217, 367.

Soward, A.M., Priest, E.R., 1977, *R. Soc. Philos. Trans. A*, 284, 369.

Spitzer, L. Jr., 1962, *Physics of fully ionized gases*, J. Wiley and Sons, New York.

Spitzer, L., Jr., 1978, *Physical processes in the interstellar medium*, J. Wiley and Sons, New York.

Spruit, H.C., Matsuda, T., Inoue, M., Sawada, K., 1987, *Monthly Not. R. Astron. Soc.*, 229, 517.

Stern, B.E., Begelman, M.C., Sikora, M., Svensson, 1995, *Monthly Not. R. Astron. Soc.*, 272, 291.

Stone, J.M., Balbus, S.A., 1996, *Astrophys. J.*, 463, 656.

Stone, J.M., Norman, M.L., 1994, *Astrophys. J.*, 433, 746.

Sunyaev, R.A., Titarchuk, L., 1980, *Astron. Astrophys.*, 86, 121.

Sunyaev, R.A., Titarchuk, L., 1985, *Astron. Astrophys.*, 86, 121.

Svensson, R., 1982, *Astrophys. J.*, 258, 321; 258, 335.

Svensson, R., 1984, *Monthly Not. R. Astron. Soc.*, 209, 175.

Taylor, G.I., 1950, *Proc. R. Soc. London A*, 201, 159.

Thompson, C., Duncan, R.C., 1994, *Astrophys. J.*, 143, 374.

Thorne, K.S., 1974, *Astrophys. J.*, 191, 507.

Thorne, K.S., Price, R.H., 1975, *Astrophys. J. Lett.*, 195, L101.

Thorne, K.S., Price, R.H., MacDonald, D.A., 1986, *Black holes: The membrane paradigm*, Yale Univ. Press, New Haven, Conn.

Toomre, A., 1964, *Astrophys. J.*, 139, 1217.

Torkelsson, U., Brandenburg, A., 1994, *Astron. Astrophys.*, 283, 677.

Tout, C.A., Pringle, J.E., 1992, *Monthly Not. R. Astron. Soc.*, 256, 269.

Tout, C.A., Pringle, J.E., 1996, *Monthly Not. R. Astron. Soc.*, 281, 219.

Treves, A., Maraschi, L., Abramowicz, M., 1988, *Publ. Astron. Soc. Pacific*, 100, 427.

Turner, M.S., Widrow, L.M., 1988, *Phys. Rev. D*, 37, 2743.

Velikov, E.P., 1959, *J. Exp. Theor. Phys.*, 36, 1398.

Vietri, M., 2003, *Astrophys. J.*, 591, 954.

Vishniac, E.T., Ryu, D., 1989, *Astrophys. J.*, 337, 917.

Wald, R.M., 1984, *General relativity*, Univ. of Chicago Press, Chicago.

Wang, Y.-M., 1995, *Astrophys. J. Lett.*, 449, L153.

Wardle, M., Königl, A., 1993, *Astrophys. J.*, 410, 218.

Weinberg, M.D., 1986, *Astrophys. J.*, 300, 93.

Whitehurst, R., King, A., 1991, *Monthly Not. R. Astron. Soc.*, 249, 25.

Wilkinson, P.N., Narayan, R., Spencer, R.E., 1994, *Monthly Not. R. Astron. Soc.*, 269, 67.

Woosley, S.E., 1993, *Astrophys. J.*, 405, 273.

Yi, I., 1995, *Astrophys. J.*, 442, 768.

Zatsepin, G.T., Kuzmin, V.A., 1966, *Sov. Phys. JETP Lett.*, 4, 78.

Zdziarski, A.A., 1985, *Astrophys. J.*, 289, 514.

Zdziarski, A.A., Coppi, P.S., Lamb, D.Q., 1990, *Astrophys. J.*, 357, 149.

Zel'dovich, Y.B., Ruzmaikin, A.A., Sokoloff, D.D., 1990, *Magnetic fields in astrophysics*, Taylor and Francis, London.

Znajek, R.L., 1978, *Monthly Not. R. Astron. Soc.*, 185, 833.

Index

Echu eo ma jañson, an hini 'oar c'hontinui
(I have sung my song; let those who know, carry on)
from Tri Martolod